Cognitive Science

Cognitive Science

An Introduction

David W. Green and others

First published 1996

2 4 6 8 10 9 7 5 3 1

Blackwell Publishers Ltd
108 Cowley Road
Oxford OX4 1JF
UK

Blackwell Publishers Inc
238 Main Street
Cambridge, Massachusetts 02142,
USA

Library of Congress Cataloging-in-Publication Data has been applied for.

ISBN 0–631–19859–8; ISBN 0–631–19861–x (pbk.)

British Library Cataloguing in Publication Data

A CIP catalogue record for this book is available from the British Library.

Commissioning Editor: Alison Mudditt
Desk Editor: Tony Grahame
Production Controller: Lisa Eaton
Text Designer: Lisa Eaton
All-round good eggs: Andrew Brockbank, Alison Dunnett, Nathalie Manners

Typeset in Baskerville 10.5 on 12.5 by Photoprint, Torquay, Devon
Printed in Great Britain by The Alden Press, Oxford.

This book is printed on acid-free paper

Contents in Brief

Contents

7 The Structure of Sentences 174

8 Meaning and Conversation 217

9 Pragmatics and the Development of Communicative Ability 244

Acknowledgments

We thank George Graham and other commentators on a prior draft of this text for their most helpful comments. We are particularly grateful to Phil Johnson-Laird for his advice, concrete suggestions and encouragement on the book as a whole. We also acknowledge the support of Uta Frith and John Marshall, John Dowell, Mark Keane, Chris McManus, Neil Smith and Herb Clark for their specific commentaries.

The authors and publishers wish to thank the following for permission to use copyright material:

Figure from H. H. Clark and D. Wilkes-Gibbs (1986), Referring as a collaborative process, *Cognition* 22, 1–39. Figure from S. Garrod and G. Doherty (1994), Conversation, co-ordination and convention: an empirical investigation of how groups establish linguistic conventions, *Cognition* 53, 181–215. Permission granted by Elsevier Science.

Material from E. Hirschman, J. G. Snodgrass, J. Mindes and K. Feenan (1990), Conceptual priming in fragment completion, *Journal of Experimental Psychology: Learning, Memory, and Cognition* 16, 634–47. Reprinted with permission from the American Psychological Association.

Figure from Philip J. Benson (1994), Visual processing of facial distinctiveness, *Perception* 23, 75–93. Reprinted with permission of Pion Limited, London.

Every effort has been made to trace all copyright holders but if any have been inadvertently overlooked the publishers will be pleased to make the necessary arrangement at the first opportunity.

Acknowledgments

List of Collaborators

Paul W. Burgess received his Ph.D. from the Institute of Neurology in London. After working with neurological patients at the Kemsley Brain Injury Unit in the UK, he moved to the National Hospital in London, and the Applied Psychology Unit in Cambridge. He is now a senior research fellow at UCL, and has published papers on the neuropsychology of neurological rehabilitation, memory disorders, schizophrenia, and executive functions.

Robyn Carston is lecturer in Linguistics at University College London. Her primary research interest is in pragmatics, in particular the development of the relevance-theoretic account of the cognitive processes involved in utterance interpretation.

Richard Cooper is a Lecturer in Psychology at Birbeck College, University of London. He studied Mathematics and Computer Science before obtaining a Ph.D. in Cognitive Science from the University of Edinburgh. His recent publications include articles in *Cognition and Artificial Intelligence.*

David Green is currently a Senior Lecturer in Psychology and Executive Director of the Centre for Cognitive Science at University College London. He has published widely in the area of language and cognition including work on word recognition and text comprehension, bilingualism and brain-damage, and reasoning. His current research interests are speech control in bilinguals and the link between cognitive models and actions.

Peter Howell is Professor of Experimental Psychology at University College London. He is co-author of the textbook *Signals and Systems for Speech and Hearing* (Academic Press, 1991) and has edited two books on musical cognition. He has authored papers on music, speech perception and production, and hearing. His main current interests are automatic recognition of speech and stuttering.

Alan Johnston is a Reader in Psychology at University College London. He has published papers on a wide range of topics in visual science including face and object perception, shape-from-shading, integration of depth cues, the geometry of topographic maps in the visual cortex, and the computational modelling of motion perception.

Daniela K. O'Neill received her Ph.D. from Stanford University. She is Assistant Professor in Developmental Psychology in the Department of Psychology at the University of Waterloo, Canada. Her major research interests are children's developing understanding of knowledge and the early pragmatic competence of infants and toddlers. She was formerly a visiting research fellow at the MRC Cognitive Development Unit in London and is continuing work begun there on the early pragmatic development of children with autism and with Down's syndrome.

John Morton is Director of the Medical Research Council's Cognitive Development Unit and Professor of Psychology at University College London. He has made theoretical and empirical contributions to a variety of cognitive areas, in word recognition, cognitive neuropsychology, memory, and human–computer interaction. More recently his particular interests have been in infant recognition of faces, the development of memory and developmental psychopathology in general. He is co-author with Mark Johnson of *Biology and Cognitive Development* (Blackwell, 1991).

David R. Shanks is Lecturer in Psychology at University College London. He received his Ph.D. from the University of Cambridge and then worked at the MRC Applied Psychology Unit in Cambridge and at the University of California, San Diego. He is the author of *The Psychology of Associative Learning* (Cambridge University Press, 1995) as well as various articles on connectionism, implicit learning, causal reasoning, and other aspects of human learning and memory.

Hans van de Koot received his Ph.D. in Linguistics from the University of Utrecht and currently teaches linguistics and computational linguistics at University College London. His research interests are in syntax and the computational properties of models of language. He has published papers in syntax and computational linguistics.

CHAPTER 1

Introduction

> The world of science, like that of art or religion, (is) a world created by the human imagination, but within very strict constraints imposed both by nature and the human brain.
>
> (Jacob, 1988, p. 306)

Outline

This chapter outlines some of the key concepts and issues in cognitive science and introduces you to the nature of this textbook. Cognitive scientists aim to understand the processes and representations underlying intelligent action in the world. They do so by building explicit models of these processes and testing predictions from them experimentally. Complete models involve at least three levels of description: the behavioral, the cognitive, and the biological. Given current knowledge, our examples of the discipline focus on the cognitive and behavioral levels. Our aim is to encourage you to participate in the debates and to provide you with some of the conceptual tools to do so.

Learning Objectives

After reading this chapter you should be able to:

- contrast everyday and scientific thinking
- characterize the discipline of cognitive science
- identify research strategies for the discipline
- understand the significance of the representational view of mind
- grasp the basic methodological requirements of the discipline
- appreciate the aims and goals of this text

Key Terms

- cognitive science
- functionalism
- information processing
- levels of description
- mental representation

Cognitive scientists are embarked on a collective and long-term enterprise to understand the mind scientifically. What does it mean to think scientifically about the mind? How does such thinking compare with everyday thinking about people? Please read the scenario in box 1.1 as we will be referring to it throughout this book with the aim of illustrating cognitive science. Hopefully, you will have no problem understanding what is going on, primarily because, like us, you have a sense of why people do what they do. As social agents, we need such a sense because we need to be able to plan our actions, predict what might happen in certain circumstances and be able to explain events in order to decide what actions to take. So, for example, Lucy, perhaps aware that her father would object to her visiting her friend, sought her mother's support. On overhearing Dad's objection, she tried to placate him with an offering of his favourite jam.

What matters to us as social agents is whether our knowledge of the social and physical world, and our methods for handling various problems, are sufficient to allow us to meet our practical concerns. We learn, or infer, the relationship between concepts and observations. Suppose, for example, the following morning, Lucy spoke to Dad without looking at him. Dad might infer that Lucy was angry with him, because when people are angry they either stare at one or look away. He might also infer that Lucy was angry because he did not relent. He based this inference on the fact that people get angry when they are thwarted. In everyday life, we do not go on to ask how such a capacity for inference is possible. In accounting for why she forgot where Granny went on holiday, Mom might say it was because she is forgetful. In everyday life, we do not go on to ask what is involved in remembering or in forgetting things. We rarely seek to explain the normal. We do not ask: how is it possible that Mom is able to read the letter? How is it possible that she was able to remember seeing the tray in the living room? We do not ask: how do we recognize a coffee pot or, assuming a dramatic staging of the scenario, distinguish Lucy's voice and face from Zara's voice and face? And yet these are most remarkable achievements! Clearly there are occasions when we do extend our accounts, such as when something unexpected happens or when things break down. Suppose Granny had a stroke and was unable to write. We would attribute her inability to the physical effects of the stroke rather than to some emotional cause. In our everyday explanations, then, we do (but perhaps not as much as we could) distinguish between what someone does and their motives or dispositions. Occasionally, we also appeal to possible physical factors. But our everyday accounts, however subtle, are particular rather than general. As part of our upbringing, we also learn certain methods for achieving our goals. Methods, for example, for coping with difficulties (e.g., ring the doctor if a fever has continued for a period of time). However, we do not generally consider whether a given method is both necessary and sufficient. It is whether or not it works that matters, for instance, given a headache, we may reach for the aspirin. The general point is this: our thinking and practical actions are determined by the goals we have in mind.

Box 1.1 The scenario

This is a fictional conversation over breakfast between two adults, Mom and Dad, their teenage daughter, Lucy, and, their two and a half year-old daughter, Zara. It is set in a kitchen/living room. The phone rings.

Mom picks up the phone, listens, and says: "I'll get her. How's the coffee?"
Lucy: "Started." Lucy goes to phone.
Zara in high chair at the table, reaches forward toward milk carton, saying: "More milk, mommy!"
Dad: "Here's some milk."
Zara turns towards the cupboard, points and says, "Mommy, – ops!"
Mom: "Hm? What darling?"
Zara: "Pops! Pops!"
Mom: "Oh, can you get her some – I'm seeing to the toast."
Lucy: "I'm eating at Jane's tonight, Mom – OK?"
Dad: "I thought it was your homework evening."
Door bell rings.
Lucy: "I'll go."
Mom frowning: "She's worked every night, John."
Dad: "She didn't work last night. She sat in front of the TV."
Lucy: "For you, from Gran I think."
Mom struggling to open the parcel.
Zara: "Gran send me bear."
Mom: "That's right! Granny sent you your bear for your birthday, last month, didn't she?"
Mom goes out to the living room.
Dad: "Toast! Lucy! It's Etna in here."
Lucy opens the windows in the kitchen.
Lucy: "I'll make some more." Lucy discovers the toaster is broken. She makes toast heating the bread in a frying pan. Zara: bangs the table "Allgone! Gone!"
Dad: "Yes, you've eaten it up . . . cleared the bowl. Lucy are you pouring the coffee? Where's the butter? Can you bring it over? And the jam."
Lucy: "Anyone seen the tray?"
Dad: "It's not over here."
Mom comes back with scissors and opens the parcel.
Mom: "It's a jumper for Zara. Look, won't she look great!" Mom puts down the jumper and starts to read the letter enclosed. "She says she has fixed up something . . . what's this? 'nolinay'? Ah! 'holiday'! Granny's writing doesn't improve. 'I've fixed up a holiday with friends who have a yacht. We're sailing to the Polish port of (she spells out the letters) SZCZECIN. I've never been sailing before.' Where did she go last year? I've completely forgotten."
Lucy: "Mom, have you seen the tray?"
Dad: "The Grand Canyon. I remember the postcard she sent us." The cat comes in, stretches, and lies down.
Mom: "I just saw it in the living room."
Lucy comes across with the toast, butter, jam, coffee pot, and mugs on the bread board.
Dad: "Ah you little monster! My favourite jam. You still have to work tonight!"
. . .

How then does scientific thinking about the mind contrast with everyday thinking about the mind? In thinking scientifically about the mind we aim not only to generate explanations that allow us to predict observable behavior but we also want these accounts to be general and to use the fewest explanatory concepts possible. As cognitive scientists, we are primarily interested in such matters as how individuals have the capacity to see the world, how they are able to plan actions in the world, and how they manage to understand one another. We also need special methods for testing accounts. These methods need to be valid (i.e., measure what they claim to measure) and to be reliable (i.e., produce data that are replicable). Cognitive scientists also make use of methods aimed at ensuring that ideas are as explicit as possible.

Here you might raise a number of objections to the very idea of a science of mind. We have written one such objection below in the form of what we call a self-assessment question or SAQ. Try to answer the question in SAQ 1.1 before reading on.

SAQ 1.1 How would you reply to someone who said that the mental worlds of human beings are too complex for scientific understanding?

The question of how the mind should be conceived is a crucial issue: one possibility is that the mind is modular. The basic notion can be understood by way of an analogy. Faced with preparing breakfast, what do people typically do? They do not try and do everything at once. Instead, the task is broken down into various component tasks: making coffee; making toast; preparing the table. Where more than one person is involved these tasks can be shared out. In the scenario, of course, Lucy seems to have drawn the short straw! The analogy to the mind is this: different mental systems or modules may process different kinds of information relevant to the achievement of a particular task. The fact that we have different organs for seeing and hearing, for instance, suggests that the mind might consist of many special mechanisms adapted to carrying out different tasks. In other words, its design is modular.

What other objections could be raised to the enterprise? One concern you might have is that mental events are private. Consider the question posed in SAQ 1.2.

SAQ 1.2 What methods can you imagine for observing or measuring mental events?

A number of methods have been developed. All mental operations take time and so we can observe their influence on the speed of response to specific probe questions just as a physicist might observe the effects of an invisible particle by its effects on other known particles in a cloud chamber. We infer processes from the reports of subjects in experiments making simple judgments (e.g., do they see a particular face as sneering or smiling?) or from the reports or protocols that individuals make during their problem solving. Other distinct methods are also possible: we can infer the nature of mental operations from the kinds of slips and errors that people make. For example, the fact that Mom intended some action (to make the toast) but did not carry it out (Mom let the toast burn) suggests that unless individuals take special precautions, some routine action (making toast) can be overridden by some more salient activity (opening a parcel from Granny). We can also infer the existence of certain mental processes and representations as a result of the performance of individuals who have suffered brain damage, perhaps as the result of a stroke. More recently, in combination with an understanding of the nature of the mental operations, we can use brain-imaging techniques that make visible areas of the brain subserving such operations. Our position is this: we see no fundamental objections to the scientific study of mind. The next step is to characterize the science of mind.

The Discipline of Cognitive Science

How then shall we define this enterprise of cognitive science? A particular scientific **discipline** considers problems of a specific range or scope and involves a set of practices and knowledge (see Kuhn, 1970; Long and Dowell, 1989). We define the scope of **cognitive science** as the interdisciplinary scientific study of mind. Its practices and knowledge derive from those of the primary contributing disciplines, which are computer science, linguistics, neuroscience, psychology, cognitive neuropsychology, and philosophy. It seeks to understand how the mind works in terms of processes operating on representations. Mind, and hence the basis of intelligent action in the world, is viewed in terms of computations or information-processes. For example, in order to read this page you must be able to go from a printed word to its meaning. Retrieving the meaning of a word is a mental process. This process must make the meaning of the word or, rather, your mental representation of its meaning, available to a further process that constructs the meaning of the sentence and so on.

Cognitive scientists aim to construct causal accounts by linking three levels of description that are used informally, and for different purposes, in everyday explanation. Following Morton and Frith (1995), we refer to these levels as the behavioral level (for example, a person's performance on specific experimental tasks); the cognitive level (i.e., the postulated cognitive or affective systems underlying the behavior); and the biological level (the nature of the brain systems as affected by genetic, environmental, and social factors that mediate the cognitive systems). Clearly in some cases, there can be a direct link between this biological level and behavior. The tics, for example, in Tourette's syndrome. But in other cases we must draw a contrast between the cognitive and the behavioral: different symptoms, for instance, might be the result of a single cognitive deficit. So, for example, the fact that an autistic child cannot engage in "pretend play" (such as when a child pretends that a banana is a telephone) and the fact that the child is socially withdrawn, may derive from a difficulty in forming a certain kind of mental representation. Box 1.2 illustrates a causal model involving both genetic and cognitive disorders.

In pursuit of understanding we build models that capture the postulated processes and representations and which, ideally, can be implemented in

Box 1.2 Levels of description

We have the option of describing events at the behavioral, cognitive, or biological levels. A complete description will include all three. Specific separation of all three is particularly useful when considering developmental disorders. In the text, we mention that autism may result from the child not being able to form a particular kind of mental representation. We also know that autism is the result of a genetic disorder. These two claims are not contradictory and can be represented conveniently in a standardized form as follows:

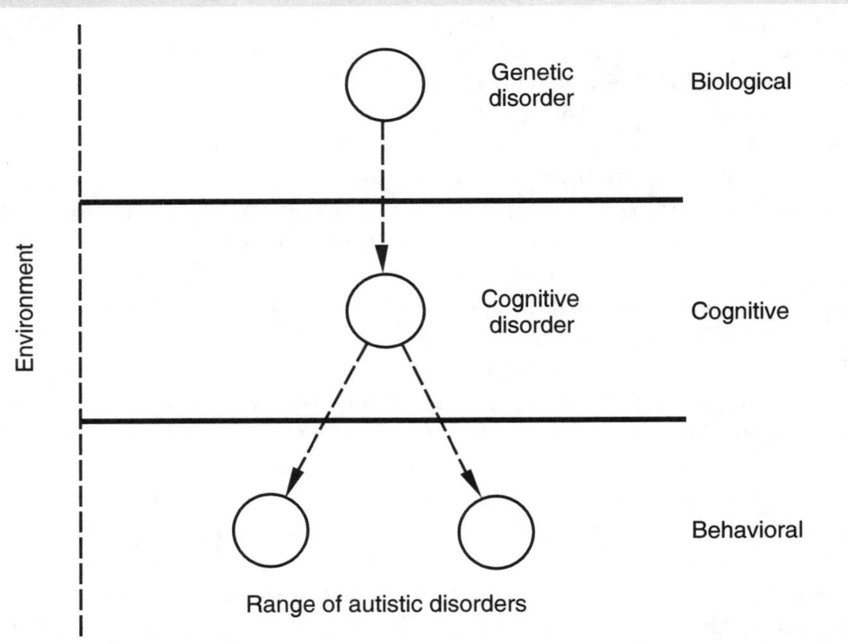

1.1 Diagrammatic representation of a causal model.

As we create more complex theories, biological or cognitive, we can make the causal model represent this complexity. Indeed we may need to unpack the biological level since the brain is organized on many different scales: systems, neural networks, neurones, synapses, and molecules.

computer programs that generate relevant aspects of the behavior in question. This is what we mean by information-processing or computational models of mind. We also test the predictions of theories by experiment with a view to determining what is necessary and sufficient for the behavior to occur.

The knowledge base of cognitive science includes theories and evidence; debates in journals and books and most importantly exemplars, that is instances of problems and their provisional solutions. This conception of the discipline of cognitive science distinguishes it from some of the work in Artificial Intelligence, which is free to develop solutions to problems (such as chess playing) unconstrained by any view of how humans think. Such work is really about technology rather than science. However, to the extent that research is oriented towards understanding the fundamental nature of intelligent action, it is relevant.

How in a nutshell can we capture the approach of cognitive science? We treat the mind as a machine whose workings we are trying to understand. What is your immediate reaction to this idea? For some people it is an incredible notion. People are not machines. Nor should they be treated as such. The approach seems to suggest that we should be interested in others only to the extent that they have utility for us. We argue that we need to change our minds about the nature of machines. A machine need not be simple. The machine approach makes clear a basic requirement: our theories must be capable of generating the behavior we hope to explain. If they can, then we have a possible explanation. Others applaud the machine approach because they take it to mean that cognitive science is embarked on a programme of reducing mind to the physics and chemistry of the components of the brain. But this is not what we mean. Reduction is not an appropriate goal because the mind is a representational system – in order to understand its workings we need to understand what it is trying to do. This is why we say that three levels of description are required for a complete causal account. Mind, from the perspective of cognitive science, is the activity of the brain at a certain level of description. It is for this reason that some researchers write of the mind/brain. When we write of the "mind" we do so with this sense in mind. Box 1.3 provides a potted history of the emergence of cognitive science.

The Mind as a Representational System

Fundamental to the endeavor of cognitive science is the notion of mental representation. What do we mean by the term? In thinking about the physical world, we do not manipulate actual bits of the world but mental entities that represent or symbolize it. Likewise, in using language we do not utter bits of the world, but words such as "coco-pops", which refer to objects in the world.

In Simon's and Newell's approach (Simon, 1969; Newell and Simon, 1972), our capacity to think, feel, communicate, and act arises because of our capacity

Box 1.3 The emergence of cognitive science

. . . cognitive science has a very long past but a relatively short history
(Gardner, 1985, p. 9)

Cognitive science emerged out of a matrix of ideas and events. One critical factor was the rejection of behaviorism. Behaviorism proposed that mind was not a proper object of scientific study – only the inputs and the outputs to an organism were legitimate objects of psychological enquiry. Perhaps, most influential in behaviorism's downfall were two papers: Lashley's demonstration that skilled behavior could not be explained in terms of associative chains, and Chomsky's proof that behaviorist approaches were inadequate to account for the structure of English sentences. Taken together these strands freed researchers to postulate mental entities and processes to account for overt behavior. These strands themselves entwined with the consequences of work carried out on skilled behavior during the Second World War (such as teaching individuals to fly planes) which showed quite clearly the impossibility of accounting for human performance without considering their mental processes.

Such an opportunity would not have resulted in a scientific discipline unless it had become possible to think about mental processes and representation. The catalyzing event was the construction in Princeton of the first digital computer. The outputs of the computer depended on the data fed into it and the nature of the program. This provided a vivid demonstration of a way to view the relationship between the mind and the brain: mental activities could be viewed as programs and the brain as the computer of the mind. The theoretical possibility of such a machine and hence one of the factors leading to its construction was the mathematical theory of computability. This established in the 1930s that a small set of operations could yield an unlimited variety of symbolic outcomes. Indeed, Turing, one of the pioneers, recognized the importance of the theory in terms of the explanation of human behavior.

From a philosophical viewpoint, such an event supported the doctrine of functionalism: it was possible to talk about the way some device was organized independently of the way it was built – its physical embodiment. Important too was the demonstration that it was possible to construct devices (such as a thermostat or a guided missile) that were self-correcting, that is sensitive to the difference between some present state and some goal state. Such work put beyond doubt the scientific legitimacy of attributing goal states to human beings. Talk about goals and talk about mechanism was compatible.

By the mid-1950s Simon and Newell had devised a program to solve problems in logic that emulated some aspects of the way people solved the problems. The 1960s and 1970s, saw an increase in the use of models (information-processing models or computational models) that postulated representations and processes to account for performance in different domains such as planning (Miller, Galanter, and Pribram, 1960); attention (Broadbent, 1958); reading (Morton, 1969); reasoning (Johnson-Laird and Wason, 1970); and consciousness (e.g., Shallice, 1972).

Information-processing models consisting of different subsystems (e.g., for recognizing a person's face or for retrieving their name) encouraged consideration of what happens when a system for processing certain kinds of information is damaged or when the connection between two such systems is broken. On the supposition that the deficits observed in brain-damaged patients reflect damage to specific processing systems, research on brain-damaged patients can bear directly on models of normal function. Indeed, such research provides some of the most direct evidence that the mind is modular in construction. Neuropsychological research is, therefore, a further important input into the matrix of the discipline. Recent advances in techniques for recording brain activity in normal individuals are likely to accelerate the process of integrating cognitive and biological levels of description.

Over the past few years there has also been a growth in different kinds of model based on the notion that the brain carries on many processes in parallel. Such research is extending earlier ideas proposed by, for example, Hebb (1949) and the work of Minsky and Papert (1969; expanded edition, 1988). It has led to a vigorous debate between those favoring symbolic computation and those favoring connectionist computation. We simply note here that symbolic computation can be carried out in parallel. Whilst we are a long way from understanding what kind of machine the brain really is, it is reasonable to believe that understanding of neural processes will help us construct biologically plausible information-processing accounts.

to process symbols. A symbol may designate information from the senses. The sight of a black, furry object may evoke the symbol CAT, designating a particular animal. A symbol may also designate another symbol. For example, the symbol CAT may designate the word *cat.* A symbol or combination of symbols may also designate an action such as feeding the cat.

Central to this approach is the idea that we can abstract from the way a symbol is represented in the brain. The precise physical nature of symbols is irrelevant to their role in governing behavior. The analogy to a computer program is helpful here: a program can be written in a relatively high- level language. It comprises a set of symbols. In order to carry out the instructions of the program it is compiled into a lower-level language, which will produce changes in terms of the states in the actual physical machine. This machine may operate in terms of silicon chips or in some other fashion. In contrast, the brain operates in terms of patterns of activity over neurons. But it is the high-level program or the functional description which is the key to understanding the performance of the machine. This is position is known as functionalism (see also box 1.2). Can we know what a computer is doing just by inspecting the patterns in the hardware?

To answer this question, consider that in writing a program we construct an abstract or "virtual machine" (Sloman, 1993). The virtual machine manipulates symbolic structures rather than physical objects. For example, a word processing package manipulates words, sentences, and paragraphs but these are not to be found in the underlying physical machine. The parts of such a virtual machine interact causally in terms of how information is processed, not in terms of physical forces or states. In a similar fashion we can understand the complexity of human performance in terms of the relationships of various virtual machines: a machine for speaking or for reading. The components of such virtual machines interact causally, but not in terms of neural impulses or patterns, even though they are implemented in a brain that does operate in this fashion. This notion is basic to our attempts to construct an information-processing or computational account of mind. We hope we have now convinced you that the cognitive level is central to a scientific account of mind, but is the view of mind as symbol-manipulation sufficient? Please read devil's advocate box 1.1, which outlines an opposing view, together with a possible resolution.

Mentalizing: Orders and Varieties of Mental Representation

Everyday thinking involves trying to understand the actions of others. In doing so individuals try to infer their intentions and this may help explain a wide range of behaviors. Why are Mom and Dad cleaning, ironing, tidying the

Devil's Advocate Box 1.1 The Chinese room argument

Searle (1980) challenged the view that mind can be understood as symbol manipulation. He considered the following thought experiment. He imagines himself sitting in a room. Slips of paper are passed in containing questions written in Chinese, a language he does not understand. His task is to formulate replies to these questions. He does so by consulting a set of rules, written in English, for manipulating symbols and writes these replies down on other slips of paper which he passes out of the room. He treats this situation as entirely analogous to that of a computer programmed to give appropriate replies in response to typed in questions. In neither case is there any understanding of meaning. Strings of symbols are input, manipulated and output. The strings of symbols only mean something to individuals who know Chinese. The meaning is extrinsic rather than intrinsic to the symbol system. Hence, Searle argued, thinking and mentalizing cannot just be symbol manipulation. Harnad (1990) refers to this as the symbol grounding problem. The question for Harnad is "how is symbol meaning to be grounded in something other than more meaningless symbols" (p. 340)? Symbols have to be connected to the world. But how? Harnad proposes a possible solution in which certain basic terms are tied to the representations of objects. Consider the meaning of "horse"; it is linked to a category of object (that captures the invariant properties of horses); likewise, the word "stripes" is linked to an appropriate perceptual representation. Presented with the definition of a zebra as "horse" and "stripes," zebra would have intrinsic meaning, because it inherits the grounding from the combination of symbols. Harnad proposes that one mechanism for extracting invariant properties uses a connectionist network. On this view, cognition as a whole is best modeled by a hybrid symbolic-connectionist system (see chapter 2, "Hybrid Models"). However some concepts, expressed in language, appear to have no basis in perception (e.g., the notion of permission or obligation). It is likely then there are innately specified primitives or subconcepts (e.g., Johnson-Laird, 1993) which underlie our ability to think about the world.

garden? Because friends will be coming to stay in their house and they want them to enjoy their stay. Social cognition requires understanding the mental states of others. On evolutionary grounds we might suppose that people are equipped to develop the capacity to represent the mental representations of others about some object or event. It helps here to distinguish different orders of representation. People's knowledge of the world, expressed in propositions about it (e.g., "dolphins live in the sea") can be tested against reality and they can then revise their understanding in the light of such tests. These are **first-order representations** (Leslie, 1987). In contrast, a proposition such as "Sailors believe that mermaids live in the sea" is not refuted by knowledge that mermaids do not exist. The belief remains intact. This is termed a second-order representation (Leslie, 1987). Equally, the child's pretence that the banana is a telephone is not refuted by pointing out that the banana is really a banana. When the child is pretending she has to insulate her pretence from reality in order for the play to go on. Similarly, people distinguish between what they know to be the case and what others' believe to be the case. The ability to form such representations is crucial to an understanding of the mental states of other people. The ability to mentalize in this fashion is also important if people are to understand the utterances of others. It was a difficulty or an inability to form such **second-order representations** that was postulated, and demonstrated, to underlie a major difficulty for those born autistic (e.g., Baron-Cohen, Leslie, and Frith, 1985).

Besides recognizing that there are different orders of representation, we also need to recognize that there are different types of representation. Perhaps this notion of different types of mental representation is best grasped by considering a simple task. Please read SAQ 1.3.

SAQ 1.3 Where is your bed relative to the door in your bedroom?

In answering this question, most individuals call up some image of the room and "inspect" that image. Mental images have properties that distinguish them from representations of the words that someone just used which might have a more sequential character or be based more on meaning: can you remember Dad's last utterance in the scenario?

Detailed models of how tasks are performed specify the steps or processes by which they yield some observed behavior. For Simon and for Newell these representations consist of symbols (e.g., our mental representation of numbers) and they are changed by the action of rule-like processes (e.g., a mental process that adds one to a previous number). For others the crucial property of mental representations (not inconsistent with the previous view) is that they correspond to the structure of situations in the world. Such a representation is called a mental model. Your mental image of your bedroom is the perceptual form of an underlying mental model. This model contains tokens representing the objects and their properties and the relations among them. Box 1.4 provides further background on mental models.

For yet other researchers, the idea that a mental representation consists of discrete symbols is too restrictive. In memory research, for instance, the traditional notion of a memory trace, refers to a continuous representation: one for each memory. Recent work using connectionist models goes further: it allows the memory for some events to be distributed across the network so there is not even a separate trace. This mode of operation is still consistent with the notion of mind as a representational system, it is just that the representation is in terms of activation patterns and weights of the connections. The representation still symbolizes. As a brief example of work in cognitive science illustrating processes and representations, box 1.5 outlines the work of Kosslyn and colleagues on the nature of mental imagery: this work moved from an intuitive analogy to the specification of mental processes and representation.

Strategies for Cognitive Science

Given the goal of cognitive science, how should research proceed? There are a number of possible strategies. We could simply develop models of interesting phenomena in the hope that we can learn something more general about the mind. For some this is as useful as setting off in a random direction in the hope that you will arrive where you want to be. You're guaranteed to arrive somewhere – it may be where you started. Alternatively, we could aim for generality from the start. Some researchers have proposed cognitive architectures that can be programmed to perform different tasks. However, to be useful such architectures (for example those proposed by Newell or by Anderson) must place constraints on how a given task can be performed, otherwise any solution will do. In fact, so far, the constraints have been rather minimal, precisely because of the need to achieve generality! The view presented in SAQ 1.4 poses a challenge to these and other strategies discussed below.

SAQ 1.4 Some have argued (e.g., Edelman, 1992) that it is only by understanding the nature of brain mechanisms that we can make good hypotheses about the workings of the mind. Can you think of arguments for and against this view?

A rather different strategy is to build simple creatures and then try to develop more complex creatures. These researchers aim to build systems that behave in the world. To date they have concentrated on such competencies as moving around and collecting objects. The proponents of this approach (e.g., Brooks,

Box 1.4 Background to the notion of a mental model

In everyday usage, we use the term model to refer to different kinds of entities and relations. On display in a shop window, you see a television set. It functions as an exemplar or model of that type of television, identical in all respects to the one you might purchase except that it is on display. An engineer might construct a scale model of a bridge which might be made of the same material as the real bridge. It is similar in critical respects to the real bridge but differs from it in scale. What of a map? It is made of different material. Nor does it look like the territory: it is a two-dimensional representation. It might contain roads marked in red and contour lines with altitude measurements. Neither type of line is found on the ground. But the map provides a model of certain critical features of the terrain, which can be used to consider possible actions and to reach certain decisions. Distances on the map correspond to distances on the terrain. In the absence of a bridge, crossing a river in a car may prove impossible. Further details may signify marsh or forest; snow-line or scree. If you know the conventions you can read the map and hence plan a route or a trip which takes account of the conditions. A map corresponds to the territory in such respects.

Craik (1943) proposed that a central function of thought was to construct such models or maps of reality. In Marr's terms this provides a description of the task of thought. Like the external models described, such mental models have a similar relation-structure to the situations they represent. The essential feature is that the model symbolizes a state of affairs. If an organism can construct a small-scale symbolic model of external reality and of its own possible actions then it is possible to try out alternatives mentally. Future possibilities can be envisaged and planned for. Of course, any plan is only a guide to action and cannot specify in detail what must be done. Actions are carried out in the here and now and are sensitive to the conditions that obtain, that is to the perceptual model of the actual state of affairs (Suchman, 1987).

Our models of the world need not be exact either. For instance, some individuals treat a thermostat as a valve. They believe that turning the dial up increases the amount of heat coming into the room in the same way as depressing the foot on the accelerator in a car increases the amount of fuel being burned. An expert's model, by contrast, treats the thermostat as a self-regulating device: if the temperature is below that desired then the heating system is turned on until the room temperatures is right. All else being equal it produces heat at a constant rate whether the temperature has fallen by 5 degrees or by 20 degrees until the right temperature is reached. An incorrect valve model of a thermostat may still lead to actions that may be energy-efficient (Kempton, 1986). Individuals with this model may set the thermostat low at night to save energy. Others with the self-regulating model may be inclined not to set it lower on the grounds that extra fuel will be needed to reheat the house in the morning.

Our model of some phenomena need not be complete either. A person using a TV set only needs to know how to switch it on, alter its brightness or sound level, or change channels. The inside can be treated as a "black box." Our models have to be sufficient for the purposes in hand. In general, they are likely to be simple representations of relevant properties (Simon, 1969).

Box 1.5 From intuitive analogy to the specification of mental processes and representations

As an example of the approach we consider the work of Kosslyn and colleagues (Kosslyn, 1980). Kosslyn was interested in mental visual imagery. One practical function of imagery is that it allows us to envisage alternative actions in the world: can we put the jam jar between the coffee pot and the sugar bowl? Kosslyn suggested a fruitful analogy: **mental images** are like displays on a cathode-ray tube generated by a computer program. This analogy distinguishes between the mental image in some visual buffer that we are aware of and the information in memory which we use to construct the image. How do we search for information in an image? Kosslyn supposed that an image can be scanned and that this scanning operation follows a straight path across the image and occurs at a constant rate. If this is so, what can we predict? Let us consider a specific task.

In one task, Kosslyn asked individuals to learn a map of a mythical island comprising various objects: a hut, a tree, a rock, a well, and a marsh. Individuals drew the map and were eventually asked to image it. On a given trial in the experiment, individuals were asked to focus on the location of one of the objects. A word was then presented which named an object at another location on the map. The task was to press a "yes" button when the named location had been reached mentally. This time interval (mental scanning time) was recorded. On half the trials the named object was not on the map. On these trials, individuals pressed the "No" button. What specific prediction can we make? Let us assume that the mental image retains the spatial structure of the external map. In this case, the time required to scan mentally between one location and another should be directly related to the distance between the two locations on the external map: the greater the distance the longer the scanning time. And this was the result. The graph of the relationship between mental scanning time and distance was a straight line.

Such a finding suggests that visual imagery is a result of a representation in a visual buffer. This buffer is limited and has two-dimensional spatial properties. Kosslyn went on to show that a variety of special procedures operate on this buffer. For example, individuals can zoom in on some part of the image filling the entire buffer with that part of the image. Individuals can also mentally rotate an image to check its correspondence with some other image. Kosslyn went on to specify the nature of these various procedures: so, for example, ROTATE takes an image as input, moves all points in a bounded region in a specified direction around a pivot and outputs the re-oriented image.

More generally, Kosslyn's account distinguishes processes concerned with storing information about the nature of objects, processes concerned with generating an image on the basis of such information and processes concerned with inspecting the resulting image. Research on brain-damaged patients suggests that each of these processes can be separately impaired. Some brain-damaged patients, for instance, cannot generate visual images despite knowledge of their visual properties.

1991) argue that it simplifies cognitive problems. Because the system, like a person, moves around in the environment, the world itself can be used as an external memory. It can remind the system which tasks still need to be performed. Given these kind of competencies, higher-level processes (such as planning) are not required to specify how an action is carried out. It remains to be seen, however, whether such systems can be scaled up to consider problems that require the use of complex knowledge.

There is one strategy that practically everyone agrees is critical to progress. It is this: in order to understand the mind, we need to understand the nature of the problem(s) it is trying to solve. Marr (1982) drew the following analogy: in order to understand how a bird flies our theory cannot be based solely on the study of feathers and wings. We also need to understand what it takes to fly: we need an understanding of aerodynamics. Only in this context, can we really appreciate the nature of feathers and wings. In developing explicit accounts of mind, Marr urged that we need to formulate a view about the nature of the task to be performed. We can then go on to specify a set of explicit procedures (i.e., algorithms) that could be used to achieve the task. We can also explore the nature of the brain systems involved (that it is the implementation of these procedures).

The work of Newell and Simon shows a clear appreciation of this need to understand the task. They expressed their understanding in terms of the notion of a problem space that captures the set of ways by which a given task could be solved. Even researchers, such as Gibson (1977 – see devil's advocate box 1.2), implacably opposed to theorizing about mental processes, acknowledged the need to understand the nature of the task.

This question of the task can be considered in a slightly different way. We are products of evolution. We can ask: what kinds of problem did the mind evolve to solve? Cosmides, Tooby, and Barkow (1992) suggest that this evolutionary strategy has a further advantage: it leads to accounts that are compatible with the data from other relevant fields such as evolutionary biology. They note that chemists do not propose theories that violate the physical principle of the conservation of energy, so cognitive scientists should envisage accounts consistent with our knowledge of evolution.

What does such a position imply with regard to the design of the human mind? Cosmides et al. argue that a crucial fact has to be recognized: the thousand or so generations since the appearance of agriculture, 20,000 years ago, is less than 1 percent of the two million years that our ancestors spent as hunter-gatherers. They claim, therefore, that the human mind fundamentally evolved in response to that way of life and not to the demands imposed by modern technological culture. Such a view also suggests a single and universal design to the human mind, despite varying levels of current technological sophistication. We should resist being too carried away by this view, since we know from work on artificial selection that even a few hundred generations can transform a species, as for instance in the differentiation of St Bernards and dachshunds from the basic dog form.

It is plausible to argue, though, for the gradual emergence of language capacity as conveying an evolutionary advantage, both because it allowed the

Devil's Advocate Box 1.2

In Gibson's ecological approach, human agents are coupled to the environment such that there may be little explicit **mental representation** of the world at all: action is somehow directly linked to perception (see also Winograd and Flores, 1986). It is undeniable that, at least in some skills, such as driving a car, we may move, as we become more expert, from a circumstance where we are aware of trying to control the car, to a circumstance where we are simply aware of driving along the road. The world appears to "afford" possibilities for action. A curve in the road "affords" turning the car wheel. But such a subjective experience is compatible with the views of Newell and Simon. Perceptual information can contact already specified symbols denoting certain actions, such as turning the car wheel. There can be perception–action links that are mediated symbolically and we may be unaware of them. We are simply aware of the products of that linkage: the position of the car on the road.

Allied to the ecological approach is the idea that individuals can perform complex tasks with little or no planning. A plan, though, is rarely a specification of a fixed set of actions. It may consist, instead, of a hierarchy of more or less detailed specifications (e.g., get money, buy food at supermarket). At a high level, it suggests an appropriate path to a goal. It might include possible actions to consider in the event of a major problem. In risky situations, thinking through possibilities, and envisaging steps to recover, is a way to reduce the risk to life and limb (if the canoe veers over to right, immediately paddle backwards).

sharing of expertise (Pinker, 1994) and the opportunity to learn about other members of the social group through gossip – a means more efficient than direct observation alone (Dunbar, 1993). Evolutionary pressures arose not simply from the physical environment but from the need to engage with other members of the social group. As noted above, human social interaction requires a capacity to represent the intentions of others and Cosmides, for instance, has proposed that we are equipped with Darwinian algorithms for detecting cheaters in social exchange.

An evolutionary point of view, then, can suggest the kinds of tasks that the mind evolved to solve. It might suggest that humans are innately endowed with specialized systems for handling different kinds of input, such as the recognition of human faces. On the other hand, the ability to read must recruit systems evolved for other purposes. Of course, given the possibility of specialized subsystems, we need to know how they work: how exactly are faces recognized? How are words read?

The key insight of this approach is that we can better understand human action by assuming that it is adapted to its environment (Brunswick, 1956; Shepard, 1981). Now it might be argued that its design should ultimately be justified in terms of an evolutionary criterion, such as the number of surviving offspring. But it is difficult to make such a measure workable in areas such as perception, memory, and categorization; though cognitive efficiency might enhance reproductive success by improving the ability of individuals to compete with other people to secure a mate and to raise offspring who will also reproduce. Shepard proposed instead that perception has been optimized through evolution to make the best possible inferences about the world, given the perceptual data. Anderson (1990) has suggested that the same might be true of memory and categorization.

Anderson focused on the principle of rationality: the cognitive system operates at all times to optimize the adaptation of the behavior of the organism. Anderson's rationalist strategy is consistent with the evolutionary strategy but, rather than concentrating on the world of hunter-gatherers, it suggests posing the problem at a more abstract level. What is the structure of natural situations? One abstract property seems evident: our natural and social worlds are to some degree probabilistic. Given certain cues, we cannot know for certain how things will turn out. The system should, therefore, be capable of updating its estimates in the light of experience. In the case of memory, the system would be behaving optimally if it made more available those memories that are more likely to be needed, that is if it mirrored the structure that exists in the environment.

Other researchers (Sperber and Wilson, 1986) have proposed that the cognitive system is designed to maximize the relevance of information processed. We are designed to allocate attention to the most relevant information and to process that information in the most relevant way. As in Anderson's view, the actual cost or effort involved in achieving a cognitive outcome is also an important consideration. When extended to the language context, this principle provides a means by which individuals can infer the meaning of an utterance from its context. It allows us, for instance, to appreciate Dad's utterance to Lucy "Toast! Lucy! It's Etna in here", even if we do not know that Etna is a volcano.

The sense in which we optimize needs some elaboration. In making decisions, for example, we are not very good at keeping all the relevant options in mind. We need to recognize that our rationality is "bounded;" we are cognitive satisficers, generally prepared to accept the first reasonable conclusion (Simon, 1969).

Furthermore, although evolution provides general constraints, how we actually think depends on what precise task we are trying to solve, and, of course, this can be shaped by the tools we are using. Tools change the nature of the tasks people perform. Task analysis seems, therefore, to be a prerequisite for effective theorizing.

Evolutionary considerations do, however, suggest that the human mind is likely to be modular, that is to comprise a set of systems specialized for different purposes. Indeed, it seems likely that decomposability is a fact of our world (true of nature and of human artifacts, such as watches, cars, houses, and computers) and mental systems. They all contain special parts. We will be exploring how some of these virtual machines or, functional systems, work. But we also need to keep in mind the fact that they generally work together. Work on this problem is only just beginning (Sloman, 1993). A solution to it may account for the apparent unity of our subjective experience. Indeed, consciousness itself may have evolved to handle a set of subsystems operating in parallel (see, for instance, Mandler, 1975; Shallice, 1991).

The Methodology of Cognitive Science

We have discussed a number of possible research strategies and have adopted the view that at least three levels of explanations are required in order to provide a full causal account. In many cases, however, the neural or biological basis of tasks is unknown and so our focus is on the cognitive level of explanation and the nature of representations.

At the core of science is the process of formulating theory. Theories rarely emerge full-blown, rather they develop from certain ideas or conjectures over time. Such development might arise as a result of the attempt to test ideas experimentally. Consider, for example, a test of the notion that individuals use visual images to solve certain kinds of problem. Waking from sleep one morning, Roger Shepard saw some forms rotating in his mind's eye. This provided an impetus to design an experiment that tested people's ability to match forms. On the assumption that individuals can mentally rotate forms then the time needed to judge whether two forms are the same or different should depend upon how many degrees they have to be rotated before a match occurs. On the other hand, if matching is based on some more abstract description, then forms which differ by a single description (e.g., arm at top or arm at the bottom) should be matched very quickly. In fact, the amount of time subjects needed to determine that the two forms matched, depended upon the number of degrees through which the forms had to be rotated. In essence, an experiment aims to distinguish between two or more proposals about the way things work. Once we have some reasonable notion of this, it then becomes possible to specify the kind of information-processes that are needed. In the case of mental rotation, there must be some spatial representation of the form and procedures that find a way to map discrepancies between them (see also box 1.5).

Experimental Methods

Part of the methodology of cognitive science, then, is the use of the experimental method (a) to identify the kinds of behavior that theory has to explain and (b) to discriminate amongst alternative theoretical proposals. It is important to recognize that a theory is always "underdetermined" by the data. That is, there are always other theoretical possibilities consistent with a given set of data. Given a model of an experimental result, how can we tell whether or not it is the correct one? In order to make distinctions between different possible models, it is necessary to include additional measures of performance. Massaro and Cowan (1993) give an example about how children add two single-digit numbers. The look-up model supposes that children of a certain age use a table which has the sum for each combination of digits. The counting model envisages that in adding 6 + 3, children choose the large number (6) and then count from this number in successive units to the value of the smaller number (6, 7, 8, 9). Both models predict that 6 + 3 will take more or less the same time as 4 + 3. But only the counting model correctly predicts that 7 + 1 will take less time than 7 + 3.

In order to progress, then, we have to develop theories that make specific predictions and to develop specific experiments to test them. Progress is achieved by proposing and testing specific conjectures: if a precise and specific experiment proves the prediction or conjecture false then the model or theory is wrong. In order to test a theory, then, it is necessary to control all factors which may affect the result. This is the reason why we can test the predictive power of a theory in the laboratory but not in a naturally varying environment. Even if a theory is found to be wrong, it may remain useful. Science is conservative and rarely discards a well-developed theory, until a much more adequate theory is in place. Furthermore, theories may have a range of convenience: Newton's laws remain useful, even if Einstein's theory of relativity offers a truer characterization.

Computational Models

Cognitive science treats the mind as a machine and thus seeks to express theoretical ideas as computational models that generate the behavior. One good reason for doing so is that such models can show that the theory is sufficiently well- specified to be programmed and hence to run on a computer that has no vested interest in ensuring a given outcome. The process of model building encourages a deep understanding of the theory. Critical to the success of the approach is a clear separation between the theory-relevant parts of the program and theory-irrelevant parts of the program. It is still relatively rare for the computational models of different theories to be submitted to comparative testing. We suggest that the comparative testing of computational models is a second critical aspect of an adequate methodology for cognitive science (see Keane, Ledgeway, and Duff, 1994). These models in turn need to be compared against the results of experimental test.

The Nature of this Textbook

How we learn about a new area depends on what we think knowing the area actually amounts to. Suppose someone thought that the goal of cognitive science was to amass facts about the mind. In this case, their effort would be best directed at memorizing as many facts as possible. Suppose, instead, they thought that cognitive science, like other sciences, is an unfolding story, a process of argument aimed at forming a justified but provisional view of reality. What should they do in this case? In this case, their effort would be best directed at understanding the debates and styles of argumentation and explanation; learning about the methods; and exploring new ways of thinking about the facts. Unlike everyday construction of knowledge, scientific knowledge is consciously constructed and refined. This process of refinement implies the need to go beyond existing beliefs and the opinions of authority figures. For example, Josephson in 1959 who was a graduate student at the time, predicted the possibility of a superconducting current between two closely spaced superconducting metals. The originator of superconductivity theory, Bardeen, disagreed. Yet the predicted effect was observed and Josephson awarded a Nobel prize.

It turns out that studies of the way in which individuals learn about science (see Reif and Larkin, 1991) indicate that many pursue inappropriate goals. They seek to memorize facts rather than understand the issues and debates. They accept knowledge acquired from textbooks uncritically and assume that science provides absolute truths. We hope this text will encourage you to become participants in the social process of constructing the science of the mind. This is the reason why each chapter includes devil's advocate boxes that put some alternative view which might itself be rebutted or circumvented. You may find objections to our arguments too. In addition, we seek to engage you by asking questions at various points. Answers to these SAQs are generally to be found in the neighboring text. Our credo is this: passion for the subject, yes! Dogma, no!

One characteristic of a discipline is a set of exemplars. This textbook provides such a set. We noted earlier that complete causal accounts require the specification of three **levels of description**: the behavioral, the cognitive, and the biological. Complete causal models are rare and so we have chosen to focus on the behavioral and cognitive levels of description and where necessary to point to the interface with the biological level. As the discipline develops, complete causal models will become more common and so a different set of exemplars may become appropriate.

We have chosen exemplars which provide a relatively broad coverage of the discipline and which also link into traditional courses in psychology. In any manageable introduction, it is impossible to cover the entire field of activity and so some topics and issues will not be covered. Each of our exemplars is treated in a separate chapter and the chapters are interlinked in various ways. Each chapter references the scenario in box 1.1. This scenario will have served its purpose if it challenges you to consider how an integrated account of mind in action is possible. The chapters are also interlinked because they reference an overlapping set of concepts and methods. Cross-referencing to certain debates and issues is also provided. As you will come across a number of new concepts we provide definitions of many of these in the glossary. First mentions of glossary items relevant to a given chapter are in bold.

Let us briefly overview the content of this text. This current chapter has the twin aims of introducing you to some of the concepts and issues in cognitive science and describing the nature of this text. Chapters 2 and 3 extend the conceptual base. Chapter 2 introduces different kinds of information-processing or computational models and stresses the need to ensure that programs embody what is critical to a theory. Chapter 3 explores in depth the nature of the notion of modularity and cognitive architecture. Chapters 1, 2, and 3 then provide an introduction to the key debates and methods. We recommend reading them once through at least and returning to them again once certain exemplar chapters have been read. The exemplar chapters have been ordered so that the more perceptual topics precede those of a more central nature: hence it makes sense to read them in this order. Chapter 4 addresses the question of how we can perceive objects and faces. Chapter 5 tackles a comparable job with respect to speech: how do we recognize words in the context of variability in the speech signal? Chapter 6 presents research on the reading of single words as well as introducing certain other theoretical tools such as developmental contingency modeling. The next three chapters are concerned with higher-level processes in language. Chapter 7 focuses on the question of how our knowledge of grammar is represented mentally and provides you with an exposition of the minimalist program. Chapter 8 considers the meaning of words and how we derive the intended meaning of utterances and coordinate our utterances in conversations. Chapter 9 extends this concern with the pragmatics of conversation by looking at the development of communicative competence – an area which is underdeveloped from a computational point of view. Chapter 10 provides an analysis of various memory systems and current efforts to model these systems. Chapter 11 considers what happens when we are faced with problems for which we have no remembered solution: it explores the idea of mental models and the acquisition of expertise. Finally, chapter 12 considers efforts to model how we manage to carry out actions and maintain control of our thoughts.

Summing-up

Cognitive science is a discipline in the process of construction. It aims to provide an explicit account of our capacity to act intelligently in the world. It does so by treating the mind as a machine consisting of different parts. Research is guided by a fundamental question: what problems is the mind designed to solve? This text presents accounts which explain behavior in terms of cognitive processes and representations, and introduces you to some exemplars of the discipline with the aim of actively involving you in the debates and issues.

Further Reading

For an enthusiastic report on the emergence of cognitive science see *The Universe Within* (Brighton: Harvester Press) by Morton Hunt (1982). A well-written and scholarly view of the emergence and development of the discipline is also provided by *The Mind's New Science* (New York: Basic Books) by H. Gardner (1985). *The Computer and the Mind* (London: Fontana Press) by P. N. Johnson-Laird (1988) provides an engaging and lucid introduction to computational issues in cognitive science. For those of you fascinated by the philosophical issues as well *The Science of the Mind*, 2nd edn (Cambridge, Mass.:, MIT Press), by O. Flanagan (1993) provides a very comprehensive and illuminating read. Many of the controversies and debates within cognitive science are colorfully presented in *Speaking Minds* (Princeton, N.J.: Princeton University Press) by P. Baumgartner and S. Payr (1995). Their interviews with twenty eminent cognitive scientists illustrate the diversity of opinion currently in the field.

CHAPTER

2 Explanation and Simulation in Cognitive Science

Outline

This chapter focuses on the structure of theories of cognitive processes and considers how computer programs may be developed to simulate the behavior of such theories. We discuss a number of approaches to simulation, each based on different assumptions about the computational basis of cognition, and give an overview of the ongoing debate between proponents of two major approaches. Also discussed is the strategy of embedding specific models within a model of the complete cognitive apparatus (a "cognitive architecture").

Learning Objectives

After reading this chapter you should be able to:

- differentiate between a flow chart and a box/arrow diagram, and know the difference between the computational assumptions underlying each notation
- discuss some problems inherent in the computational modeling of cognitive phenomena
- compare and contrast several symbolic models
- explain the various mechanisms used in connectionist models
- describe the fundamental differences between symbolic and connectionist models
- appreciate the relative strengths and weaknesses of symbolic and connectionist models, and discuss how these are addressed by hybrid models
- debate the role of cognitive architectures in computational modeling

Key Terms

- Cognitive architecture
- Cognitive theories
- Computational modeling
- Connectionism
- Hybrid models
- Parallel Distributed Processing
- Symbol systems

Introduction

One of the central tools identified in chapter 1 that is available to the cognitive scientist is the construction of **computational models**, that is, the use of **simulation**, in the development and testing of theories of cognitive processes. Thus, if we have a theory of how Zara acquires language, or why Lucy decides to use the bread board as a tray, one thing we might do with that theory is implement it as a computational model and execute that model to (hopefully) simulate Zara's acquisition of language, or Lucy's problem-solving skills. Simulation ensures that our theories are fully explicit, and allows us to be confident of the predictions or consequences of complex theories. But what sort of theories are amenable to simulation, and how do we go about producing such a beast? What techniques might a cognitive scientist use in conducting the simulation? Are there special considerations to which simulation in cognitive science (as opposed to simulation in, for example, weather forecasting) must attend? These are the questions we will attempt to answer in this chapter. We begin though by looking at cognitive scientific theories, for these are the things we wish to implement. What, then, does a cognitive scientific theory look like, and what background assumptions do such theories make?

Theories and Diagrams

Cognitive scientists have used a number of different methods for presenting their theories. One of the most common approaches is to present a theory as some form of diagram consisting of boxes joined by arrows and annotated with a natural language description (see figure 2.1 for two examples). Frequently, such annotated diagrams are developed into computational implementations, but the diagrams themselves are an important way of expressing theories of cognitive processes.

The diagrams in figure 2.1 actually belie two very different approaches to theorizing as the assumptions underlying each are radically different. We refer to figure 2.1(a) as a **box/arrow diagram**. In this form of diagram, boxes are used to represent postulated cognitive processes and arrows between those boxes used to represent communication, or flow of information, between those processes. These sort of diagrams are to be distinguished from **flow charts**, see figure 2.1(b), which are often used to express procedures or algorithms. In flow charts, the boxes represent operations or decisions rather than processes and the arrows represent flow of control. Flow charts may be used entirely descriptively, to describe a series of operations performed during some process, but generally the implication behind them is that some general-purpose processor (like a desk-top computer) carries out the operations represented by the boxes in sequence, just like a computer program. Indeed, a major use of flow charts is in program development. Box/arrow diagrams, in contrast, treat each process as a functionally distinct information processing "module" (see chapter 3), and there is no commitment to a single controlling process. Arrows in these diagrams indicate possible routes for information flow between modules (rather than routes for the flow of control), and as such the lack of an arrow between two boxes, indicating that the boxes don't directly communicate, means just as much as its presence. Note that neither notation makes any necessary commitment to the underlying structures that implement the various boxes (they may or may not, for example, be distinct regions of the brain: see Mehler, Morton and Juszuck, 1984).

Although the distinction between the two forms of diagram is important, it is often ignored. This may be because it is sometimes difficult to distinguish between the two. Theorists often use such diagrams without explicitly stating their assumptions, and even when it is clear that they intend a diagram to be interpreted as a box/arrow diagram rather than a flow chart, other assump-

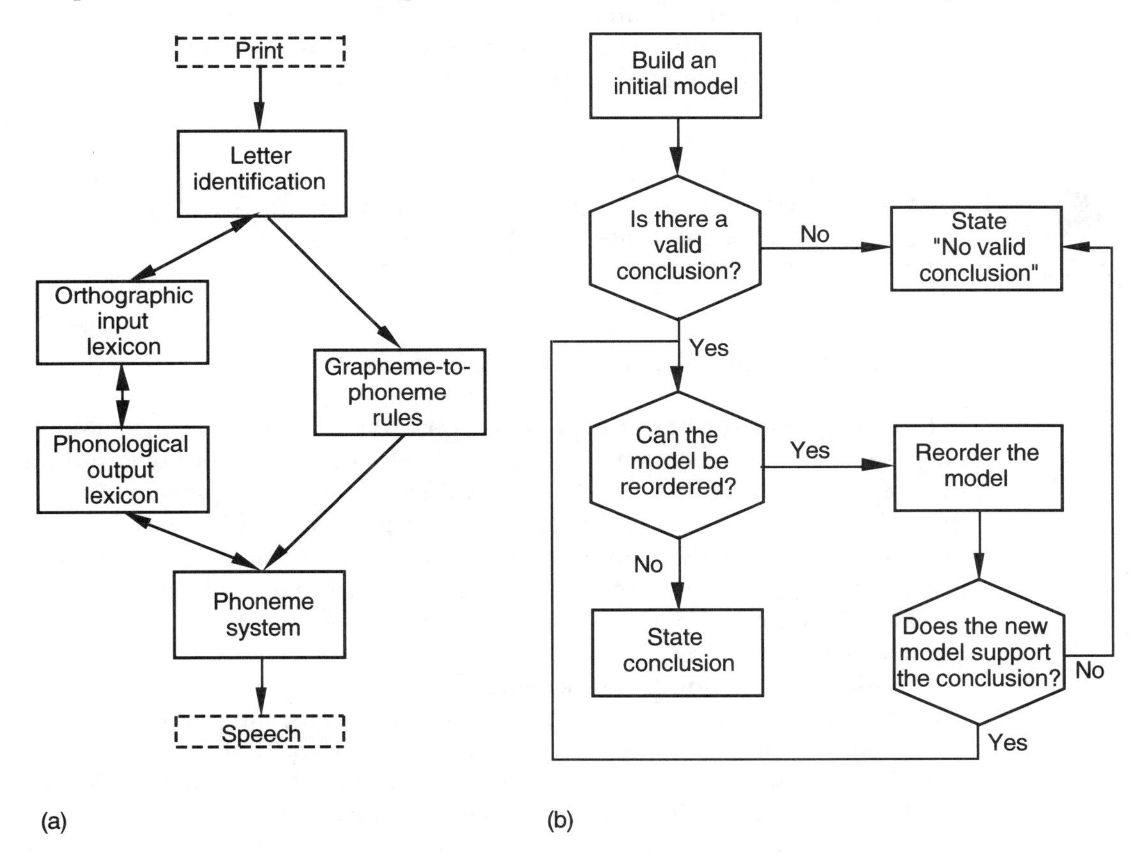

2.1 (a) A box/arrow diagram for reading (see ch. 6); (b) A flow chart for model based reasoning (see ch. 11).

tions may remain implicit. Thus, it may be necessary to learn the assumptions of a particular theorist from his/her accompanying descriptions. One question that must be answered concerns whether the relationship between two connected boxes represents a stage relationship or a cascade relationship. In a stage relationship all the operations of the first process are completed before the second one takes over. In a cascade relationship, partial data may be passed through to the second process before the first has completely finished its task, allowing the second process to begin working on incomplete data.

Once we have a diagrammatic representation of a theory, and we are confident of the intended interpretation of the diagram, we can think about how to implement the theory in terms of a computational model. For this, it is necessary to fill in the boxes with appropriate computational processes. There is little commitment within the diagrammatic tradition as to how this should be achieved. In the following sections, we will consider a number of different approaches to computation and it will be seen that any of these may be adopted for this task. Approaches may even be mixed in implementing a single diagram. All that is important, from the box/arrow perspective, is that this fleshing out should preserve the pattern of connections (and disconnections) expressed by the original diagram, for this is the theoretical content of the notation.

SAQ 2.1 Draw flow charts for the various processes involved in the breakfast scenario (e.g., preparing coffee, speaking on the telephone, making toast).

How does Simulation Assist Explanation?

Before we go any further, it is perhaps appropriate to consider in detail the purpose of simulation. Often the theories developed by cognitive scientists are so complex that it is difficult to be sure of their precise consequences. Even so, these theories are intended to explain some aspect of behavior, and, typically, to make predictions. Simulation then – the encoding of theories as computer programs, the execution of those programs, and the comparison of the resulting behavior with empirical data – is one approach to verifying the claimed behavior.

The use of simulation is not peculiar to cognitive science. Simulation plays a key role in the study of all complex dynamic systems, ranging from animal population modeling to weather forecasting, and the mind is certainly a complex dynamic system. Cognitive science is not, however, amenable to a standard, run of the mill, application of generic simulation techniques. Special techniques are required to address problems specific to the field. Two issues of paramount importance concern the relation between theory and implementation and the interpretation of simulation results. The first of these arises from the abstract nature of psychological theorizing. Psychological theories are rarely specified in the detail required for computational implementation. The second, that of the interpretation of results, arises from the fact that simulations rarely cover perceptual, cognitive, and motor processes. Simulations generally only concern limited, central, aspects of cognition, and these aspects are often only indirectly evident in outward behavior. In order to interpret simulation results as behavior it is generally necessary to make assumptions about the processes not covered by the simulation.

The abstract nature of psychological theorizing may present a problem for computational simulation, but the act of developing a computer simulation can also be used to help isolate poorly developed aspects of a theory. Because simulation requires a complete specification of a theory, it is not possible for a theorist to implicitly assume certain properties of theoretical mechanisms or to neglect to completely specify some critical process. Thus, the act of implementing a theory as a computer program can force a degree of precision which is sometimes lacking in cognitive psychology.

Simulation may also be used in a more experimental way in theory development by first implementing an initial version of a theory and then adjusting the implementation to produce a simulation which captures all the subtleties and nuances of the empirical data. In its basic form this is not a particularly sound way of developing a theory – an arbitrary program that replicates some aspect of behavior is only instructive if we know how or why the program replicates the behavior – but if the adjustments are systematic and theory driven, such that successive simulations may be seen as successively better approximations to a final theory, with each approximation generating further testable predictions, then the empirical approach to theory development can be genuinely productive.

Symbol Systems

Many "higher-level" cognitive functions, such as problem solving, reasoning, and language, appear to involve the explicit manipulation of symbols. Problem solving, as it occurs in playing a game of chess or making a shopping plan, for example, seems to involve the mental representation of relevant aspects of the world, or partial "models" of the world, and the systematic manipulation of those models. Thus when playing chess, a novice generally considers various moves by imagining how the board would look if he/she were to make those moves. People's abilities in domains such as game playing and planning have lead to the development of computational models of the underlying cognitive processes, models which make explicit use of discrete symbolic representations. The domains of problem solving, reasoning, and language are all discussed in some detail in later chapters. Here, we consider first the nature of symbolic representations, and then a number of different symbolic computing machines which are often used in symbolic modeling.

Representation in Symbolic Models

Central to the enterprise of symbolic modeling is the assumption that there exist such things as **mental representations** and that those mental representations are structured, semantically interpretable, objects. For a representation to be structured and semantically interpretable, that representation must consist of symbols which have parts, and the parts must have independent meanings which contribute to the meaning of the symbols which contain them. The symbol "34", for example, has parts (the symbols "3" and "4"), and the meaning of "34" is a function of the meaning of "3" in the tens position and "4" in the units position. The arabic representation of numbers, then, is a structured, semantically interpretable representation.

Fodor and Pylyshyn (1988) give several reasons for supposing that structured mental representations play a pivotal role in cognition. Firstly, they cite the productivity of thought. Even with only a finite number of basic concepts we can mentally represent an infinite number of thoughts. How can this be, unless thoughts are structured, semantically interpretable objects? Secondly, they note that cognitive representations are **systematic**. If we can have the thoughts that Zara is hungry and that Mom is tired, then we can also have the thoughts that Mom is hungry and Zara is tired. We can systematically exchange like terms in mental representations to yield new mental representations. Thirdly, they point out that cognitive representations are **compositional**. The meaning of a thought is a function of the meaning of its parts (and their mode of combination). Thus, the thought that Zara is tired is a function of the meaning of the thought "Zara" and the meaning of the thought "is tired." Finally, they argue that inference is rational. If we know that A and B is true, then we may infer that A is true, irrespective of what A may stand for.

Models of Symbolic Computation

Given symbolic (i.e., structured, semantically interpretable) representations, how might we manipulate them? The answer from computer science is: with a computing machine. A wide variety of such devices have been developed for different tasks. What they have in common is a **representational system** (for expressing input and output), a processing strategy, and (sometimes) a set of predefined machine operations (a machine language). To clarify these concepts, we now consider a number of computing machines that have been used for symbolic modeling within cognitive science.

Finite state automata

The simplest interesting symbolic computing devices are **finite state automata** (FSA). These machines take as input a finite sequence of symbols and classify that input on the basis of its structure. In doing this they make use of a state, which is effectively a memory for a single item of information. The machine processes its input in a cyclic fashion, taking one input symbol on each cycle and using this symbol and its current state to generate a new state (i.e., alter its memory). It does this by looking up a table of state transitions, which show for each possible state and input symbol, the machine's resultant state. Thus, as

Box 2.1 Propositional knowledge representation

One of the central areas of Artificial Intelligence is that of the representation of knowledge. In order to develop computer programs which process information, it is necessary to specify information in a precise and consistent manner. A number of knowledge representation schemes have been developed, and in general different schemes are appropriate for different applications. Here we consider just one scheme, that of **propositional representations**.

Within this scheme, the central idea is to express all information in terms of properties of objects or relations between objects. Thus, we might represent the state of a chess board by a series of propositions of the form:

```
position(white-pawn, c3).
position(black-rook, d4).
```

A representation of a more complex situation, such as that of our breakfast scenario, would involve a number of different relations. To start with, we might have:

```
is-a(t1, tray).
is-a(d1, door).
behind(t1, d1).
is-a(j1, jar).
is-a(j2, jar).
contains(j1, jam).
on(j1, t1).
contains(j2, peanut-butter).
```

There are two important things you should note about this example. Firstly, we've used the symbols "t1", "j1", etc. for objects. If we had used a proposition like "contains(jar, jam)," we wouldn't have been able to distinguish between the two jars. Secondly, you should notice that different relations hold between different types of things. Thus, "is-a" holds between an object and an object-class, whereas "behind" holds between two objects.

SAQ 2.2 Imagine the scene in our kitchen at a particular instant during breakfast. Try to describe the scene in words, and then convert that description into propositional terms by completing the above list.

Computer programming languages (most notably Lisp and Prolog) now exist which allow the direct use and manipulation of these sorts of propositional representations.

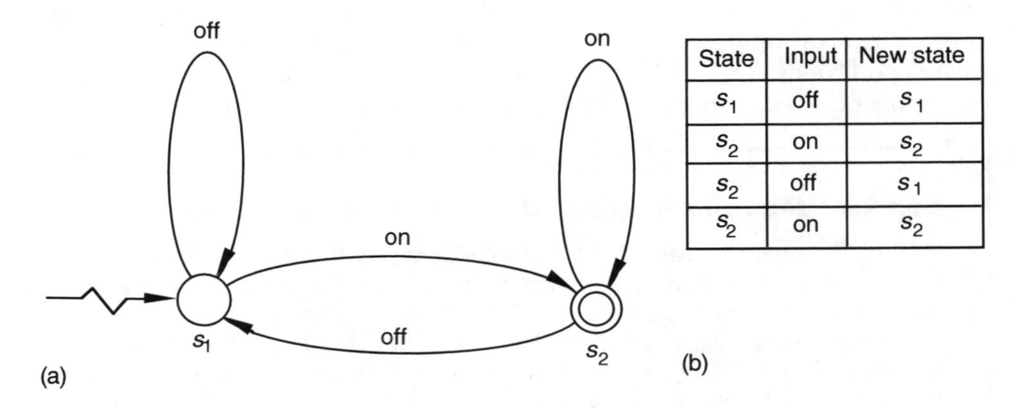

State	Input	New state
s_1	off	s_1
s_2	on	s_2
s_2	off	s_1
s_2	on	s_2

2.2 (a) A Finite State Automaton; (b) Its state transition table.

processing continues, the machine progresses through a sequence of states while using up its input. Certain states are termed "accept" states, and if, when all the input has been used up, the machine is in such a state, then the machine is said to have classified the input sequence as acceptable. Otherwise the machine is said to have rejected the input sequence.

Figure 2.2(a) shows a diagrammatic representation of a finite state automaton. An alternative representation would be in terms of a state transition table, as shown in figure 2.2(b). In figure 2.2(a), states (s_1 and s_2) are represented by circles, with the initial state being tagged by a zig-zag arrow, and the accept state (which is the only other state in this case) being marked by a double circle. State transitions are shown by labeled arrows between states, with the label indicating the input symbol under which the corresponding state transition is allowed. This particular finite state automaton takes as input a sequence of **on** and **off** messages, and accepts those sequences which end with **on**.

SAQ 2.3 What is the sequence of states (i.e., s_1, s_2) that the FSA in figure 2.2 goes through as it processes the sequence "on off off on on on off off off on."

Push down automata

If a finite state automaton is in a particular state, and it encounters a certain symbol, its state will always change to that specified in its state transition table. A **push down automaton** (PDA) has, in addition to a state, an infinite (though limited in access) memory, and the resultant state can be dependent on the contents of this memory, as well as the current state and input symbol. The memory of a push down automaton is organized in the form of a **stack**. A stack (or push down stack) is a storage device with a "top-most" element. Elements may be added to the top of the stack (or "pushed" onto the stack) at will, but the stack may only be accessed from the top. Thus, only one stack element is available at any time, and recovering an element that has been buried in the stack may require "popping" several items off the stack.

SAQ 2.4 A stack may seem like an artificial device, but examples of stacks in everyday life abound. A tray dispenser in a canteen, is one common example. Try to think of a few other examples of stacks.

The addition of a stack to a finite state automaton gives a push down automaton substantially greater computational power, in that it increases the range of tasks of which the automaton is capable. Checking that brackets in a text match (i.e., that there are equal opening and closing brackets, and that no closing bracket appears before its corresponding opening bracket), for example, is one task which requires (at least) a push down automaton. A finite state automaton is incapable of doing the necessary counting of brackets.

Turing machines

A stack is a very limited memory structure. Although infinite, only the topmost element is ever available. A far more power computational device (indeed, the most powerful computational device) is a **Turing machine**. A Turing machine (conceived of by Alan Turing) has an infinite memory in the form of a tape. This tape can be read off, or written onto, and all locations on the tape are open for inspection at any time. Like the previous devices, a Turing machine has a state (which may be any one of a finite set) and a state transition function, but the state transition function is dependent on the contents of the tape location which is currently being examined, and specifies, in addition to a resultant state, an instruction to move the tape to the left or right (or keep it still), and an instruction to overwrite the tape's current symbol with some other symbol.

Turing machines are of enormous significance in the theory of computing. It can be shown, for example, that any task which may be specified in terms of a procedure that is known to terminate (i.e., an **algorithm**) can be performed by a Turing machine with an appropriate state transition function. Universal Turing machines, machines which emulate any other Turing machine given a tape with a description of another Turing machine and its input, can even be designed. As such, many philosophical debates within cognitive science have centered around the theory of Turing machines. Is the human brain Turing powerful (i.e., can it also compute any function, given an appropriate algorithm)? Does the design of the machine which executes a program matter (given that a Turing machine can execute any algorithm)? If not, is there any reason why we should not work in terms of Turing machines (rather than less powerful designs)?

Von Neumann machines

Although Turing machines are able to compute any function, they are not particularly useful as general computing machines. Standard modern computers are generally based on a design by John von Neumann. Strictly speaking such machines are less powerful than Turing machines because they have finite memory, though in practice this is not usually a limitation. What makes a von Neumann machine useful is that it has an area of memory in which it can store a program. The program is a sequence of machine instructions (like "add the contents of one memory to those of another memory"). This ability to store and follow a program means that the machine can be easily programmed to do virtually any task, and this has been the key to the machine's success.

Processing in a von Neumann machine is based on a fetch-instruction/execute-instruction cycle. The machine has a special memory location which points to the next instruction, and on each cycle the machine fetches the current instruction, moves the instruction pointer to the next instruction, and then executes the instruction it has just retrieved.

Production systems

Though useful as general computing devices, few would propose that cognitive processes are the result of a von Neumann machine executing a program.

Box 2.2 Automata and language

As described above, the input to an automaton is a sequence of symbols, and all that an automaton does is accept or reject its input sequence. If the symbols are words (of English, for example), then automata can be designed to accept only those word sequences which are grammatical sentences (of English). This process of classification, when accompanied by a process of building a description of the relations between words in a sentence, is known as **parsing**.

Finite state automata are clearly insufficiently powerful for the task of correctly classifying English sentences. Although they can be designed to accept simple sentences, difficulties arise with sentences with embedded sentences (such as "Dad believes that Zara likes coco-pops," where "Zara likes coco-pops" is itself a sentence). In general, center recursive constructions (e.g., sentences with embedded sentences) require at least the power of a push down automaton, and one of the major debates in computational linguistics during the 1970s and 1980s concerned whether any natural language required more powerful computational machinery. The rationale behind the debate was that if all natural languages could be parsed by appropriate push down automata (and other non-natural languages can be invented which require more powerful computational machinery to parse them), then this would seem to suggest that the human language processing apparatus has the computational power of a push down automaton, and not, for example, a Turing machine. It is now generally accepted that there are natural languages which require more than push down automaton power, and the debate has shifted to other areas of language complexity.

Production systems are an alternative class of machine which have been proposed as the computational basis of cognition. These machines were first introduced to psychology by Newell and Simon (1972) in their work on problem solving, and achieved the height of their popularity in the late 1970s and early 1980s.

A production system is a cyclic processor with two main memory structures: a long term memory containing rules (or productions) and a working memory containing a (symbolic) representation of the system's current state. The rules specify the conditions under which external actions (e.g., add sugar to the coffee) and changes to working memory (e.g., add to working memory the information that sugar has been added to the coffee, and remove the goal to sweeten the coffee) are justified. Rules are intended to embody encapsulated pieces of knowledge, such as "adding sugar to the coffee will achieve the goal of sweetening the coffee." Different production systems use different notations, but in a vanilla flavor production system this rule might be generalized and written as:

IF goal(sweeten(X)) and available(sugar) THEN action(add(sugar, X)) and retract(goal(sweeten(X)))

In ordinary English, this says that if (in working memory) the goal is to sweeten something (X) and sugar is available, then carrying out the action of adding sugar to X will achieve the goal (and hence the goal should be removed, or retracted, from working memory).

A production system's processing cycle consists of two phases. In the "recognize" phase, the processor finds all rules in long-term memory whose conditions are satisfied by the elements in working memory. In the "act" phase, the system chooses one matching rule and executes its right-hand side, thus changing working memory and/or performing some action. The execution of a rule will generally result in further rules becoming applicable, and in this way a production system can, given an appropriate set of condition-action rules (and an initial state), exhibit complex sequences of behavior.

There are two principal issues in the design of a generic production system: the pattern matching mechanism and the conflict resolution strategy. Pattern matching is the process that determines which rules are applicable given the state of working memory. Pattern matchers differ in the complexity of the conditions which they allow. The simplest matchers only allow tests of the form **if A and B and ...** where each element of the condition must match an element of working memory (that is, conjunctive conditions). More powerful pattern matchers may allow conditions of the form **if A or B** (that is, disjunctive conditions) or **if not A** (that is, negated conditions) and complex combinations of such conditions.

Conflict resolution is the mechanism that determines which rule will apply when the pattern matcher finds more than one possible rule during the recognize phase. One approach is to favor specific rules over general rules.

This strategy views rules with many conditions as special case rules, whereas rules with few conditions are seen as general, or default, rules, to be used only when special cases do not apply. A second strategy is to prioritize rules by attaching numbers to them. A rule with a higher priority may then be favored over one with a lower priority. (See chapter 11 for an example of the use of a production system in an explanation of expertise.)

SAQ 2.5 Symbolic modeling seems appropriate for domains such as problem solving, where production systems have been quite successful. Are there any tasks or domains for which you think symbolic modeling would be inappropriate?

Connectionism and Parallel Distributed Processing

The roots of symbolic modeling lie in the theory of computation developed by John von Neumann (1947) and Alan Turing (1950), and embodied in virtually all existing electronic computers. A radically different approach to computation is employed by models that compute through the simultaneous operation of numerous simple communicating processing devices. This form of processing, often termed **connectionism**, takes its inspiration from neurophysiology, with the processing devices (often called units, nodes, or artificial neurons) corresponding to simplified neurons and the communication channels (often called connections or links) corresponding to simplified synapses.

A connectionist network is a set nodes connected in some fashion. Nodes within the network have an **activation level** – a number (usually either between 0 and 1, or between –1 and +1) which varies as the node becomes more or less active. The nodes interact via the flow of activation along the connections, which are generally directed (so activation flows from one node to another, but not back) and also have a number attached to them. This number, the weight of the connection, determines the strength and nature of the interaction between two nodes. The weight may be either positive or negative. If there is a connection from node A to node B and the weight of that connection is positive, then node A will tend to excite node B when node A is active (thus causing node B to become active). If the weight is negative, node A will tend to inhibit node B when node A is active (thus causing node B to become inactive). The weight is thus a measure of the correlation between two nodes, but in general any node will be simultaneously excited and inhibited by many other nodes, with the activation of each node being given by some function of the weighted sum of its inputs. It is therefore difficult (and usually not informative) to consider individual nodes in isolation.

One branch of connectionism which has received a great deal of attention within cognitive science is **Parallel Distributed Processing** (PDP). Indeed, the terms are often taken as synonymous. The distinction between the two forms of computation rests on their representational assumptions. Connectionist networks in general make no assumptions about representation. On the one hand, nodes in such networks may have well-defined interpretations (e.g., a specific node might be active if and only if an input stimulus is the color red, so the node can be interpreted as meaning "redness"). Such representations are often termed local representations. On the other hand, representations may be distributed over many nodes (such that redness cannot be determined by examining a single node) and any one node may participate in many different representations (so one node might be active when either red or orange is present) with interpretation only being possible at the level of whole sets of nodes (so differentiating between red and orange will require looking at the activation distribution over a set of nodes). These representations are termed global or distributed representations. PDP, then, is that branch of connectionism that involves distributed (and not local) representations. Such representations are radically different from those employed by more traditional symbol systems because they lack properties such as systematicity and compositionality.

Classes of Network

There are many different classes of network that fit the above description of connectionism, and in order to tie the description down it is instructive to consider some of these classes in some detail. We consider three in particular: **associative networks**; **feed-forward networks**; and **recurrent networks**. Each of these networks has been used in the development of a number of models of cognitive processes. We also consider how networks may be joined together to produce modular systems.

Associative networks

Associative networks are designed to recognize and reconstruct patterns. Typically, an activation pattern is presented to some subset of the units, and activation is allowed to propagate to other units. The activation pattern of the network as a whole generally settles into a stable state, as the network reconstructs one of the patterns that it has previously experienced.

Within an associative network, nodes whose activations are positively correlated are connected by strong, positively weighted connections. Nodes whose activations are negatively correlated are connected by strong, negatively weighted connections. Nodes whose activation is uncorrelated are connected, if at all, by only weak connections.

One way of viewing an associative network is as a complex landscape, consisting of hills and valleys, with the network's activation pattern at any time being a position in this landscape. Every valley corresponds to a pattern that the network can recognize. Giving the network an input is like placing a ball somewhere in the landscape. As it rolls to the bottom of a valley, the network settles into a stable state, and the valley that it ends up in will correspond to the pattern that the network reconstructs for the corresponding input.

Feed-forward networks

Feed-forward networks do not reconstruct patterns. Instead, they map from one domain to another. In a feed-forward network nodes are grouped into "layers" and activation spreads through the layers in sequence. The simplest such network has just two layers, an **input layer**, at which a representation of the stimulus is presented, and an **output layer**, at which a representation of the response is generated. A layer of activation levels is often referred to as an activation vector, so these networks can be viewed as transforming an input vector to an output vector.

Two-layered feed-forward networks are strictly limited in their computational power. Certain mappings cannot be encoded by such networks, and for these functions it is necessary to chain several networks together. This is equivalent to adding further **hidden layers** between the input and output layers. In a three-layer network (see figure 2.3), input units feed activations to a set of so-called hidden units, which in turn feed activations to the output units. Three layers are sufficient for most purposes.

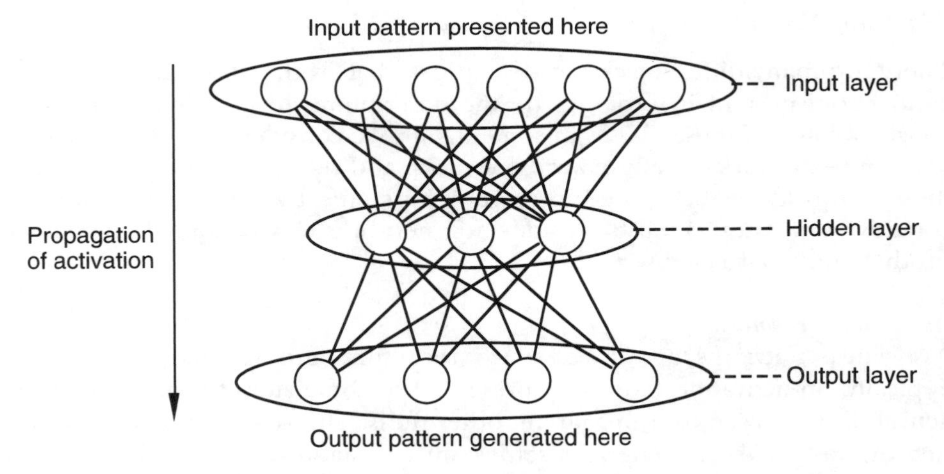

2.3 A three-layer feed-forward network.

Recurrent networks

Standard feed-forward networks are appropriate for mappings where the output is dependent only on the current input. However, they are not able to encode mappings which are dependent not just on the current input, but also on previous inputs. The crux of the problem is that feed-forward networks have no memory. Jordan (1986), and later Elman (1990), demonstrated that such mappings could be performed by adding further units to encode context, with feedback links from either the output units or the hidden units to these

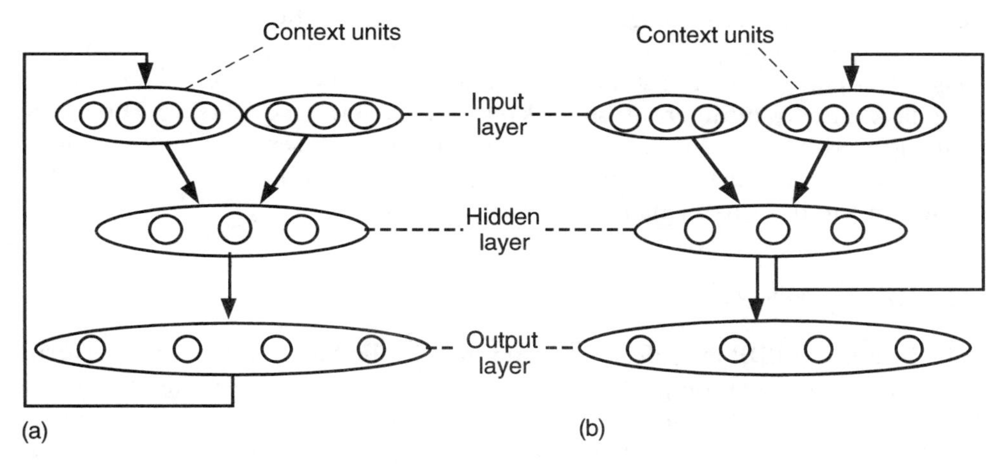

2.4 Recurrent networks of (a) The Jordon type; (b) The Elman type (connections are shown as single arrows between layers).

context units and feed-forward links from the context units into the hidden units. In these networks, known as recurrent networks, context units act as a form of memory for the network's state, and allow the output of the network to be sensitive to its state on the previous input cycle.

Figure 2.4(a) depicts a recurrent network with feedback from the output layer. On each processing cycle, the existing context units feed, together with the input layer, into the hidden layer. In addition, the existing output vector is copied to the context units. The output of the network is thus dependent on the current input and the previous output.

In figure 2.4(b), the recurrent connections come from the hidden layer. Processing is analogous (with the activation vector of the hidden layer being copied to the context units on each cycle), but the hidden layer is able to encode more information than the output layer. In these networks, the hidden units are used to encode not just an intermediate stage in the input/output mapping, but also information about the current position in the output sequence. Consequently, many more hidden units are typically required than in the case where recurrent connections originate only from output units, but it can be arranged that the output of the network is a complex function of the complete input history (rather than just the current and immediately previous inputs).

Recurrent networks have been successfully applied to a number of problem domains, perhaps the most interesting of which is the syntactic analysis of natural language (Elman, 1990; Miikkulainen, 1993). Elman (1990), for example, used such networks to predict the syntactic category of the next word given a partial sentence.

Modular networks

It may be unclear how the networks of the form discussed here relate to psychological theorizing, which, as discussed above, often involves the construction of box/arrow diagrams. One answer lies in the use of networks (of various types) to flesh out the boxes. Thus, Miikkulainen (1993), for example, has developed a model of story understanding based on the decomposition of the task into a number of subtasks (such as sentence parsing and story parsing), with each subtask being performed by a separate connectionist network. The networks communicate by passing activation vectors along paths which in the box/arrow diagram correspond to arrows.

Properties of Networks

The interest in PDP models lies not so much in the fact that they can perform associations and mappings, but firstly in the fact that there are algorithms by which such networks may learn to perform such associations and mappings (and generalize the performance in the process) and secondly in the way that such networks breakdown when damaged.

Box 2.3 An example network: the past tense of verbs

Consider the function that maps the sound of verbs in their base form to the sound of their past tense form (see Rumelhart and McClelland, 1986). This function will map "jump" to "jumped," "kick" to "kicked" and "bounce" to "bounced," but also "give" to "gave," "go" to "went," "teach" to "taught," and so on. So the general rule is to add "ed", but there are a number of exceptions. Within the symbolic paradigm, the obvious way to perform this function would be to list the exceptions and apply the general rule whenever the stimulus isn't listed.

Within the connectionist paradigm, we begin by representing the elements of the input and output domains by feature vectors. Given the nature of the problem, some kind of phonological representation seems appropriate. A simple approach would be to limit our inputs to words of, say, at most 10 phonemes long (and our outputs to perhaps 12 phonemes), and represent each phoneme (English has around 40 of them) in each position by a feature, the value of which will be 1 if the word does have that phoneme in that position and 0 otherwise. Each input word may then be represented by a vector of 410 binary digits, with the first 41 features corresponding to the first phoneme (assuming 40 different phonemes plus one empty phoneme for short words), the second 41 to the second phoneme and so on. Similarly, each output word will be represented by a vector of 492 binary digits (12 positions with 41 phonemes per position). The simplest network which may perform the mapping is a two-layered feed-forward network with one node for each input feature and one node for each output feature (so in this case we have 902 nodes). If we then link each input node to each output node, and weight those links appropriately (see below, where we discuss a process known as learning or training which allows appropriate weights to be determined), then the network will, when presented with a feature vector of an input word at its input nodes, develop (an approximation to) the feature vector of the past tense of that word at its output nodes.

Learning

Learning (or training) within a connectionist network typically involves presenting the network with a series of patterns (the training set) and adjusting the weights on the connections in such a way that those patterns are somehow encoded in the network's weights. Most algorithms involve only small adjustments to the weights on each trial, but require that the complete training set be presented numerous times (often in the order of thousands) before a reasonable degree of learning is achieved.

Different learning algorithms are appropriate for different classes of network. In the case of simple associative networks, learning may be achieved by presenting an activation pattern to the network and adjusting the weight of each connection by an amount proportional to the correlation between the activity of the corresponding nodes. Thus, if for a given pattern two nodes are active, their connection weight should be increased. Similarly, if two nodes are inactive, their connection weight should be increased. The weights between active and inactive nodes, on the other hand, should be decreased. This simple form of learning, known as Hebbian learning, is both highly effective and biologically plausible. However, it is only appropriate for associative networks.

Networks which map input patterns to output patterns (such as feed-forward and recurrent networks) require more complex learning algorithms. The most common approach, backpropagation of error, involves presenting the network with an input pattern, and then comparing the output that is actually produced with the output that should have been produced. The difference between these two patterns is then calculated, and propagated backward (from the output layer) through the network (towards the input layer), with small weight adjustments being made at each step, so that when that pattern is next presented, the network's output will more closely match the correct output. Unlike Hebbian learning, learning by backpropagation is supervised, in that an explicit mechanism is required to compare the output produced by the network with the required output, and propagate the errors back along the connections.

Once trained, PDP networks often exhibit psychologically interesting properties such as interference and generalization. Unlike a symbol system, the capacity to encode information in a network of a fixed size is flexible. If the network has too few units to encode the complete training set (that is, the training exemplars saturate the network), then similar patterns in the network will tend to interfere, and the performance of the network (in terms of the accuracy of its mapping) will be less than optimal. The reverse side of this, however, is that unlike in the symbolic case, where performance would fail if a rule is absent, in the PDP case the network will, if over-saturated, approximate appropriate performance. Networks are also often able to extend the mapping defined by their training exemplars to unseen inputs. This property of generalization was demonstrated by the past-tense verb learning network of

Rumelhart and McClelland (1986), which was able to generate "wept" as the past tense for "weep" and "bid" as the past tense for "bid", despite these verbs not being in the training set. This was possible because the novel verbs were similar to verbs in the training set (e.g., weep is similar to sleep), and the network was able to generalize to the new verbs.

Damage and the PDP modeling of neuropsychological syndromes
A further property of PDP networks, desirable from the perspective of modeling cognitive functioning, is that when damaged (through either the removal of some portion of connections, or the introduction of random noise into the propagation of activation), networks tend to produce behavior similar to that exhibited by patients with various forms of neurological damage. Most notably, and unlike symbol systems, PDP networks degrade gracefully. When a PDP network is damaged it is generally still able to function, though the functioning is somewhat impaired. Hinton and Shallice (1991), and later Plaut and Shallice (1993), for example, conducted extensive simulations of deep dyslexia, a deficit in reading where patients produce various classes of error, including visual errors (reading "cat" as "cot") and semantic errors (reading "cat" as "dog"). A full discussion of the deficit, which typically arises as a result of localized brain damage, is given in chapter 6. The research of Hinton, Shallice, and Plaut involved constructing various networks to perform the reading task (actually the task of mapping from an orthographic representation of a word to its phonological representation), training the networks with a suitable training corpus, and then damaging the network in a variety of ways. Certain classes of network, when damaged, produced errors which were qualitatively similar to those produced by various patients.

SAQ 2.6 In which domains do you think connectionist/PDP modeling is most appropriate? Are there domains for which you think connectionist modeling would be inappropriate?

Symbolic Network Models

Parallel distributed processing is characterized by two features: its use of interconnected units and its use of feature-based distributed representations. Distributed representations require interconnected units because a model without interconnected units could not support such representations, but the opposite is not true: models may have interconnected units without using distributed representations. Indeed, many computational models within cognitive science take this form. Strictly speaking some of these models might be referred to as being connectionist but not PDP. Here we treat them as a separate (and heterogeneous) class.

Semantic Networks

A popular approach to modeling people's knowledge of word meaning and/or concepts is the **semantic network**. In a semantic network, nodes are used to represent high-level symbolic concepts, such as "cat," "Tom" and "chase," and connections or links between those nodes are used to represent interrelationships between those concepts (such as the inclusion relation between "Tom" and "cat", or the agent relation that might hold between "Tom" and "chase" if we know that Tom is chasing something). The network can be used to make inferences by following links between concepts. Thus, from the links between feathers, bird, and animal in the network shown in figure 2.5, we may infer that some animals have feathers. Similarly, we may infer that Tweety has feathers, because Tweety is a bird and birds have feathers.

Semantic networks have enjoyed considerable popularity, and there are many variations on the basic model. Generally these variations relate to the inheritance mechanisms that may operate between nodes and the way such mechanisms can be specified so as to allow special cases (such as "penguins don't fly") to interact with default rules (such as "all birds fly"). A survey of

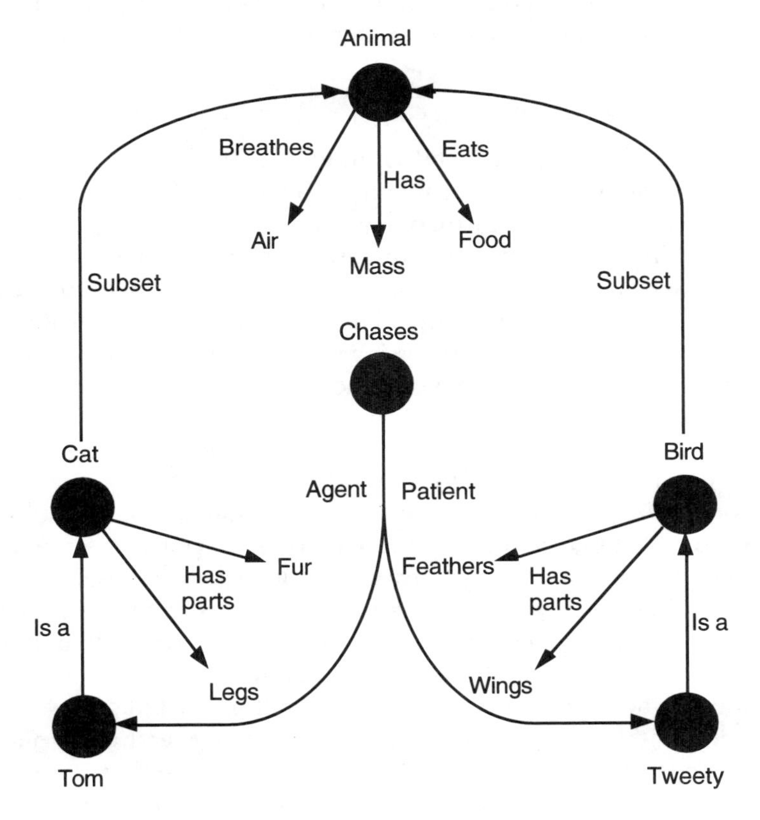

2.5 A semantic network (adapted from Rumelhart and Norman, 1985).

some of the research in this tradition can be found in Rumelhart and Norman (1985).

Production Systems with Spreading Activation

Although production systems are often seen as the prototypical symbolic model, Anderson has developed a series of hybrid production systems (ACT: Anderson, 1976; ACT*: Anderson, 1983; ACT-R: Anderson, 1993) in which continuous activation values are associated with discrete symbols (working memory elements). Within ACT-R, the most recent of Anderson's production systems, elements are never removed from the system's working memory. Instead, when they are originally entered into working memory they are given a high activation, and as time progresses their activation decays. The "act" cycle therefore can only add elements to working memory or reactivate existing elements.

The activation of working memory elements triggers productions, and on each cycle the production that is triggered most strongly is fired. This means that productions can fire even if not all of their conditions are matched (that is, several highly active working memory elements may trigger a production even if no working memory element matches one of its conditions).

Interactive Activation Networks

One common use of symbolic network models is to capture weighted interactions between high-level (i.e., symbolic) concepts. This is achieved in an **interactive activation network** by mapping the high-level concepts onto nodes, and then working in much the same way as in standard connectionist modeling: by associating an activation value with each node, associating a weight with each connection between nodes, and specifying rules for the interaction of the activations. Processing in these models then consists of allowing activation to spread between the symbolic nodes until either the activation of some node exceeds a threshold or the system reaches a stable state. Models of this sort allow subtle interactions between numerous factors to be weighed up and sorted out into a single activation value.

Such models are able to produce discrete behavior by including some mechanism for selecting the most active symbol. Behavior is then determined by this symbol. Simply associating activation values with symbols, however, does not guarantee that there will be a single most active symbol. Several symbols could end up being equally active. This problem is avoided in interactive activation models by the inclusion of competitive processes that operate between nodes. These processes take the form of further links between nodes. Typically, competition is incorporated into an interactive activation network by adding self-excitatory links (positively weighted feedback links from each node to itself), and lateral inhibitory links (negatively weighted links between competing nodes); see figure 2.6(a). The excitatory links encourage nodes to be active, but the inhibitory links ensure that dominant nodes over-power less active nodes. The most active node will receive the most self activation and the least lateral inhibition (assuming that the activation passed along each link is proportional to the activation of the node from which the link starts), and, with appropriate weightings, will force down the activations of all competing nodes, even those whose initial activation was only slightly less than the dominant node; see figure 2.6(b). While in the simple case this elaborate competitive process yields the node that was initially most active as the winner, this need not be the case when there are multiple asynchronous activation sources influencing each node. As in standard connectionist models, the activation of each node is then given by the complex interaction of a number of activation sources. Competition provides just two more sources of activation which ensure that the system reaches a state where only one node is significantly active. McClelland and Rumelhart's model of letter perception (see box 2.4) is a classic example of how such a network may work, illustrating the complex interaction of data-driven and expectation-driven processes that operate in this apparently simple domain.

SAQ 2.7 A number of different classes of computational model have been discussed in the last three sections. Attempt to draw a semantic network showing the interrelationships between them. For example, production system models form a subclass of symbolic models. Also show on your network the properties, such as ability to learn and type of representation, which members of the various classes exhibit.

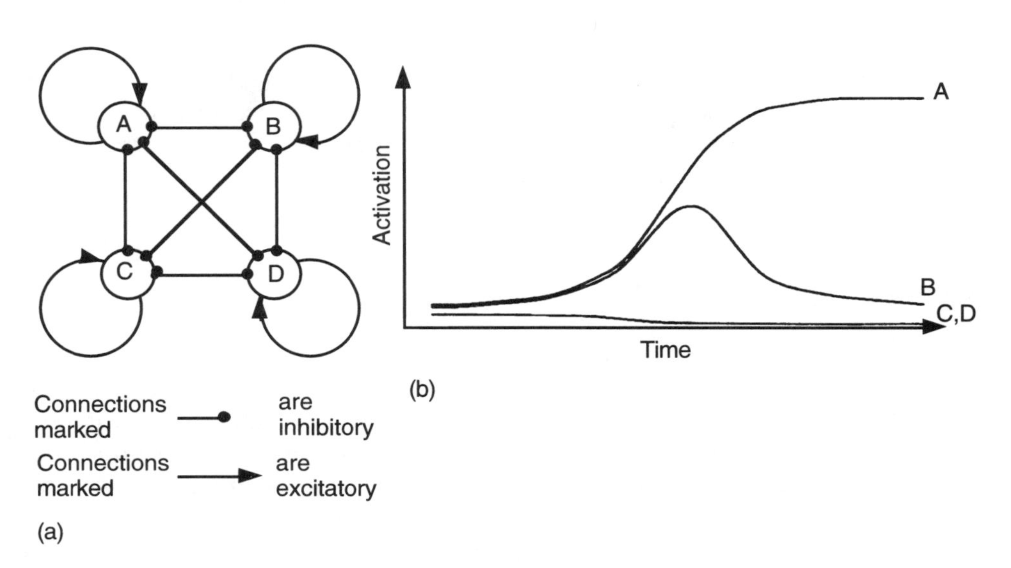

2.6 (a) Four nodes in an interactive activation network; (b) The activation profiles of the nodes in (a) as competition proceeds.

Box 2.4 An interactive activation model of letter perception

McClelland and Rumelhart (1981) present a three-level interactive activation model of letter and word perception. The system takes as input a representation consisting of a set of visual features, and maps this to a word, via a representation of the letters contained in the input. Each level (feature, letter, and word) consists of a network of nodes, with nodes in the levels corresponding to distinct elements of the corresponding domain (so each node in the letter network corresponds to a distinct letter).

Feature nodes receive activation directly from the system's input. These nodes are connected to the letter nodes, and excite those letter nodes compatible with them while inhibiting those letters nodes with which they conflict. So the feature node corresponding to a vertical stroke will inhibit the letter node corresponding to "o" because "o" is not compatible with a vertical stroke. Letter nodes also inhibit each other, so that only one letter node tends to be active at a time (i.e., the letter level is an interactive activation network). Letter nodes for each position in a word feed into the third layer – the word nodes. This layer is also an interactive activation network, with inhibitory connections between words such that only one word tends to be active at a time. Finally, there are feedback connections from the words to the letters, such that active words tend to excite letters within them and inhibit other letters.

The feed-forward connections from one level to the next and inhibitory connections within each level result in the appropriate word node becoming active when the system is presented with a featural representation of the word. More interestingly, when presented with a corrupted input representation, say a word with one letter blacked-out, the system is able to construct a guess at the input word. This is because, when a word is partially excited, feedback connections from the word level to the letter level excite all the letters which make up that word. This excitation then feeds forward again to the word level, further boosting the activation of the word. It is through positive feedback loops such as this that the system is able to generate an appropriate guess. Feedback loops represent a general mechanism by which interactive activation models are able to combine top-down or expectation-driven information (in this case about possible words) and bottom-up or data-driven information (in this case about features present in the stimulus).

Symbols or Connections?

Connectionist modeling, or at least the concepts behind connectionist networks, has a long history. McCulloch and Pitts first investigated networks of simple processing units in 1943. This work aroused substantial interest, especially when, during the 1950s, learning mechanisms for the simple processing units were developed. Interest waned, however, after Minsky and Papert showed in 1969 that there were certain things that simple processing units could not learn. The current interest in connectionism dates to the early 1980s, when learning mechanisms for complex networks of units were discovered and shown not to be subject to the earlier failings. Since then, an intense debate – concerning connectionist modeling, symbolic modeling, their relation, and their role in cognitive science – has been raging.

The Symbolic/Connectionist Debate

The symbolic/connectionist debate is sustained in part by the apparently complementary properties of connectionist and symbolic models, in part by the explanatory force of models couched in each technology, and in part by the way in which advocates of each school have presented symbolic and connectionist technologies as being stark alternatives.

We have seen that connectionist networks appear to have a number of psychologically interesting properties. They exhibit interference, generalization, and graceful degradation. Symbolic systems generally have difficulty accounting for these properties of cognition, but are, on the other hand, able to account for the productivity of thought in terms of systematic, compositional representations. The connectionist school objects that symbolic modeling assumes systematic, compositional representations (together with processes such as dynamic memory allocation for the construction and manipulation of stacks), without making any attempt to account for how those representations might be encoded within the brain. The symbolic school, in turn, claims that such representations are necessary for an account of all but the simplest cognitive processes (Fodor and Pylyshyn, 1988). The response from the connectionist camp has been to engineer connectionist methods for representing structured information. Pollack (1990), for example, has developed a method for encoding recursive data structures (such as stacks) in connectionist terms. One interesting property of such encodings is that they inherit the necessary symbolic properties of systematicity and compositionality, and the connectionist properties of graceful degradation, interference, etc.

A further line of attack on the symbolic camp comes from those who assert that symbolic modeling assumes explicit sequencing of instructions based on a preconception of the brain as a von Neumann-type machine. If the brain is not a von-Neumann type machine (if it does not function in terms of a fetch-instruction/execute-instruction cycle), then basing explanation on such algorithms, it is claimed, is misguided. This objection is sustained by the claim that connectionist models are neurophysiologically more plausible than symbolic models, because connectionist units share many properties with real neurons. This is not to say that proponents of connectionism in general wish to equate the units within a connectionist network with individual neurons in the brain. Smolensky (1988), for example, argues that connectionist modeling operates at what he terms the subsymbolic level – a level above that of the neurophysiological implementation but below the symbolic level.

The debate over the explanatory force of each class of model has tended to be highly subjective. Those of a symbolic persuasion have tended to level one of two criticisms at connectionist models. Firstly, processing in connectionist networks tends to be opaque in the sense that once a network has been trained, it is difficult to see how it is working. It is effectively a "black-box." As such, it offers nothing in terms of explanation. Secondly, it is often claimed that connectionist models address the implementational level, and not the algorithmic level. It is the algorithmic level, such cognitive scientists claim, at which cognitive scientific explanation should be pitched. One connectionist retort to these concerns is based on the concept of emergence. Complex systems exhibit emergent behavior: behavior that results from the interaction of the subsystems and which could not have been predicted from a knowledge of how the individual subsystems act in isolation. If symbolic rule-based behavior is an emergent phenomenon, then attempting to explain it in symbolic, algorithmic terms will fail – it may be describable in those terms, but it will only be explicable in terms of the complex interaction of processing subsystems.

There are further issues. What, for example, is the true nature of rules. Much of behavior may appear to be rule-governed, but is this because we are following explicit (symbolic) rules (like production systems), or does rule-governed behavior emerge from lower-level processes? Connectionist networks are able to exhibit, and indeed learn, rule-governed behavior without recourse to explicit rules. Networks excel at abstracting implicit "rules" from a set of examples (as in the case of learning the past tense of verbs). Furthermore, the rules so derived are flexible in that they generalize to novel situations and produce sensible behavior when all of their preconditions are not satisfied. Connectionists take people's ability to learn flexible, implicit, rules as evidence for their approach. However, the situation is not clear cut. Current network learning techniques (such as backpropagation) are implausible on both practical and neurophysiological grounds: they generally require thousands of presentations of a set of training examples (when we are capable of learning from a single exemplar), and, though networks themselves are neurophysiologically inspired, most current learning mechanisms are generally regarded as neurophysiologically implausible. Furthermore, irrespective of the plausibility or otherwise of existing network learning algorithms, what a network learns is often intimately related to the representations chosen for its inputs and outputs, and details of the network's connectivity (how many hidden units it has, for example). At present there appear to be few principles to guide these assumptions.

The debate between the symbolic and connectionist approaches is fierce, but there are those who point to the complementary successes of each approach, and argue that we should simply accept that cognitive science has two different technologies, technologies that are appropriate for different tasks or domains. There appears to be a fit between tasks and technologies, such that a connectionist approach is suitable for some tasks (generally low-level, primarily perceptual tasks), and a symbolic technology is appropriate for other tasks (generally higher-level, sequential reasoning tasks). This notion of fit between a model and a task is widely accepted, at least implicitly, within the connectionist school, where as we have seen a number of substantially different types of network are now common, and the decision about which network to apply to what task is based on task-specific properties. Nevertheless, hardline adherents of either school tend to be dogmatic, arguing that their preferred approach can account for all (interesting) aspects of cognition, and the debate remains very active. Perhaps the most level-headed assessment is that of Minsky and Papert (1988: p. xv) who assert:

> it never makes any sense to choose either of those two views as one's only model of the mind. Both are partial and manifestly useful views of a reality of which science is still far from a comprehensive understanding.

Hybrid Models

There are many who concur with Minsky and Papert's assertion that the symbolic/connectionist debate is unfounded, and that both symbolic and connectionist approaches may contribute to the simulation of cognitive processes. Some advocates of this position have gone so far as to attempt to combine symbolic and connectionist techniques within **hybrid models**. These models are based on the twin assumptions that (a) neither symbolic nor connectionist techniques are inherently flawed, and (b) the techniques of both forms of modeling are compatible.

Despite the apparent potential of such hybrid models in addressing the issues raised by the debate between the symbolic and connectionist schools, the combination of the approaches is not straightforward, and it is instructive to consider the theoretical possibilities. Firstly, hybrid models might be constructed by joining together separate symbolic and connectionist submodels. The behavior of these "physically hybrid models" is a complex function of the interaction of the behavior of the various components. Secondly, "non-physically hybrid models" may be developed in which there is only a single physical system (i.e., just one symbolic system, or just one connectionist system, but not one of each), but hybridness is obtained by describing the system in both symbolic and connectionist terms. This approach effectively operationalizes the levels of description argument in chapter 1. In particular, the connectionist part of such a model is generally identified with the neurophysiological level, with the symbolic part being identified with the cognitive level.

Physically hybrid models

As noted above, there is sometimes a certain fit between a task and an approach to modeling that task, be it symbolic or connectionist. Likewise, there are tasks which seem not to be well suited to either connectionist or symbolic approaches alone, but which seem to naturally decompose into subtasks, some of which are suited to symbolic modeling and others of which are suited to connectionist modeling. The task of reading aloud sequences of letters (including words and nonwords) is one such example (see chapter 6). In this task, well defined rules (grapheme–phoneme correspondences) appear to be employed when reading nonwords and words with a regular pronunciation, but exemplar-based processes appear to be operating in the reading of words with irregular pronunciations. Tasks such as this are particularly suited to physically hybrid modeling, where separate submodels, employing whichever approach – symbolic or connectionist – are most appropriate, and can be built for the separate subtasks. These models may then be combined into a single, physically hybrid model, producing a model of the complete task.

Two issues are central to physically hybrid modeling: interfacing, or the method of communication between symbolic and connectionist subsystems, and modularity, or the decomposition of a task into subtasks. In general, interfacing is kept to a minimum. This is consistent with the view of subsystems as independently functioning modules whose internal representations are private to those modules, but even so it is necessary to consider how the results of symbolic and connectionist subprocesses should be combined within the model as a whole. One popular approach is the use of a symbolic network, such as an interactive activation network, where nodes have a symbolic interpretation and receive activation from both symbolic and connectionist subsystems. Competitive influences may then act on the combined outputs of both systems to weigh up the contribution of each subsystem to the behavior of the system as a whole.

Non-physically hybrid models

Although physically hybrid models make use of both symbolic and connectionist techniques, such models do not necessarily inherit the desirable properties of each technology. There is no guarantee that the model as a whole will inherit graceful degradation from its connectionist submodels and rule-based behavior from its symbolic submodels. Non-physically hybrid models address this issue by tying the symbolic and connectionist technologies together more closely than in the physically hybrid case. In these models, a single system is described or interpreted in both symbolic and connectionist terms. The central issue in this form of hybrid modeling is therefore one of levels of description.

The approach is exemplified by the connectionist production system of Touretzky and Hinton (1988). This system was developed in an attempt to demonstrate that connectionist networks were capable of higher-level rule-governed behavior. The system was not intended to do any particular task, but rather was intended, like most production systems, to be programmable – Touretzky and Hinton were only concerned with providing production system (i.e., rule-based) functionality in connectionist terms. The system they built was modular in that it contained subnetworks corresponding to the various subcomponents of any production system (a working memory, a production rule memory, etc.), and each subcomponent could be described in connectionist terms, or (approximately) in symbolic terms. Of particular interest was the fact that the production system gained characteristic connectionist properties in virtue of being implemented in a connectionist technology. The system as a whole inherited a degree of damage resistance, and a working memory capacity dependent on the similarity of items stored.

SAQ 2.8 Based on your answers to SAQ 2.5 and SAQ 2.6, can you think of some domains where physically hybrid modeling would be appropriate?

Architectures

The modeling techniques considered above have generally been used for developing computational models of specific cognitive processes (such as working memory) or specific tasks (such as reading). Some theorists have argued that this single domain approach is flawed, given that (1) most real cognitive tasks actually involve the interaction of many cognitive processes (reading aloud, for example, involves processes of perception and articulation, as well as a variety of more central cognitive processes; see chapter 6), and (2) much of our cognitive apparatus is likely to be redeployed across a variety of tasks (writing and speaking, for example, may both draw on the same mechanism for converting thought to words). These theorists have instead argued that we need to look at the mind as a complete entity, and develop complete, or "unified" theories of cognition. This requires a theory of the **cognitive architecture**. We take up the issue of cognitive architectures in more detail in chapter 3. Here we focus on exactly what constitutes such an architecture.

In computer science an architecture is the fixed processing structure underlying the design of a computing machine. The hardware of any particular computer embodies a specific architecture, and it is the architecture which gives the computer various processing capabilities and characteristics. A cognitive architecture is, similarly, an arrangement of functional components (e.g., processors and buffers) together with a processing strategy.

For a computer to achieve a given task, it must be supplied with a program. The architecture is just the basis for executing the program. In the same way, a cognitive architecture is devoid of any task-specific knowledge. Indeed, an architecture can be seen to be the result of abstracting task-specific knowledge from the control structures which use that knowledge to produce intelligent behavior. Thus in the case of ACT-R (Anderson, 1993) or Soar (Newell, 1990), two architectures based on production system technology, appropriate productions encoding task specific knowledge are required if the architecture is to carry out a particular task.

An architecture is not, however, just a framework within which to build cognitive models for given tasks or domains by providing the requisite knowledge. Connectionism, as described above, is a framework. It provides a set of concepts and techniques which may be put together in various ways to construct a computational model. An architecture is more than this. It is a fixed structure (ignoring changes due to aging) – a fixed arrangement of the pieces that allow intelligent behavior. This is not to say that there cannot be connectionist architectures. Indeed, both symbolic (e.g., Newell, 1990) and connectionist (e.g., Schneider and Detweiler, 1987) architectures have been proposed.

There is a further parallel between computer architectures and cognitive architectures: different architectures make different algorithms easy (see Pylyshyn, 1991). One architecture might be very efficient at certain operations, but inefficient at others, whereas a second architecture may have the opposite characteristics. This gives architectures a kind of "fingerprint." We can differentiate between architectures on the basis of the way they use the same knowledge in performing the same task.

There is also a major difference between computer and cognitive architectures. It is that as a complete theory of the processing structures involved in cognition, a cognitive architecture must include mechanisms for learning, that is, ways of acquiring knowledge. Computer architectures do not, as yet, learn.

Though it is not a necessary characteristic of architectures, most cognitive architectures current in the literature are based on a unitary processing mechanism and a unitary learning mechanism. That is, all aspects of behavior are explained in terms of one processing mechanism operating with different knowledge, and all varieties of learning are explained in terms of a single mechanism for acquiring knowledge. These unitary assumptions mean that the processing mechanism must abstract away from all possible tasks. A unitary architecture is therefore likely to be a general-purpose processor capable of computing any computable function (i.e., architectures are generally Turing powerful). Despite the "fingerprint" effect discussed above, such general purpose computing machines can have only a limited impact within specific domains: different theories of single domains can be "implemented" within the same architecture by giving that architecture different knowledge. This has led to architectures such as Soar being criticized by, for example, Cooper and Shallice (1995), who suggest that modular architectures – architectures where different modules are dedicated to different subfunctions – may be a more fruitful avenue for research.

Summing-up

This chapter has discussed a central issue in cognitive science: the role of computational simulation in cognitive scientific explanation. We began by looking at the ways in which theorists present their theories, and noting that the complexity of typical theories of cognitive processes requires that simulation techniques be used to determine their consequences. We then surveyed some of the computational techniques available to the cognitive scientist. Apart from purely symbolic and connectionist approaches, we also considered symbolic approaches which include the use of networks. The advantages and disadvantages of each approach were debated, and hybrid methods were presented as a possible resolution to this debate. We also considered cognitive architectures, models of the complete cognitive processing apparatus within which models of particular tasks or domains may be embedded.

Further Reading

A number of books present collections of readings in various areas of computational modeling. One worthy example which includes chapters on both symbolic and connectionist approaches is *The Simulation of Human Intelligence*, by D. Broadbent (1993), Basil Blackwell, Oxford. An excellent presentation of the theory of computing machines as it relates to natural language is given in *Mathematical Methods in Linguistics*, by B. Partee, A. ter Meulen and R. Wall (1990), Kluwer Academic Publishers, Dordrecht, The Netherlands. A number of classic papers within the symbolic tradition are included in *Issues in Cognitive Modeling*, by A. M. Aitkenhead and J. M. Slack (1985), Lawrence Erlbaum Associates, Hove. An extensive discussion of many of the issues surrounding the development of symbolic architectures is contained in *Unified Theories of Cognition*, by A. Newell (1990), Harvard University Press, Cambridge, MA. Architectures (of all varieties) are discussed in *Architectures for Intelligence*, a collection edited by K. van Lehn (1991), Lawrence Erlbaum Associates, Hove. A very readable introduction to connectionist modeling and the issues surrounding connectionism in cognitive science is contained in *Connectionism and the Mind*, by W. Bechtel and A. Abrahamsen (1991), Basil Blackwell, Oxford. The technical details are well covered in *Introduction to the Theory of Neural Computation*, by J. A. Hertz, A. S. Krogh, and R. G. Palmer (1991), Addison-Wesley, Reading, MA. The two volumes of McClelland, Rumelhart and the PDP Research Group should also not be overlooked: *Parallel Distributed Processing, Volume 1: Foundations*, by D. E. Rumelhart and J. L. McClelland (1986): MIT Press, Cambridge, MA; and *Parallel Distributed Processing, Volume 2: Psychological and Biological Models*, by J. L. McClelland and D. E. Rumelhart (1986): MIT Press, Cambridge, MA.

Exercises

1 What sorts of tasks/domains from the scenario do you think would be most amenable to symbolic modeling? For each task, what sort of symbolic computing machine do you think would be most appropriate?
2 What sorts of tasks/domains from the scenario do you think would be most amenable to connectionist modeling? For each task, what sort of network do you think would be most appropriate?

CHAPTER 3

The Architecture of the Mind: Modularity and Modularization

Outline

In this chapter we focus on the overall organization of the mind, rather than on the properties of any particular mental capacities which are the subject of later chapters. The central question here concerns the extent to which the mind can be seen as an homogeneous whole, employing general all-purpose computational strategies, and the extent to which it has to be understood as a constellation of special-purpose processors with quite idiosyncratic computational properties. Naturally, this question raises a host of sub-questions. If some, at least, of our mental activities are the result of special-purpose processors, how fine-grained are they? For instance, within visual perception are there distinct systems for color perception and shape perception, are there distinct systems for perceiving tables, cats and human faces? If the adult human mind is structured, is it so from birth or does it acquire its specialist skills through its interaction with the environment? Does an evolutionary perspective on human cognition favour a unitary or compartmentalized view of the mind? And, relatedly, what does research on non-human animal behavior and the cognitive capacities underlying it suggest about the human mind?

You might like to consider the plausibility of the following analogy before embarking on the chapter and then come back to reconsider it at the end:

> The mind is probably more like a Swiss army knife than an all-purpose blade: competent in so many situations because it has a large number of components – bottle opener, cork-screw, knife, toothpick, scissors – each of which is well designed for solving a different problem. (Cosmides and Tooby 1994a)

Learning Objectives

After reading this chapter you should be able to:

- discuss the concept of cognitive architecture

- discuss the characteristics of a modular architecture as opposed to a general or nonmodular architecture
- present the case for a modular account of perceptual capacities
- discuss whether the human language ability is a mental module
- outline the frame problem and some of the debate around it
- bring to bear evolutionary (phylogenetic) considerations in discussing these issues
- bring to bear developmental (ontogenetic) considerations in discussing these issues

Key Terms

- computation
- constructivism
- domain-specificity
- implicit vs. explicit knowledge
- informational encapsulation
- modularity
- nativism
- reflex

Introduction

As for virtually any everyday interaction, the members of the family having breakfast together in our scenario in chapter 1 are engaged in a vast range of mental activities. They are processing many visual and auditory stimuli and so perceiving objects and sounds. They are coordinating their perceptions and their actions in reaching for objects and moving about the room. All of them are both producing and comprehending utterances, mostly spoken but some written. Their memories are actively engaged, both in recognizing the people and objects around them and in recalling (or trying to recall) past actions and states of affairs. They are tackling the range of problems presented by their various goals, from making breakfast and opening a parcel, to negotiating their relationships with each other.

We will consider, on the one hand, whether some of these different mental capacities are specialized and independent of others and, on the other hand, whether some fall together as the achievements of a more general unitary intelligence. The issue here is the nature of our **cognitive architecture**; that is, the structure or design of the human mind. Although the plasticity and creativity of human behavior are striking, the cognitive processes that underlie this flexibility are bounded by our cognitive make-up and by our experience. Architectural constraints on our behavior are a result of the way our minds are constructed as distinct from those constraints that arise from habit, learning, or features of our individual experience; it is these cognitive constants or fixed points that we are interested in discerning here. In a bid to separate out architectural, from other, constraints on human cognitive function Pylyshyn

(1991) applies a kind of test: for some cognitive regularity, does this regularity hold necessarily or can it be made to go away by changing subjects' beliefs or goals? If a change in function is brought about by a change in such **representations** then the original regularity cannot be (solely) a property of architecture, which should remain invariant across changing goals and knowledge. That is, cognitive architecture is said to be "cognitively impenetrable." A **cognitively penetrable** process, on the other hand, (i.e., one that can be caused to change tack by changes in belief and goal) does not reflect architectural properties of the mind. We will put this test into practice in the next section in assessing whether or not certain aspects of human perception are a function of mental architecture.

Cognitive (or functional) architecture should not be equated with physical architecture, though there may be some important correlations between them. The organizational level of cognitive architecture is the level at which the mind is representational and where the representations correspond to beliefs, goals, memories, percepts, etc. (see Pylyshyn, 1991, p. 191); in terms of Marr's three levels of explanation, outlined in chapter 1, this is the **algorithmic** and representational level. A particular cognitive architecture may be implemented or realized in various physical forms (biological or non-biological).

Perceptual Processes: Computational Reflexes

First, consider some standard cases of simple **reflexes**: the knee-jerk, the eye-blink, and the sneeze. On the whole, all that's needed to evoke a reflex is presentation of the appropriate eliciting stimulus, a sharp tap to the tendon below the kneecap, the close approach of some foreign body to the eye, for instance. Reflexes are largely unaffected by the beliefs and goals of the behaving organism: " . . . suppose you know perfectly well that under no conceivable circumstance would I stick my finger in your eye. Suppose that this belief of yours is both explicit and deeply felt . . . Still, if I jab my finger near enough to your eye, and fast enough, you'll blink . . . The blink reflex has no access to what you know about my character or, for that matter, to any other of your beliefs, utilities or expectations. For this reason the blink reflex is often produced when sober reflection would show it to be uncalled for . . . " (Fodor, 1983, p. 71). In other words, reflexes are the paradigm case of cognitively impenetrable behaviors.

Now consider intelligent activities like deciding what to buy mother for Christmas, making plans for the holidays, or musing on the meaning of life. These thinking cases seem to be the opposite of reflexes in all respects. Reflexes are fast, automatic, unaffected by beliefs and goals, and built into the organism (innate); decision making and planning are relatively slow, non-automatic (they are largely under the organism's voluntary control), not innate and, of course, highly sensitive to beliefs, goals, hopes, etc. (i.e., cognitively penetrable).

Crucially, these intelligent activities are cognitive (often referred to as "higher cognitive processes") and viewed from a cognitive science perspective (as opposed to a physical perspective) involve **computations** over representations, while the reflexes mentioned can be described only from a physical or neural perspective. But could there be such a thing as a cognitive reflex and what would such a thing be? It would be a computational response which is fast, automatic, innately specified, and immune to the beliefs and goals of the organism. In a series of writings, Fodor (1983, 1985, 1986) has proposed that there are indeed such reflexive (as opposed to reflective) mental processes; they are the perceptual processes, including language perception.

This thesis can be convincingly motivated by considering some cases of illusions. The most famous is the Muller–Lyer visual illusion shown in figure 3.1. The line with the inward pointing arrowheads appears longer than the line with the outward pointing arrowheads. Even once people are fully assured that the lines are of the same length, perhaps having measured them themselves, they see one as longer than the other. Knowledge of the facts of the matter does not affect perception. And this immunity of the illusion to corrective knowledge, knowledge which would, after all, make for a more accurate perception, generalizes to many other cases of visual illusion (see, for instance, Rock, 1984).

The following kinesthetic illusion of self-motion is familiar: when sitting in an unmoving train carriage and seeing a train on the adjacent track pulling out of the station you feel that your carriage is moving. You may know full well that your train is not moving, perhaps that you are going to be held up there

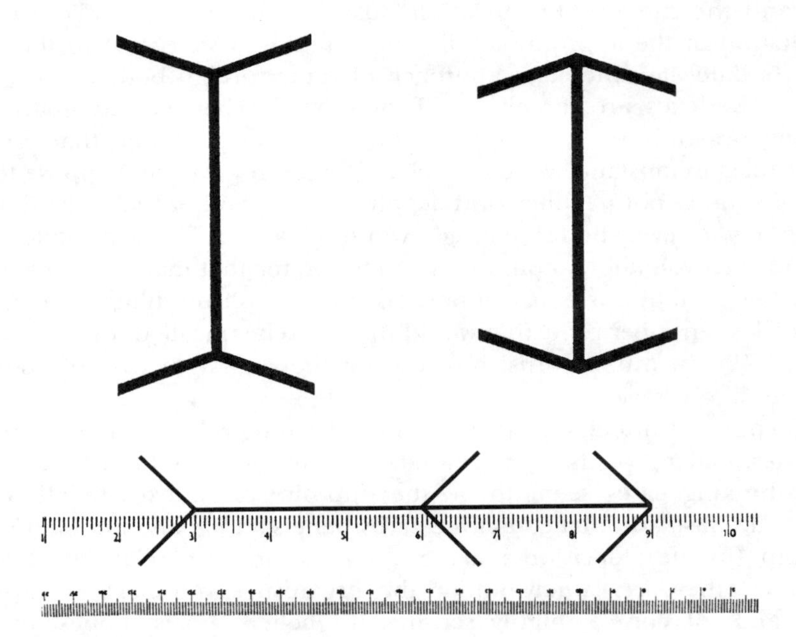

3.1 The Muller–Lyer Illusion.

for some time, and yet none of this interferes with the illusory perception of movement. The virtually opposite illusion may arise too, the perception that objects in front of you are moving, though they are not: if you push your eyeball with your finger (as opposed to the usual, internal, way of initiating eye movement) you perceive movement of, say, the print on the page in front of you (Fodor, 1983, p. 67). Again, you know that it is your eye rather than the scene before your eyes that is moving but this knowledge does not inform your perception.

There is much less work on auditory, than visual, illusions, but these too exist and are equally immune to knowledge of the facts (see Deutsch, 1975, 1983). And in language perception there are a range of illusions; for instance, it seems that if we hear a group of phonemes followed by a cough which is followed by another group of phonemes, we cannot avoid perceiving a word, if replacing the cough by a phoneme would make a word. For example, /infor(cough)ation/ is perceived as /information/ with a cough in the background and no amount of informing otherwise alters the perception. Experiments have shown that subjects systematically mislocate a click in a stream of speech as falling at the boundary of a major grammatical category, such as a clause, and, again, factual beliefs do not seem to correct this misperception.

In their immunity to beliefs and information, active in the organism's overall cognitive system (their cognitive impenetrability), perceptual processes are similar to reflexes. They are also fast, automatic, and innately specified, like reflexes, according to Fodor. However, they are computational and to that extent are like the thoughtful processes of decision making and planning. A group of cognitive psychologists, known as the New Look Cognitivists (e.g., Bruner, 1957; Bruner and Klein, 1960; Gregory, 1970), at whom Fodor's analogy with reflexes is aimed, emphasized the intelligence of perception and thus its similarity with higher cognitive processes. What they were impressed by was the disparity between the information contained in the **proximal stimuli** impinging on the sensory receptors and the information in the perceptual responses. For instance, as you walk around a table with a square top, light reflected from the tabletop hits your retina, projecting upon it a two-dimensional trapezoid of changing dimensions. However, you do not perceive an ever-deforming, two-dimensional trapezoid. Instead, you perform computations of some sort on these data to construct a "percept" of a stable, three-dimensional, square tabletop. This sort of argument is an instantiation of a general argument for the intelligence of perceptual systems known as "poverty of the stimulus." The argument applies equally to the case of language: from a certain vibrating of the tympanic membrane in the ear we may perceive that someone has uttered the sentence "It will rain tomorrow" and this perception is stable across quite a range of pronunciations, voice pitches, and extraneous noise (see chapter 5). It seems evident that there must be some system of computational principles in the mind of the organism, which operates on the meagre and variable information in the stimulus to transform it into the perception of a structured, meaningful sentence. The poverty of the stimulus argument has been employed to considerable effect at another levels in discussing the acquisition, or growth, of mental capacities, such as language, in the individual (Chomsky, 1986a, 1988). Here the conclusion of the argument is that we must start life with mental systems which are already highly structured and specific, so that they can mature into adult perceptual and linguistic systems on the basis of the minimal and unstable input that the environment provides.

SAQ 3.1 If a square is projected onto a screen and then its sides are expanded and contracted in a regular fashion, viewers report seeing a square of constant size that recedes from the viewer, then approaches the viewer, then recedes, etc. Consider the sequence of images this stimulus projects onto the viewer's retina. Is the perception reported the only one compatible with those images? If not, what does this indicate about the perceptual processes operating on the input?

The Cognitivists, then, were impressed by the considerable powers of computational elaboration that human perceptual systems must have, and a lot of very important work was done to determine the sorts of principles at work in visual and language perception. Perceptual processes could not be reduced to reflexes because reflexes are dumb and perception is intelligent. This perspective was, of course, a healthy reaction to the earlier **behaviorist** position, which eschewed mental processes and attempted to account for intelligent behavior in terms of learned (**conditioned**) reflexes. The cognitive architecture of the behaviorist is a single, undifferentiated, **domain-general**, all-purpose learning system, on the basis of which the organism's experiences determine the formation of conditioned reflexes. On this view, whatever structure the mind may ultimately have is simply a reflection of the structure of the environment. There is no place for specific computational principles underlying particular mental capacities; there are no systems of inbuilt knowledge; and no learning devices beyond the general **associative** laws (primarily the law of contiguity), which determine the development of the conditioned stimulus-response units (reflexes), the form which all psychological functioning is supposed to take.

The cognitivist reaction to this was the starting place of current cognitive science; the sort of explanations of mental function that cognitive science looks for, such as computational descriptions of the task and accounts in terms of particular representations and algorithms (see chapter 1), rest on the cognitivist premise that explanatory accounts of mental function have to focus on the complex processes that mediate between observable stimulus and observable behavioral response. Fodor endorses all this but claims that what is missing from this picture of "intelligent" perception, which is taken to be essentially the same as the higher cognitive functions of problem solving and decision making, is any account of the unwavering refusal of perceptual processes to heed certain items of relevant information in arriving at their conclusions. The persistence and unshakability of perceptual illusions, such as those mentioned above, are strong indications that, in one respect at least, perception is more like a dumb reflex than an intelligent thoughtful process. Fodor's point is that the cognitivists have failed to distinguish two ways in which a process can be intelligent/dumb. Reflexes are dumb in that they are non-computational (i.e., they involve 'straight-through' connections) and they are encapsulated from the beliefs and wishes of the organism (that is, they are cognitively impenetrable). Thinking, planning, and fixing beliefs (including making decisions) are intelligent processes in that they are computational (inferential) and they take account of beliefs and wishes. The essence of Fodor's position on perception is that "although it is smart like cognition in that it is typically inferential, it is nevertheless dumb like reflexes in that it is typically encapsulated" (Fodor, 1985, p. 2).

Ramachandran (1985, p. 99) makes the same point from a different perspective when he says of the view that perception is "intelligent":

> This is an example of what I call the "animistic fallacy" – the tendency to **homunculise** our perceptual mechanisms and to attribute our own thoughts, motives, and cognitive capacities to what are essentially low-level rote mechanisms. While it is undoubtedly true that perception often mimics intelligent behaviour, it may be intelligent only in the same sense that a beaver may be said to have knowledge of architecture just because it builds elaborate dams. The distinction here is between **implicit** and **explicit** knowledge. Implicit knowledge (like the beaver's knowledge of the workings of dams) is available only for a single purpose, whereas explicit knowledge is available for a much wider behavioural repertoire (e.g., teaching architecture, creating new architectural designs, etc.).

This important distinction between the implicit knowledge embodied in a computational reflex, like visual perception, and the explicit knowledge of a flexible, thoughtful, creative cognitive system will be taken up again in a later section.

SAQ 3.2 Experiments seem to indicate that when we hear a word like, for instance, "bread," the word "butter" is made immediately more accessible, as opposed to a word like "tree" which is not associated with "bread." Apparently, this is so even when "butter" is completely irrelevant, so that its activation serves no purpose, say in the context of an utterance about taking some stale bread to feed the ducks. If these results can be sustained (there is disagreement about them), what do they indicate about the processes involved in word perception?

Devil's Advocate Box 3.1

Is Fodor really right that perception is fixed, rigid, and impenetrable? Surely our perceptual experiences can be modulated at least to some extent by our beliefs. Churchland (1988) takes this line against Fodor. He discusses several ways in which perceptual experiences do seem to be under the control of top-down processes. For instance, the perception of ambiguous figures like the Necker cube (figure 3.2) can be brought under control with a little practice, so that from perceiving it in one way you can make yourself perceive it in reverse by just telling yourself to do so.

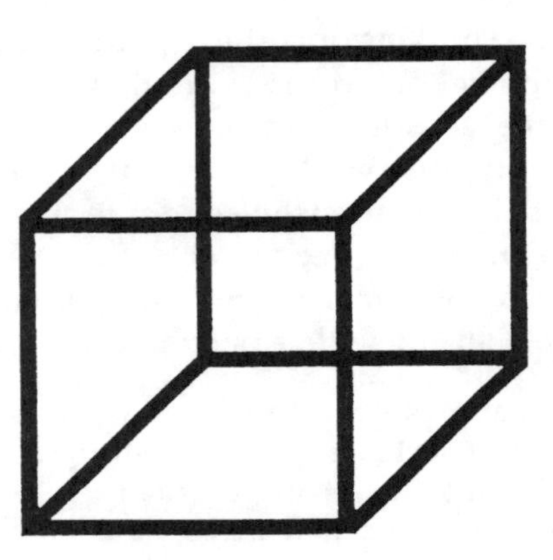

3.2 The Necker Cube.

Even more compelling are cases in which perception can be radically changed by experience. The best examples of this come from case studies of people adapting to "inverting" lenses, optical devices that turn the world upside down or the wrong way round. The effect of wearing such lenses is initially extremely disorienting, but, with sufficient exposure, people do ultimately manage to adapt to the new visual input so that they are able to recoordinate their vision with the rest of their sensory and motor systems and so navigate their way smoothly about the world and reach out in the right direction to grasp objects, etc. Eventually, people lose their perception that things are upside down – things appear just as they were before they put on the inverting lenses. So it seems that some of the supposedly fixed computational assumptions the perceptual system makes are open to revision. Churchland considers it quite possible that there are no assumptions about the world which are immune to revision given enough exposure to a world in which those assumptions do not hold.

See Fodor's defense of his views against Churchland's points (Fodor, 1988).

Mental Modules

Fodor's basic "functional taxonomy of psychological processes" distinguishes **transducers**, **input systems**, and **central systems**. The transducers (sense organs) are interpreted in traditional fashion as providing (sense) modality-specific representations of proximal stimulus configurations, "the distribution of stimulations at the 'surfaces' (as it were) of the organism." They are not computational; their function is simply to translate stimulus information into a format in which perceptual computation can be performed. This is where cognitive psychology begins; the perceptual "input" systems, as Fodor calls them, do their constrained computational work on the stimulus and deliver representations, which "are most naturally interpreted as characterizing the arrangements of things in the world" (Fodor, 1983, p. 42). In other words, the function of these input systems is as an interface between **proximal stimuli** and the identification of **distal stimuli**: they deliver up representations of the world to the higher thought processes of the central systems. The central processors are concerned with the fixation of belief (which includes decision making) and the planning of intelligent action. In so doing, they may use information from any and all of the input systems and integrate it with existing beliefs and wishes. Fodor's thesis is that the mind has a mixed architecture: the input systems, and probably output systems too (concerned with processes of motor control and speech production, for example), are specialized dedicated systems with highly restricted communication with other cognitive systems, while the central information-integrating, belief-fixing systems are general and may be influenced by information from just about any source. We will take up the nonmodularity of the central cognitive systems in a later section. Here let's look more closely at the properties of a mental module.

Modular mental systems are **domain-specific**, that is, they take as their input a limited and eccentric set of stimuli, the computations they perform are generally idiosyncratic and the conclusions they reach (their output hypotheses) concern only a limited range of distal properties. Fodor (1983, p. 47) emphasizes that within such broad domains as vision, audition and language perception there are subdomains which are probably analyzed by specialized subsystems, for example, within vision: edge detection, shape analysis, color perception; within language: affix recognition, grammatical analysis, thematic relation assignment. Certain relatively higher-level perceptual achievements of particular importance to human life, such as the recognition of human faces and voices, might constitute specific domains served by specialized computational systems dedicated to them.

Domain-specificity is an essential property of a modular faculty of mind. It distinguishes this way of conceiving of mental architecture as structured, from another according to which the divisions crosscut domains: here there are such general faculties as memory, attention, judgment, sensibility, and perception. For instance, memory for the properties of objects, for faces, for one's native language, for what one did yesterday is a single system; similarly, the faculty which is able to exercise judgment on moral issues is the same as that which makes aesthetic, practical, and other judgments. The modularity thesis, on the other hand, takes the position that "memory" for visual perceptual principles and for grammar may be quite distinct and that visual and linguistic judgments depend on principles that have little, if anything, to do with each other.

That a modular system works on a delimited domain is probably not enough to distinguish it from nonmodular systems. Hirschfeld and Gelman (1994), who take a domain-specific perspective on cognition, say: "Yet modular and domain-specific approaches also contrast in significant ways. The principal difference is the former's emphasis on specificity in functional cognitive architecture and the latter's focus on specialization for specific types of knowledge." It seems safe to say that while domain-specificity is a necessary property of a module there may be specific, fairly self-contained, domains of human knowledge that are not served by their own exclusive architectural unit. A case might be the knowledge and ability involved in the efficient driving of a motor car. This is a fairly circumscribed domain and yet there is little temptation to think of it as a mental module in Fodor's sense.

SAQ 3.3 Consider the following interesting phenomenon, known as the McGurk effect. Subjects receive simultaneous visual and auditory input. The visual input consists of a talking face, making the lip and other articulatory movements for a particular consonant–vowel syllable, say, [baba] or [gaga]. The auditory input which is dubbed over this is, say, [gaga] or [baba]. Subjects are asked to report what they hear. The results are as follows: (1) when the vocal movements are those for [gaga] and the auditory stimulus is [baba] they report hearing [dada]; (2) when the vocal movements are for [baba] and the auditory stimulus is [gaga] they report hearing [gabga] or [bagba]. It is clear from this that speech perception is affected by input from a visual domain. Does this indicate that the mental system which processes speech is not domain-specific? If not, why not?

The second essential property of an architectural module is the one that the illusions so nicely demonstrate: **informational encapsulation** (or cognitive impenetrability). Modular systems do not, cannot, draw upon the full range of the organism's knowledge, beliefs, and desires. They are restricted to their own proprietary knowledge system (or data base), which is brought to bear on stimuli from the domain that triggers their functioning. Consider again the McGurk effect of SAQ 3.3 in the light of the following remark by McGurk and Macdonald (1976, p. 747): "We ourselves have experienced these effects on many hundreds of trials; they do not habituate over time, *despite objective knowledge of the illusion involved.* By merely closing the eyes, a previously heard

[da] becomes [ba] only to revert to [da] when the eyes are open again." In other words, the effect is clearly informationally encapsulated.

It might seem that encapsulation is an unhelpful property for a cognitive system to have since it gives rise to illusions (i.e., wrong perceptions). It will be argued in a later section that the encapsulation of perceptual systems does in fact generally promote our survival. To put it briefly, " . . . a condition for the *reliability* of perception, at least for a fallible organism, is that it generally sees what's there, not what it wants or expects to be there. Organisms that don't do so become deceased" (Fodor, 1983, p. 68).

SAQ 3.4 Across a range of experimental tasks subjects are much quicker to react to (utter, comprehend, or make judgments concerning) the word "nurse" than, say, "clown" in the context of the following sentence: "The doctor handed the blood sample to the ———." What might this indicate about the information available to the processes responsible for recognizing words? How does this bear on the question of whether the language system is informationally encapsulated or not?

A third property of perceptual input systems is that they are mandatory (assuming the transducers are working, of course). We cannot help but see an object in our visual field as a three-dimensional object rather than a two-dimensional array of varying colors and lines. The fourth property which is their speediness follows directly from the other three. Their domain-specificity makes possible (though does not require) the evolution of a dedicated circumscribed computational architecture; their mandatoriness means there is no process of decision making involved in setting their operations in motion

Devil's Advocate Box 3.2

The idea that perceptual systems are modular is called into question by people whose sensory perception is "synesthetic." That is, when presented with a stimulus in one sensory domain they have a perceptual experience not only in that domain but also in some other. For example, on hearing certain sounds or words some people see particular colors and/or shapes moving about; others when tasting a food feel a shape (rounded, pointed) in their hand, etc. It seems that the input systems of these people are not modular or, at least, are much less encapsulated than Fodor claims.

Does the existence of synesthetic perception really threaten the modularity thesis? Can you think of ways in which the two could be reconciled?

Cytowic (1989) and Baron-Cohen et al. (1993) give very different accounts of this intriguing phenomenon.

and their encapsulation ensures that there is only a delimited and fixed set of information that they need take account of. Input processing is indeed very fast: "To all intents and purposes ... object recognition and sentence recognition are instantaneous processes" (Marshall, 1984, p. 221).

Among the further modular properties input systems have, Fodor lists the following: (5) relatively shallow output representations, which are the result of the encapsulated perceptual processes. These provide input to the higher cognitive processes, which integrate them with information from other sources. Just how shallow or developed output representations are is an important issue; in the case of vision, they could be anything from reports on size, shape and color, to three-dimensional models of objects and scenes, to conceptual reports ("that is a chair"); (6) the **ontogeny** (growth, development) of an input system seems to proceed at the same pace and to pass through the same sequence of stages across the species; (7) each system has its characteristic and specific breakdowns and a breakdown in one system may leave others entirely intact; (8) these systems are generally associated with fixed neural architecture, which may, but need not, entail localization in the brain (see Farah, 1994). There is strong evidence from neurobiology that visual perception has these last two properties: a complex division of labor in visual perception "is manifested anatomically in discrete cortical areas and subregions of areas specialized for particular visual functions; it is manifested pathologically in an inability to acquire knowledge about some aspect of the visual world when the relevant machinery is specifically compromised." (Zeki, 1992, p. 43). He goes on to give examples of damage to specific regions of the cortex that result in people losing the ability to see just one aspect of the visual world, such as color, form, or motion. Similarly, Moscovitch (1992) has developed a neuropsychological model of memory in which the modular view is very evident; he distinguishes various types of memory such as (a) module-internal memory systems for faces, objects, words; (b) a sensori-motor procedural memory; (c) a domain-general central system memory. He correlates these memory types with particular brain regions and switches easily between cognitive and neural vocabulary, very much in accordance with Fodor's fixed neural architecture property; for instance, he writes of the "cortical modules" (i.e., perceptual modules) and the "hippocampal component," a mental module which mediates encoding, storage, and retrieval of episodic memories (see chapter 10, "Learning and Memory").

The last three properties mentioned here are all consequences of the fact that mental modules, as Fodor uses the term, are innately prespecified. An analogy with bodily organs is apt. The human body is both an integrated whole and a complex of different, innately determined physical organs: the heart, the liver, the limbs, etc. The same goes for the human mind which is (partially at least) a complex of mental organs. As in anatomical study of the human body, the map of mental architecture must reflect the discontinuities and joints of these structurally distinct systems.

SAQ 3.5 Consider whether the ability to play chess competently is likely to involve a specialized faculty. Which of the properties of modular systems just reviewed does chess-playing have and which does it not have?

Language Module

The putative modularity of the language processing system is generally felt to be in greater need of demonstration than that of the perceptual systems, which are more directly linked to input from sensory transducers. Without doubt, a distinction has to be made between language perception and utterance interpretation; no one would claim that recovering the full significance of an utterance could be achieved by an automatic modular system (see chapters 8 and 9). However, once such a distinction is in place, the idea that language perception is modular has to be taken seriously. A quick run through the checklist of modular properties is suggestive. Language perception is generally fast and automatic; you can't help hearing an utterance of a sentence in your native language as an utterance of a sentence rather than, say, a stream of sound and, like object recognition, sentence recognition is virtually an instantaneous process. The time taken to recover semantic content from a spoken sentence is apparently comparable to the time taken to make one of the simplest voluntary responses, the two-choice reaction (e.g., pushing a button when, say, the right-hand light goes on). As Marshall (1984, p. 221) puts it: "Comprehension follows the speech wave as quickly as the relationship between the acoustic signal and the phonological percept allows." Given the complexity of the processes mediating between the acoustic signal and the lexical, syntactic, and logical structure of the sentence, this strongly suggests that language perception is a blindly operating, dedicated system.

It looks very much as if the system responsible for speech perception is innately specified, including its receptivity to the visual input provided by the movement of articulatory organs, as demonstrated by the McGurk effect discussed above. Three-month-old babies presented with the sound [bababa] show a marked preference for a film showing an actor whose mouth movements coincide with [bababa] rather than one showing the same face mouthing [gagaga] (see Kuhl and Meltzoff, 1982). But the innateness of the language capacity in humans seems to go much deeper than this. There is now a large body of work which indicates that at least the core elements of the lexical and syntactic structure of a person's native language are acquired in an orderly and rapid way which cannot be accounted for by the variable, and inevitably "underdetermining," evidence provided by the environment. Furthermore, studies of children born deaf but with hearing parents who do not know sign language, have shown that these children develop a visuo-manual system that observes several of the fundamental constraints of natural language (Goldin-Meadow and Feldman, 1979). The extraordinary thing about this is that these children have received effectively none of the linguistic input that is available to hearing children or to deaf children with signing parents. Strong "poverty of the stimulus" arguments lead us to the conclusion that linguistic knowledge is in large part innately specified (Chomsky, 1986; Crain, 1991).

There are now a number of well-documented cases of **"double dissociations"** between linguistic and other cognitive abilities. These include cases of fluent well-formed sentence production despite other marked cognitive deficits (see, e.g., the case of Laura in Yamada, 1990) and the converse case of serious linguistic impairment with normal ability in other cognitive spheres (see, e.g., the cases of linguistic **aphasia** discussed by Shallice, 1988). Just as with the perceptual systems, there is strong evidence for fixed neurological structures for language, which are located in the left hemisphere of the brain; moreover, the evidence is that these are **cross-modal**, in that they are the same for deaf people for whom the linguistic stimulus comes as visual sign rather than auditory speech (Coltheart, Sartori, and Job, 1986; Gazzaniga, 1989; Pinker, 1994, chapter 10).

Recently the fascinating case of a linguistic **"savant"** has been investigated; he is a man who can produce, comprehend, translate and communicate in up to twenty languages, though he scores well below average in nonverbal intelligence tests and is unable to perform the daily tasks involved in looking after himself (Smith and Tsimpli, 1991, 1995). This dissociation of mental functions might appear to support the modularity of language. However, there are savants with a single outstanding ability in mathematics or drawing or music, coupled with broad intellectual deficit. These are all clearly cases of an enhanced capacity which is domain-specific, but they do not so obviously have the other properties of modular systems that Fodor cites, so that whether the existence of a specifically linguistic savant provides evidence for a language module cannot be assumed without further argument.

Let us consider now whether the two central properties of a Fodorian module apply to language: domain-specificity and informational encapsulation (together with the related issue of the output of the language input system). The evidence for the domain-specificity of the language processor is strong: the mechanisms which perform phonetic analysis appear to be triggered into operation only by the acoustic stimuli of verbal utterances and to be distinct from the perceptual mechanisms that analyze nonspeech sounds (Fodor, 1983, p. 48–51). As we have seen, the McGurk effect shows that the speech processor admits input from two modalities (auditory and visual). However, it is a very restricted and eccentric domain of visual stimuli: just the movements of the organs involved in articulation; seeing any other part of the person's body moving or even facial movements unconnected to speech such as smiles, frowns and twitches, has no effect on speech perception. Furthermore, this cross-modal interaction is specific to the domain of language. A motion picture of a bouncing ball accompanied by a sound track does not affect the auditory perception of music or of a voice-over (see Fodor, 1983, p. 132). The domain-specificity of language processing appears to go much deeper (further into the organism, as it were) than phonetic analysis. If current work in generative grammar is correct the principles that account for intuitions of syntactic well-formedness and which determine the structure of an utterance in comprehension are peculiar to linguistic analysis just as the principles of visual perception are specialized for that particular task. See, for instance, the principles of economy of derivation, such as "Procrastinate," discussed in chapter 7, "The Structure of Sentences."

Finally, let us move to the highly contentious issue of the informational encapsulation of the language processor. Most people are agreed that low-level linguistic processes like phonetic analysis are stimulus-driven **(bottom-up)** and that high-level processes, such as inferring an ironical interpretation (see chapter 8, "Meaning and Conversation"), which depend on assessing a speaker's communicative intentions, are highly context-dependent, employing information from a range of sources **(top-down)**. It is the (many) processes in between that provoke debate, in particular the processes of word recognition and structural analysis **(parsing)**. Closely related to the encapsulation issue is the question of the output of the autonomous language processor: the highest level of representation which is cognitively impenetrable. Fodor claims that "the grammatical and logical structure of an utterance is uniquely determined (or, more precisely, uniquely determined up to ambiguity) by its phonetic constituency; and its phonetic constituency is uniquely determined in turn by certain of its acoustic properties ... " (Fodor, 1983, p. 89). Others maintain that top-down information is brought to bear much earlier on in linguistic processing and, if it makes sense at all to postulate an output representation from purely stimulus-driven analysis, it must be at a much lower level (see, for instance, Marslen-Wilson and Tyler 1987).

Parsing is basically a reflex, according to Fodor, that is, it is context-free and immune to beliefs, hopes, expectations. A cognitive penetrability test for this hypothesis can be constructed by embedding structurally ambiguous phrases, such as "visiting relatives" and "shaking hands," which have a verbal understanding and an adjectival understanding, in clearly disambiguating contexts:

> Verbal:
> If you want a cheap holiday, visiting relatives ...
> As a traditional way of gaining votes, shaking hands ...
>
> Adjectival:
> If you have a spare bedroom, visiting relatives ...
> If you're trying to thread a needle, shaking hands ...

There are at least two views on how the parsing process may work which are compatible with Fodor's view that it is encapsulated. Either both of the possible structural analyses are assigned immediately in all cases, and then a choice is made between them on the basis of higher-level contextual information (given in the preceding clause), or the analyses are constructed one at a time, but in a fixed order which takes no heed of context, say, first the adjectival and, then, if and only if this subsequently proves incompatible with context, the verbal analysis. However, if the process is open to top-down influence (i.e., it is cognitively penetrable) a single structural analysis compatible with available contextual information will be made at the outset, so just the verbal analysis in the first two examples above and just the adjectival analysis in the second two examples. This debate between the modularists, on the one hand, and the "strong interactionists," on the other, has spawned a huge programme of experimentation which is ongoing. Although the issue remains unresolved, there is certainly no shortage of results that seem to indicate that syntactic analysis is largely encapsulated, admitting only the very limited feedback from central systems of rejection of a particular analysis so as to prompt the system to try again (see, for instance, Ferreira and Clifton, 1986; Altmann and Steedman, 1988).

Another area of linguistic analysis in which the issue of whether contextual knowledge (say, knowledge about the topic of conversation) plays a predictive role is that of word recognition, those processes responsible for "deciding" which words of the language the acoustic stream represents. The case for cognitive penetration may seem quite well supported here; recall the SAQ 3.4 above, where the point was made that there is a lot of experimental evidence that subjects react more quickly to (i.e., identify) a word which is clearly related to the earlier material in a sentence than to a word which is not so related. For example, subjects are, significantly quicker to respond to "ants" than "mats" in the context of a sentence such as the following:

The woman was upset to find spiders, cockroaches, and ...

It seems then that central systems information (that, say, spiders, cockroaches, and ants are all of a kind, namely insects) is being fed into the lexical processor and affects the ease with which words are recognized. Fodor's response to this is that there is a crucial distinction that the "interactionists" are failing to make here, a distinction between lexical associations and propositional beliefs and judgments. This is the difference between having an associative connection between lexical items like "nurse" and "doctor," or "spider," "cockroach" and "ant," on the one hand, and having the knowledge/belief that doctors and nurses work together, that spiders, ants, etc. are types of insect, on the other hand. He postulates a lexical network, which consists of fixed associative connections between lexical items, as part of the proprietary data base of the language module (see Fodor, 1983, 1986). These connections may appear to be functioning like intelligent judgments but they are not; they give rise to dumb reflex-like responses: someone says "salt," you say "pepper," not because you are bringing to bear your belief that salt (in the world) is standardly accompanied by pepper, but because there is an associative connection between the phonological forms for them in your lexical network. On this view, then, the differential results for word recognition in the context of the sentence above are not due to penetration by higher-level knowledge but to a fast automatic activation of words immediately connected to each other in the network. Fodor's idea here is similar to Ramachandran's view cited above that the apparent "intelligence" of visual perception is a case of partial mimicry by the dumb perceptual routines of more thoughtful higher-level processes. The facilitated recognition of associated words is not a product of intelligent evaluation but of a module-internal lexical network which partially mimics our beliefs about how things connect up in the world.

So there are two hypotheses here: the encapsulation hypothesis and the interactionist hypothesis, both of which are compatible with the word recognition results just discussed. Is there any way to choose between them? Recall the point made in SAQ 3.2 above that recognition of the word "bread," for instance, facilitates recognition of the word "butter" as opposed to, say, "hammer," even when it is evident from the context that "butter" is completely irrelevant. This looks more like a dumb associative connection than an intelligent judgment made on the basis of available higher-level information. This result has been demonstrated experimentally by Swinney (1979) and Seidenberg et al. (1982), among others, in testing subjects' processing of ambiguous words such as "coach," "pen," "ring," "bug." Their results show pretty unequivocally that there is automatic facilitation of words related to the "bus" sense of "coach," for example, even when everything in the context dictates that only the "teacher" sense is appropriate, as in a case such as the following:

> The young tennis player wanted to talk to his coach.

This does strongly suggest that there is an initial set of encapsulated (cognitively impenetrable) processes of word identification, prior to the disambiguation process which is, of course, entirely dependent on considerations of context and relevance (for further discussion see Marshall, 1984; Marslen-Wilson and Tyler, 1987; and, for recent work which casts doubt on the results cited here, Moss and Marslen-Wilson, 1993).

SAQ 3.6 Consider the processes involved in determining the reference of "she," "it," and "there," necessary in comprehending an utterance of "She put it over there." What sort of knowledge is necessary for this? Are these modular (encapsulated) processes or unencapsulated higher-level cognitive processes?

In the next section we will consider briefly Fodor's view that the "higher" cognitive processes (such thoughtful processes as planning, decision making, fixing beliefs) are nonmodular.

The Central Systems and the Frame Problem

Establishing beliefs about the way things are in the world and making decisions regarding what to do in given situations are processes that can involve considerable weighing up of existing evidence and of likely outcomes. Such processes may be conscious or unconscious; they may differ greatly in the amount of time and effort one is prepared to put into them (deciding where to sit on the train to work will generally be quick and untaxing, while deciding whether or not to emigrate to New Zealand may be a lengthy and difficult process). Essentially, though, these all involve processes of forming a hypothesis ("I'll sit in that seat," "I'll emigrate to NZ") and seeing whether it is supported or not by relevant and available information. This process of hypothesis formation and evaluation is known as **non-demonstrative inference**; it is not demonstrative in that the conclusion finally reached does not follow in any absolute way from the facts considered; it is not provable in the way that a deductively inferred conclusion is (from the premises "Today is Tuesday" and "If today is Tuesday then I have a psychology lecture" the conclusion "I have a psychology lecture today" follows necessarily and must be true if the premises are true).

It is the job of central processes to integrate the deliverances of the modular input systems with each other and with relevant stored knowledge, in arriving at well-evidenced beliefs about the world and in making informed decisions. Such processes are domain-general and unencapsulated, that is, nonmodular; the information they draw on in any given instance of belief fixation may involve several knowledge domains and there are no built in constraints on how much information may be consulted before the process is terminated. For instance, the emigration decision may depend on evidence from any or all of your perceptual systems (the beautiful landscapes you saw, the fresh air you smelled, the good food you tasted, etc., when you were there on a visit), linguistically communicated information on job prospects, cost of living, standard of living, etc., and your stored mental representations concerning the living standards, governmental policies and environmental conditions of the country you currently live in.

All of this seems rather obvious, but the problem with these sorts of processes, according to Fodor, is that very little is known about how they work and there is no immediate prospect of achieving any deeper grasp of them. The difficult questions are: (a) how is a hypothesis (potential belief or decision) arrived at? (b) How is the relevant evidence marshalled? (c) How is the degree of support (or confirmation) of the hypothesis by the evidence determined? We do not know, says Fodor, and most likely cannot know, precisely because of the global, unencapsulated nature of these processes, that is, the absence of any architectural constraints on their functioning. He makes an analogy here with the unsuccessful bid to understand how science works. Science proceeds by forming hypotheses and looking at how well they are supported by the available evidence. Of course, this is a collective and very long-term endeavor and usually involves far more rigorous testing than our everyday nonscientific belief fixing, but the processes involved are, he claims, essentially the same. What we find in scientific practice is that for any given hypothesis about how some aspect of the universe works, there is no non-arbitrary way of delimiting the relevant data which may have a bearing on its correctness (i.e., it is an entirely unencapsulated process). For instance, confirmation of medical diagnoses may depend not only on established medical knowledge but also on information about the physical and social environment, or, to take a more central example for cognitive science, in accounting for the underlying structure of human language, the relevant data may concern both cognitive and physiological aspects of the human makeup and a wide range of different facts, physical, chemical, biological, about the typical human environment. As Fodor puts it "In principle, our botany constrains our astronomy, if only we could think of ways to make them connect" (Fodor, 1983, p. 105). Even less well understood than the processes of scientific hypothesis confirmation are the processes involved in coming up with scientific hypotheses. These seem to involve the quite mysterious capacity for "analogical reasoning." For instance, what was known about the structure of the solar system was taken over to model the structure of the atom; closer to home, facts about the symbol manipulating properties of computers and the relation of these to their physical hardware were applied to the far less accessible domain of human cognition, and have played a central role in the emergence of the discipline of cognitive science. There is no saying in advance where to look for good ideas when embarking on some scientific enquiry; good scientists often follow their hunches, they make a connection they haven't seen before while lying in the bath or they wake up after a good night's sleep with a productive new idea. There seems to be nothing systematic in the process. Some loose notion of unconscious analogical reasoning is probably involved but saying that is doing no more than labeling the problem. The point is we do not understand where good scientific ideas come from and no more do we understand where good (insightful, original, creative) ideas in everyday life come from (but see chapter 11 for an account of analogical problem solving).

SAQ 3.7 What important differences are there between scientific thinking and your own ordinary day-to-day thinking and decision making? How sound is Fodor's analogy between the two, in your view?

The problem presented by the global nature of central thinking is starkly apparent in attempts within **artificial intelligence** to get machines to perform even the simplest of belief-fixing or decision-making tasks that human beings generally perform quite effortlessly. The specific form of the **frame problem**, as this is known, is usually expressed as the problem of getting the robot to revise or update its beliefs (knowledge representations) so as to appropriately accommodate the changes brought about in the world by its own actions. Fodor gives the following example of a simple action which could apply equally to robots or humans: turning on the refrigerator. "When I turn my refrigerator on, certain of my beliefs about the refrigerator and about other things become candidates for getting updated. For example, now that the refrigerator is on, I believe that putting the legumes in the vegetable compartment will keep them cool and crisp ... Similarly, now that the refrigerator is on, I believe that when the door is opened the light in the fridge will go on, that my electricity meter will run slightly faster than it did before, that a certain humming noise that wakes me up in the night is the result of the fridge being on, etc." (Fodor, 1987, p. 30). That is, on the one hand, this simple action has a huge range of side effects and requires considerable updating of beliefs. On the other hand, it must also be the case that a great many of the system's beliefs, in fact most of them, are not candidates for updating as a result of plugging in the fridge: the belief that 2 + 2 = 4, the belief that the French word for airplane is "avion," the belief that mother likes chocolates, etc.

The point is that appropriate belief update is generally not a problem in practice for humans – we just do it and usually without much deliberation – while for the robot designers there is no nonarbitrary way of framing that subpart of the robot's data base that is to be adjusted. Of course, there are various bits of gerrymandering that can be done to tag the data likely to need revising as a result of some particular action the robot is programmed to perform (see papers in Pylyshyn, 1987); this is obviously a practical solution in many cases where what matters is just that some very delimited task be performed and some delimited set of representations be adjusted accordingly. But, clearly, this way of being artificially intelligent does not reflect natural human intelligence where belief revision is an unencapsulated process which is performed effortlessly and appropriately in response to any of thousands upon thousands of actions of varying degrees of complexity. The problem, then, for cognitive science is why the frame problem is, in fact, not a problem in practice for human cognition; it is, in short, the problem of describing and explaining how unencapsulated domain-general processes work.

SAQ 3.8 Consider Lucy's action of opening the window in the scenario. Which beliefs of hers will inevitably be updated as a result? What are some of the (many) beliefs she has that won't be affected at all?

Fodor's conclusion, then, is that cognitive scientists would do well to concentrate their attention on coming to grips with how our perceptual, linguistic and motor capacities work and give up on the central thought processes. "The condition for successful science is that nature should have joints to carve it at: relatively simple subsystems which can be artificially isolated and which behave, in isolation, in something like the way that they behave in situ" (Fodor, 1983, p. 128). In other words, the scientific study of the mind is likely to be limited to those aspects of the mind which are modular. You will have seen from the table of contents of this book that we do consider some topics that lie outside those that Fodor considers tractable (for instance, how conversations work in chapter 8 and general problem solving in chapter 11); we leave it to you to judge whether or not we should have heeded Fodor's warning.

In the next section we will consider a number of views on why the mind should be modular, at least to some degree. These will not include the point just made that modularity is conducive to scientific study! It seems rather unlikely that our minds have come to be arranged in any particular way so as to make it easier for us to investigate and understand our own cognitive capacities. What most of the arguments for the advantageous nature of modularity have in common is a "design" stance; that is, they are of the following form: "in constructing a cognitive system with the capacities of the human mind, a modular arrangement of parts makes good design sense because ... " However, if Fodor is right about the central systems, the opposite question also has to be addressed: why should the mind be, at least partially, nonmodular?

How Modular Should the Mind Be?

Why should any complex system be modular, that is, made up of components which are largely autonomous, having fixed and limited connections to other components? A frequent response to this question is to cite Simon's (1962) parable about the two watchmakers, Hora and Tempus. They both made fine watches of 1,000 parts. Tempus constructed his watches in such a way that "if he had one partly assembled and had to put it down – to answer the phone say – it immediately fell to pieces and had to be reassembled from the elements." By contrast, Hora who made watches no less complex, designed them " ... so that he could put together subassemblies of about ten elements each. Ten of these subassemblies, again, could be put together into a larger subassembly; and a system of ten of the latter subassemblies constituted the whole watch." The economic viability of the two watchmakers obviously differs: " ... when Hora had to put down a partly assembled watch in order to answer the phone, he lost only a small part of his work, and he assembled his watches in only a fraction of the man-hours it took Tempus" (Simon, 1962, p. 470).

Simon goes on to apply his parable to the process of biological evolution, pointing out that the approach of Tempus simply could not mirror the process of **natural selection** by the environment, the process which, as Darwin has shown us, accounts for all existing forms of life. The assembling over time of complex organic systems, for instance the human mind/brain, proceeds in a piecemeal fashion, adaptation by adaptation, solving the problems posed by changing environments as they arise, one by one. The evolutionary force of natural selection is quite different from both Tempus and Hora in that it does not have their foresight and purpose (it has no ultimate design in mind); as Dawkins (1986) has put it, nature is a "blind watchmaker." However, natural

Devil's Advocate Box 3.3

Are our central systems really as unstructured and unencapsulated as Fodor claims? Does the long-term collaborative enterprise of scientific theorizing really provide the best model for everyday processes of thinking and decision making? Should cognitive scientists really abandon any attempt to understand the higher cognitive processes of the human mind?

Sperber and Wilson (1986, 1995) are doubtful about all these propositions. They point out that a great deal of ordinary thinking differs markedly from scientific thinking in that it involves spontaneous, instantaneous, and unconscious inferences. The slow, careful, self-conscious processes of scientific thinking and testing may involve quite different inferential techniques from those we all employ when making mundane spontaneous inferences about, say, whether we should take an umbrella with us, how to avoid a wavering cyclist when driving, or what our neighbor intends to communicate when she points out that our tree hasn't been pruned for a long time. Sperber and Wilson focus on the last of these, that is, on the non-demonstrative inferential process of comprehending utterances. This is unencapsulated like scientific theorizing in that it may take its premises from virtually any knowledge domain. For instance, in the scenario, Dad's utterance "It's Etna in here" communicates that the kitchen is unacceptably smoky, and that Lucy should rescue the toast and open the window. To arrive at this interpretation, in addition to linguistic decoding of the sentence used, Lucy must use perceptually derived information in assigning a location to "here," and stored general knowledge about the volcano Etna and her personal knowledge of her father and his likely expectations of her.

Sperber and Wilson's account of how these rapid spontaneous inferential processes work depends on a few very intuitive assumptions about cognition: that our stored knowledge is differentially accessible and that the relative accessibility of bits of knowledge is shifting all the time and can be altered by, for instance, the prompting provided by the concepts encoded in an utterance; that the less accessible a piece of knowledge is the more the cognitive effort involved in retrieving it; that cognition is least-effort oriented so that there are clear limits on how much effort anyone will expend in deriving relevant information. On this basis they develop an account of the spontaneous, rapid, non-demonstrative inferential processes of comprehension. The communicative principle of relevance, which provides the necessary constraint on the inferential processes, is discussed in chapters 8 and 9. The point in the present context is that this principle of relevance provides an account of why people are not prone to the frame problem in arriving at a confirmed hypothesis about a speaker's communicative intention. With luck the discoveries in the realm of utterance comprehension may carry over to other varieties of central systems thinking. Fodor may be unduly pessimistic about the prospects for the study of higher cognitive processes.

selection is just like Hora in that it depends on stable intermediate forms on which the next adaptation may build.

The evolutionary argument for modularity is compelling and it is abundantly supported by **ethological** investigations of the cognitive mechanisms underlying the behavior of non-human species. For instance, various species of bird have distinct modules underlying their singing activity, their navigational capacity and their food-caching, each apparently served by its own memory which is incompatible in its functioning with the memory of the other systems (see Sherry and Schacter, 1987). To cite just one more example, the spatial location system of rats has been shown to rely entirely on geometric information and to be encapsulated from other types of information, such as color, texture and smell, which the rat is capable of perceiving independently (see Gallistel and Cheng, 1985).

Marr (1976) provides further argument for what he explicitly calls "the principle of modular design" whereby any large computation "should be split up and implemented as a collection of small sub-parts that are as nearly independent of one another as the overall task allows" (Marr, 1976). The advantage of this is that making a change in one place need not affect other places. A nonmodular design, on the other hand, means that the system becomes "extremely difficult to debug or to improve, whether by a human designer or in the course of natural evolution, because a small change to improve one part has to be accompanied by many simultaneous compensating changes elsewhere" (Marr, 1976). The principle of modular design is popular in computer programming and artificial intelligence for this sort of reason. There are many practical reasons for making programs modular: a module can be, as it were, extracted whole from one complex program and zipped onto other programs; a malfunction (bug) can be more readily isolated and repaired if the overall program is modular. So we seem to have here a quite general argument for the modularity of complex systems of which the human mind is the most striking example.

SAQ 3.9 The McGurk illusion shows that speech perception is cross-modal in that it responds not only to a particular type of acoustic stimulus but also to a particular type of visual stimulus. Why might the cognitive subsystem responsible for speech analysis have evolved in this particular way?

These considerations would seem to mediate rather strongly against any conception of the mind as a general all-purpose problem solver. There is no support here for any sort of uniform cognitive architecture, such as that which the behaviorist approach to human intelligence entailed, or those which have been proposed more recently by some cognitive scientists with a strong artificial intelligence leaning, such as the SOAR architecture of Newell (1990, 1992) discussed in the previous chapter.

SAQ 3.10 Most people find it quite natural to think of the human body as consisting of distinct organs each designed or adapted to solve a particular problem: to pump blood, to detoxify blood, to digest food, etc. Can you conceive of a unitary general account of the body? What would it be like?

By now you may be convinced that cognitive architecture must be entirely modular. This position was in fact proposed two centuries ago (note that this was well before the theory of evolution), by the great anatomist and phrenologist, Franz Joseph Gall, and has more recently been endorsed by Cosmides and Tooby (1994a, 1994b) who are strong advocates of an evolutionary or design perspective in cognitive science. As we've seen, though, this is not Fodor's view: input (perceptual) systems and, doubtless, certain output (motor) systems are modular, but central processes of thought and belief fixation are nonmodular. Let's try to see why he takes this position and assess whether he has good grounds for it. First, what he has to say about the modularity of the input systems meshes well with the points made by Simon and Marr above. He takes a **teleological** (design-oriented) stance; that is, he looks at the function of perception in our lives and argues that given that function it ought to be modular in structure. The function of perception is to work out the way things are in the local environment on the basis of the effects these things (distal stimuli) have on various sense organs; "it [perception] is built to detect what is right here, right now – what is available, for example, for eating or being eaten by" (Fodor, 1989). It is therefore understandable that perception should be performed by fast, mandatory, encapsulated systems, enabling immediate reaction, and that err, if at all, on the side of false positives. So, for instance, perception may jump to conclusions of a potential danger when there is none (a false positive), but that is a small price to pay for consistently getting it right when the distal stimulus is in fact a threat to survival.

The property of informational encapsulation is crucial here; an efficient survival- promoting visual perceptual system should not retrieve from memory everything it knows about tigers in the process of deciding that some distal stimulus is indeed a tiger. Even if retrieval processes operate in parallel and are very fast, there would still be the problem of deciding for each bit of tiger-related information how much support (degree of confirmation) it gives to the hypothesis that there is a tiger in the visual field. What the blind, dumb processes of perception buy is a certain objectivity: even if the last thing you expect to encounter at your local shopping precinct is a tiger, it is as well that perception ignores expectations and beliefs, and does its identificatory job quickly and directly.

Let us move now to what seems in the present context to be the more contentious aspect of Fodor's thesis, that is, the alleged nonmodularity of the higher-level problem-solving, decision-making systems. He summons a teleological argument here too: "If the perceptual mechanisms are indeed local, stupid and extremely nervous, it is teleologically sensible to have the picture of the world that they present tempered, reanalyzed, and ... above all *integrated* by slower, better informed, more conservative, and more holistic cognitive systems. The purposes of survival are, after all, *sometimes* subserved by knowing the truth ... Nature has contrived to have it both ways, to get the best out of fast dumb systems and slow contemplative ones, by simply refusing to choose between them" Fodor (1985, p. 4). The idea is that modular perceptual systems, geared as they are towards quick self-preservatory reaction, are shortsighted and irrational. Humans are fitted up with other cognitive systems which are not prey to such defects, systems which are not encapsulated but which can draw on an indefinite range of information and so enable us to discover deeper regularities about the way the world works; "I believe we are devices built to find out what's true" (Fodor, 1986, p. 18).

It may be "teleologically sensible" to have domain-general, unencapsulated systems, that can take the long-term view, as it were, but is it evolutionarily plausible? Fodor sees these systems as a late evolutionary development, preceded by the modular input systems: "Cognitive evolution would thus have been in the direction of gradually freeing certain sorts of problem-solving systems from the constraints under which input analyzers labor – hence of producing, as a relatively late achievement, the comparatively domain-free inferential capacities which apparently mediate the higher flights of cognition" (Fodor, 1983, p. 43). Sperber (1994a) takes issue with this idea, pointing out that it requires a process of *de*modularization, a process that results in the merging of modules, and that the evolutionary story in terms of natural selection does not provide any means by which this could take place. It may well be advantageous "to trade a few domain-specific inferential micromodules for an advanced all-purpose macrointelligence ... But evolution does not offer such starkly contrasted choices. The available alternatives at any one time are all small departures from the existing state" (Sperber, 1994a, p. 44). Natural selection is, at most, very shortsighted and seizes opportunistically on immediate solutions to immediate problems. While Fodor's conception of human central thought processes may not be wrong it receives little support from evolutionary considerations. Sperber favors a different view of the central systems, on which they too are virtually entirely modular though with rather complex interconnections mediating the flow of information among the discrete modules. This, he says, is the kind of cognitive organization that evolutionary considerations lead us to expect in a species which relies heavily on its cognitive abilities for its survival.

In this section we have taken a **phylogenetic** (evolutionary) perspective on human cognitive architecture. In the next section we move to an **ontogenetic** perspective and consider what the study of the neonate mind and its subsequent development indicates about human cognitive architecture.

Progressive Modularization and Going Beyond Modularity

A central issue within the study of cognitive development is the initial or core cognitive architecture which the child brings to the developmental process; that is, the mechanisms which form the basis for cognitive development rather than its outcome. On one view, baby architecture is essentially homogeneous, and any articulation within adult cognitive architecture is the result of developmental processes which gradually **assemble** and differentiate systems. The alternative view is that mental architecture is heterogeneous with specialized subsystems from the beginning. Fodor's innate perceptual and language modules, of course, fall within the latter view. The developmental psychologist Leslie (1988, 1994) argues for even greater modularity within core architecture, pointing out the great benefits for cognitive development if certain central subsystems, such as the child's **theory of mind** (the ability to attribute states of mind, like beliefs and intentions, to other people) and theory of physical objects, are embodied in modular mechanisms. However, the focus of this section is, as the phrase "progressive modularization" in the title indicates, a representative of the first view.

Karmiloff-Smith (1992, 1994) argues against Fodor's thesis of innately specified mental modules as entailing a too rigid view of the mind. She makes a case for processes of progressive modularization over time of a wide range of mental abilities including, but going way beyond, the perceptual and linguistic capacities to which Fodor attributes modularity. She also argues for a complementary set of processes of what she calls "**representational redescription**," which involve several cycles of recoding existing knowledge so that it becomes progressively more available to other knowledge systems, that is, less and less encapsulated.

Before we look at these processes, we need a little background. Karmiloff-Smith is attempting to steer a middle way between the strong nativism and modularity of theorists like Fodor and Chomsky, on the one hand, and the anti-nativism and anti-modularity of the great pioneer of developmental psychology, Piaget. Piaget's position is known as **epigenetic constructivism**: cognitive development is the outcome of a self-organizing system that is structured and shaped by its interaction with the environment. The mind of the new born is essentially unstructured and knowledge-free; it is equipped with just three domain-general processes which, in conjunction with a few innate reflexes such as sucking, looking, and grasping are all that the child brings to the developmental process. The environment supplies the rest. These three self-regulatory processes are the complementary processes of **accommodation** (whereby adjustments are made to the cognitive system to accommodate incoming data), **assimilation** (whereby incoming data is made to fit into the existing cognitive system), and the process of **equilibration** (whereby a structure settles into a relatively steady state so that it can accept and process data without further change). The child acts on the environment, initially just employing the few sensorimotor reflexes at its disposal, and the environment in its turn has a huge role to play in the gradual emergence of structure in the mind. Piaget's view of infant architecture bears strong similarity to that of the behaviorists': it is a uniform, knowledge-empty, all-purpose learning system with a few domain-general processes that account for all aspects of subsequent development, whether perceptual, linguistic or higher-level thought. The main difference between Piaget and the behaviorists is his view that infants are active participants in assembling their information structures rather than merely passive recipients and storers.

Karmiloff-Smith departs from Piaget in acknowledging innate domain-specific predispositions to attend to particular stimuli in the environment and, for a few domains, including language, innately given skeletal knowledge structures. In short, she supports some degree of **Cartesian nativism** (that is, the view that beliefs or knowledge may be innate), but not **architectural nativism** (that is, the view that the performance mechanisms that put knowledge systems to work are innate). She departs from Fodor in that she regards the encapsulation (hence the modularity) of various mental capacities to be a result of developmental processes rather than a genetic given. In her view, the claim that modules are innately prespecified does not comport with the known plasticity of the infant mind; she cites in this regard the way in which what would be the auditory processing region of the brain of a hearing child is taken over by visual processing in a deaf child. Another example is the emergence of language processing circuitry in the right hemisphere of the brain if the areas of the left hemisphere usually responsible for language have been extensively damaged at a very early stage in a child's life.

Furthermore, Karmiloff-Smith considers the standard nativist argument regarding the poverty of environmental input to be overstated: "we are discovering that environments have far more structure than was originally thought" (1994, p. 698). In her view, as well as the innate predispositions and knowledge that govern every child's growth there are important constraints placed by those species-typical aspects of the environment within which the child develops. It is the interaction of both common innate and common environmental constraints that accounts for the common paths of development across individuals of many cognitive subsystems; abnormal development may be the result of either abnormality in the innate endowment or in the environment.

Two points must be noted about this idea of developmental modularization. First, it is applied to a wide range of cognitive capacities, many of which would fall within the central systems rather than the input systems for Fodor. As well as language, she discusses the modularization of such systems as knowledge of the physical properties (weight, mass, rigidity, etc.) of objects, the mathematical ability to manipulate numbers, and the ability to attribute psychological states, such as beliefs, desires and intentions, to other people, in order to explain or predict their behavior (theory of mind). She also discusses the automatization and effective encapsulation of skills that may be learned later in life, such as piano playing and touch-typing. Second, as she acknowledges, this idea that a modular architecture is the result of gradual development rather than being innately specified weakens Fodor's criteria for a module; these "modules" are not hard-wired nor of fixed neural architecture. But, she emphasizes, the outcome of progressive modularization is domain-specific, fast, mandatory cognitive systems, which are relatively encapsulated. The encapsulation criterion seems to be somewhat weakened for this notion of an assembled "module": you will see that this is necessary if you try applying the cognitive (im)penetrability test to a learned automatized skill or to a central system capacity such as attributing intentions or beliefs to a fellow human being.

SAQ 3.11 Consider the ability to read fluently. Would this be a module by Fodor's criteria? Would it be a module by Karmiloff-Smith's criteria? If your answer is "no" in either case, which criteria does it not meet?

Let us move now to the second of the two types of domain-general process that Karmiloff-Smith sees as central to the development of the human mind: representational redescription. In line with the ethological studies mentioned earlier, she points out that as regards the modularity of our cognitive architecture we are like every other species. What distinguishes us is our capacity to recode existing knowledge into different types of representations making that knowledge accessible to different parts of the mind and so able to interact with other knowledge domains. Consider the beaver's innately specified capacity to build dams; this is a set of encapsulated procedures which prompts particular set behaviors. The beaver can't do anything else with this knowledge; it can't use it to modify the design of its dams or to build some other sort of structure or to reflect on the ultimate futility of dam building. This is implicit procedural knowledge which stays implicit, encapsulated, inaccessible. Human children, on the other hand, as well as developing fast, automatically activated, encapsulated procedures for using language, for doing arithmetic, and for interacting with the physical world, become able to think about their knowledge in these domains, to adapt it to new situations, to make links across different domains, and ultimately to talk about it. It is this internal (and, presumably, innately given) capacity to extract information from implicit knowledge systems, and reformulate and relocate it that underlies the flexibility and creativity of the central systems of the human mind.

Karmiloff-Smith claims that there are at least two levels of redescription, that is, of making knowledge progressively more explicit and more available for interaction with other knowledge domains; the first is a level at which generalizations or rules are induced from a bunch of procedures, the second is a level of reformatting that makes this sort of knowledge available to consciousness and perhaps verbal report. Evidence for the first of these levels, at which the reformulated knowledge is still inaccessible to consciousness, comes from a well-known phase in the development of a range of domains when children formulate over-general rules and their behavior on some task degenerates as a result. One of the many examples of this is the phase in which children form incorrect past tenses such as "putted," "goed" and "runned," having produced the correct irregular past tenses at a younger age. Evidence for the second level is relatively straightforward since we are able to experience our own conscious thinking about, for instance, language or gravity or the number zero.

The upshot of these internally driven redescriptive processes is that any one knowledge domain may be represented several times over in the mind: as implicit knowledge in a fast, automatic, encapsulated system and as explicit, relatively unencapsulated knowledge which is available for integration with other knowledge domains. In this way, Karmiloff-Smith is able to give an account of how the human mind "goes beyond modularity."

We cannot attempt here a full assessment of this "modularization and redescription" framework. Let us, however, consider how well it is supported by what we know about the input systems, perception and language, Fodor's chosen cases of innately specified modules. With regard to language, Karmiloff-Smith expresses doubt that there is a ready-made module, part of the infant's core cognitive architecture. She says "Attention biases and some innate predispositions could lead the child to focus on linguistically relevant input and, with time, to build up linguistic representations that are domain-specific. Since we process language very rapidly, the system might *with time* close itself off from other influences – i.e., become relatively modularized" (Karmiloff-Smith, 1992, p. 36). However, the aspects of language she focuses on in demonstrating both modularization and redescription are, from the point of view of current work in linguistics (see chapter 7), rather peripheral: the development of inflectional endings, such as those for the past tense or plural, of determiner systems ("a," "the," "my," etc.) and of personal pronoun systems. As Smith (1994) points out, there is a striking contrast between these aspects of language development that support the modularization plus redescription thesis, on the one hand, and the development of the syntactic system, on the other. Correct sentence structure seems to unfold effortlessly, without relevant environmental evidence and without the sort of developmental phase of mistakes and overgeneralization which typifies the acquisition of, for instance, the past tense affix. Furthermore, the knowledge that this calls on remains **tacit**, forever inaccessible to conscious introspection. In other words, this core linguistic knowledge offers no support for either of the general developmental processes of modularization or redescription.

SAQ 3.12 Consider now the perceptual modules. Do you think that children come spontaneously to reflect and comment on the principles that underlie their visual or auditory capacities, or is perceptual knowledge in humans, like the dam-building instinct in beavers, forever implicit and inaccessible?

Karmiloff-Smith's thesis does not seem to present a challenge to the central Fodorian claim concerning the innately specified modularity of the input systems. We seem to be dealing with (at least) three rather different kinds of mental systems here: (a) Fodorian modules (perception and language); (b) domain-specific knowledge systems with some degree of innate specification (theory of mind, number theory, physical theory); (c) automatized skills for which the mind is not directly programmed and which require a considerable learning effort. The latter two may fall within the range of Karmiloff-Smith's "modularization and redescription" framework, provided we accept a broader concept of a mental module than Fodor does.

Summing-up

There seems to a broad consensus nowadays that the cognitive architecture of adult humans is, at least partly, modular. In Fodor's view the perceptual and linguistic processors are modular mechanisms while the bulk of higher-level central processing is non-modular. In seeking an account in terms of natural selection of how complex cognitive systems could have evolved we find support for a more thoroughgoing modularity of mind. This architectural issue arises again when we consider the mind of the neonate and its subsequent development. Here there are at least two views: (a) the initial architecture has modular perceptual and linguistic (and perhaps other) systems, which greatly facilitate the processes of acquiring knowledge about the world; (b) the initial architecture is not modular and the modularity of the adult mind is the result of a developmental process of gradual modularization.

We have to conclude that the term "module" is being used in two rather different ways. While both uses take domain-specificity to be a central feature of a mental module, they differ on a number of other criteria. Fodor's modules are innate, richly specified, neurally fixed, and primitive. The view that modular organization of the mind emerges as a result of developmental processes allows for "modules" that are neither innately given nor neurally fixed and that are effectively assembled through interaction with the environment. As for encapsulation (or cognitive impenetrability), while this is the essence of modularity for Fodor, it seems to be a more relative notion on the second view.

Two last points. First, now might be a good time to go back to the quote at the beginning of this chapter and reconsider whether it is better to think of the mind as analogous to a Swiss army knife or to an all-purpose blade. Second, in the chapters that follow, the term "modularity" will arise fairly frequently, not always accompanied by a gloss saying whether the innate Fodorian or the constructed variety is intended. It will be as well to bear in mind the two ways in which the concept is understood and it could be seen as an ongoing exercise to clarify for yourself which sense seems to be the appropriate one (if either!) in any given instance.

Further Reading

For an excellent, quick and comprehensive account of Fodor's view of mental architecture as partly modular and partly nonmodular see Garfield's introduction to the volume of papers on (and against) modularity that he edited: J. Garfield (ed.) (1987) *Modularity in Knowledge Representation and Natural Language Understanding*, Cambridge, MA: MIT Press. The most accessible of Fodor's own writing in this area is J. Fodor (1986) The modularity of mind, in Z. Pylyshyn and W. Demopoulos (eds), *Meaning and Cognitive Structure*, 3–18, Norwood, NJ: Ablex. However, the basic and essential source for this chapter is the short monograph, J. Fodor (1983) *The Modularity of Mind*, Cambridge, MA: MIT Press.

The volume edited by L. Hirschfeld and S. Gelman (1994) *Mapping the Mind: Domain Specificity in Cognition and Culture*, Cambridge: Cambridge University Press, contains several papers which take Fodor's mental architecture as their point of departure; we recommend Sperber's, Cosmides and Tooby's and Leslie's as well as the introductory overview by Hirschfeld and Gelman.

A. Karmiloff-Smith (1992) *Beyond Modularity: A Developmental Perspective on Cognitive Science*, Cambridge, MA: MIT Press, is a very useful and interesting book; it compares the Fodorian view with the Piagetian view, it sets the modularity issue in a developmental context and it attempts to establish a compromise position involving both modularization and demodularization.

CHAPTER 4

Surfaces, Objects, and Faces

Outline

As discussed in chapter 3, we can think of the visual system as a modular component of the cognitive system which can itself be subdivided into separate modules each dealing with a particular aspect of visual experience. The aim of this chapter is to provide an understanding of the issues and problems in the areas of object perception and object recognition, and to illustrate some of the theoretical approaches that have been developed to account for our perception and understanding of the nature of objects. The first section considers how we should evaluate a theory of object perception. The next section outlines problems with feature matching approaches to recognition and the requirement for structural descriptions. We then address the problem of object constancy. How can we recognize an object from a variety of viewpoints? Object constancy is not perfect and so we also consider evidence that performance in object recognition tasks does, under some circumstances, depend upon viewpoint. Objects are made up of parts, and strategies for partitioning objects into their constituent parts are explored. Faces are an important class of object that have been subject to more careful study than other classes of objects and therefore we look at the specific problems addressed by theories of face recognition in some detail. Finally we consider the problem of describing the global shape of objects.

Learning Objectives

After reading this chapter you should be able to:

- distinguish between template matching, feature matching, and structural description approaches to recognition
- describe methods for representing the shape of surfaces
- appreciate the problem of object constancy and describe approaches to solving this problem
- describe factors affecting face recognition, discuss how we represent faces and discuss the processes which constrain the shape of faces
- appreciate the problem of the representation of global shape

Key Terms

- 2½D sketch
- global shape
- local shape
- object constancy
- primitives
- structural description

Introduction

When Lucy says, "Anyone seen the tray?", she is inviting all the members of her family to look for the item. First they are expected to search their memories. Perhaps they placed it down somewhere, or perhaps they spotted it earlier without taking much interest, or had tripped over it coming into the kitchen. If they have no recent memory of seeing the tray, the next thing she expects them to do is scan the cluttered environment of the kitchen; the floor, the work surfaces, the table, starting with places they would expect to see the tray and then searching places one might not expect to see a tray.

When one of the family catches sight of the tray there is no certain way of knowing exactly what it will look like. How the tray will look to each member of the family will depend upon the position of the individual observer relative to the tray and how the tray is positioned in 3D space. Because each member of the family has a different view of the tray the retinal images formed in the eyes of each person will be different. There may be other objects in the way which partially block the view. The exact pattern of light and dark will depend upon the illumination at that place in the room, whether there is light from the window or from the electric lights in the ceiling. There may be shadows from other objects falling on the tray, breaking up the contours. Dad may have never seen that particular tray before, even though it was bought three months before.

At least trays are rigid objects. They do not change their shape over time. If Lucy had asked, "Has anyone seen the cat?", there would be an additional difficulty. The family would not know if the cat was curled up in a ball, sitting with its tail curled as in a child's drawing, or standing, stretching with its back arched and its tail high. Nevertheless, when one of the group sees the cat, and points it out, under most circumstances, none of the others would have any difficulty in recognizing that it is a cat, seeing its location and whether the patterns of light and dark on its coat result from surface markings or shadows.

A cognitive science approach to this problem starts with a close analysis of the task the visual system has to perform. Of course we know that the family have to search for the object, form an appreciation of the object from the pattern of light falling on the eye, and identify it as a cat, but an analysis at this level would not be sufficient to tell a "man from Mars" how it is done or to design a computer program for a new robot visual system which we would like to fit to an artificial breakfast chef. However, cognitive science expects that an answer of sufficient detail to allow implementation in an artificial system is both possible and necessary for a complete understanding of object recognition. It is important to note that the aim here is to understand cognition, and particularly human cognition, not to build an artificial visual system which may or may not have to struggle with the problems faced by the human visual system.

The questions that a cognitive scientist has to ask about the activity in the scenario in relation to how the family see and interact with objects are:

- How do we characterize the problem of object perception?
- What form does the solution to the problem take?
- What intermediate representations are involved between the images on the retinae and the solution?
- What mechanisms link these representations?
- How do we evaluate a theory of object recognition?

We will address some of these questions shortly, but before we consider specific theories of object perception we should think about how we might evaluate a theory of object recognition once we have it.

Marr and Nishihara's Criteria

How should we evaluate a theory of object recognition? Marr and Nishihara (1978) considered three sets of criteria with which to assess the design of an object recognition system; (a) accessibility, (b) scope and uniqueness, and (c) sensitivity and stability. We consider each in turn. *Accessibility* is essentially an engineering criterion. We need to evaluate how easy it would be to derive a particular object description from the input image or pairs of images on the retinae. Marr and Nishihara used this criterion to argue that the recovery of the object description should proceed directly from a representation of **features** present in the image, such as object boundaries and contours rather than derived properties which might encode something of the three-dimensional surface structure or volumetric properties of the object. There is some merit in this and it follows on from Marr's more general advocacy of modularity in the design of biological systems. We may want to have many routes to an object level description. Of course what is easy for current algorithms in computer vision, running on the most recent serial hardware, has only passing relevance to what might be easy for a biological system which has evolved through many simpler variants, and where the computational architecture is predominately parallel rather than serial. What is important for a biological system is whether the methods it adopts are effective and robust rather than whether they are easy to implement.

SAQ 4.1 How easy do you think it would be to recognize an object from a cast shadow? Are shadow puppets designed in any particular way?

Scope is a statement of the class of objects that are to be included within the particular representational framework. The set of terms that you would wish to use to provide an efficient description of faces may be different from the set that you would wish to use to most efficiently describe household furnishings or citrus fruits.

The question here is one of generality. Should we decide to use a descriptive term that is sufficiently general that it can be used for all objects or should we design a set of descriptors which are specific to a particular object class? *Uniqueness* addresses the problem of object identity. In order to say whether

Devil's Advocate Box 4.1 Shape primitives

4.1 Objects which share some properties of facial features can easily be organised to give the impression of a face.

We tend to think that the primitives of a shape description should be appropriate and specific to the object we are attempting to describe, but it is quite possible to build an object, here a face, out of quite inappropriate objects. The figure shows a face built from the contents of a cooked breakfast.

two objects are the same, even though the images they project on the retinae may be different (because of transformations in scale or **viewpoint** or because the objects have altered over time), we have to compare the descriptions we have derived from the images or compare the current description with a description recovered from memory. If the descriptions are the same then the objects can be considered to be the same even though the orientation from which we view the object may well be different. Now if the system can come up with two different descriptions given the same object we have the problem of how we can tell that the two descriptions refer to the same object. It is therefore necessary, in Marr's view, that any particular object should always give rise to the same description. In fact perceptual psychologists and artists have been cunningly inventing objects over the years, like the Necker cube and Rubin's vase, which do give rise to more than one plausible object description as was discussed in chapter 3. Inevitably what is seen is a reversal in the interpretation of the object from time to time. It would appear that even if two descriptions are possible only one of the alternatives is available at any given time.

The most important of Marr and Nishihara's criteria are *stability* and *sensitivity*. Any system for object recognition should allow us to say that two objects are the same in some sense and also allow us to discriminate between them. A horse and a cow share certain properties (figure 4.2). They both have four legs, a body, a head, and a tail. The body parts articulate in similar ways. We can easily see the similarities. However, a tipster at a racetrack should not only be able to note the difference between a horse and a cow, but he should also be able to discern the fine differences in gait and stature between one thoroughbred and another if he is to make a considered judgment upon

4.2 A scheme for representing objects must be able to allow for generalization across similar objects and discrimination between objects.

which to place his bet. We need to develop a descriptive representational system that allows us both to generalize and to discriminate.

A stable description is one in which minor changes in the object do not affect the derived representation. Partial descriptions such as, "John is composed of a head, a body, two arms and two legs," are stable descriptions. Stable descriptions support generalization across instances. The description of John is always true irrespective of his current posture. A sensitive description is a detailed description which changes to reflect any minor change in the object, like a curl of the lip or a narrowing of the eyes. Complete descriptions, which could involve details of the position of each hair on the head, are sensitive descriptions. Sensitive descriptions support discrimination. This flexibility is an essential requirement for an effective system for the representation of objects. For example, by setting a criterion similar objects can be grouped and others excluded. Repeated application of this operation would allow the ordering of objects along a dimension, a fundamental requirement for a system that wishes to make relative judgments, such as, our chair is more similar to John's chair than it is to Mary's chair. Now we have some criteria in mind with which we can begin to evaluate theories of object recognition we can turn to strategies. Most theories of object recognition draw on one or more of the strategies outlined in the next section.

Templates, Features, and Structural Descriptions

In the scenario all the family members are asked if they have seen the tray but when they search the scene for a tray how do they know whether they have found the item in question? If they had a clear picture in mind of how the tray would look then they need only match up the object with their visual **template**. However, as we have seen from the previous discussion of the scenario, the problem is in the premise. Recognition by templates is characterized by a matching process involving information which is represented in the form of images. Template matching works well in situations in which the material to be discriminated and classified falls into a small number of uniquely defined categories. Perhaps the best example of successful template matching systems are automatic character readers for information on bank cheques. Each of the characters is designed to have a unique pattern of light and dark. There is a sense that all systems for recognition involve a process of matching, but templates are too rigid a way of holding information about an object. The problem is that there are many simple ways in which one can transform an object to make a template quite useless. Transformations that are continuous, like magnifications or rotations, provide particular difficulties. In many cases we only need to shrink or rotate the object to disrupt the match, and if we had a template for all rotations and sizes of objects we would soon run out of storage space in any conceivable physical device. If we want to retain the idea of an image-based match we have to develop a much more sophisticated

Box 4.1 Representing the information content of images

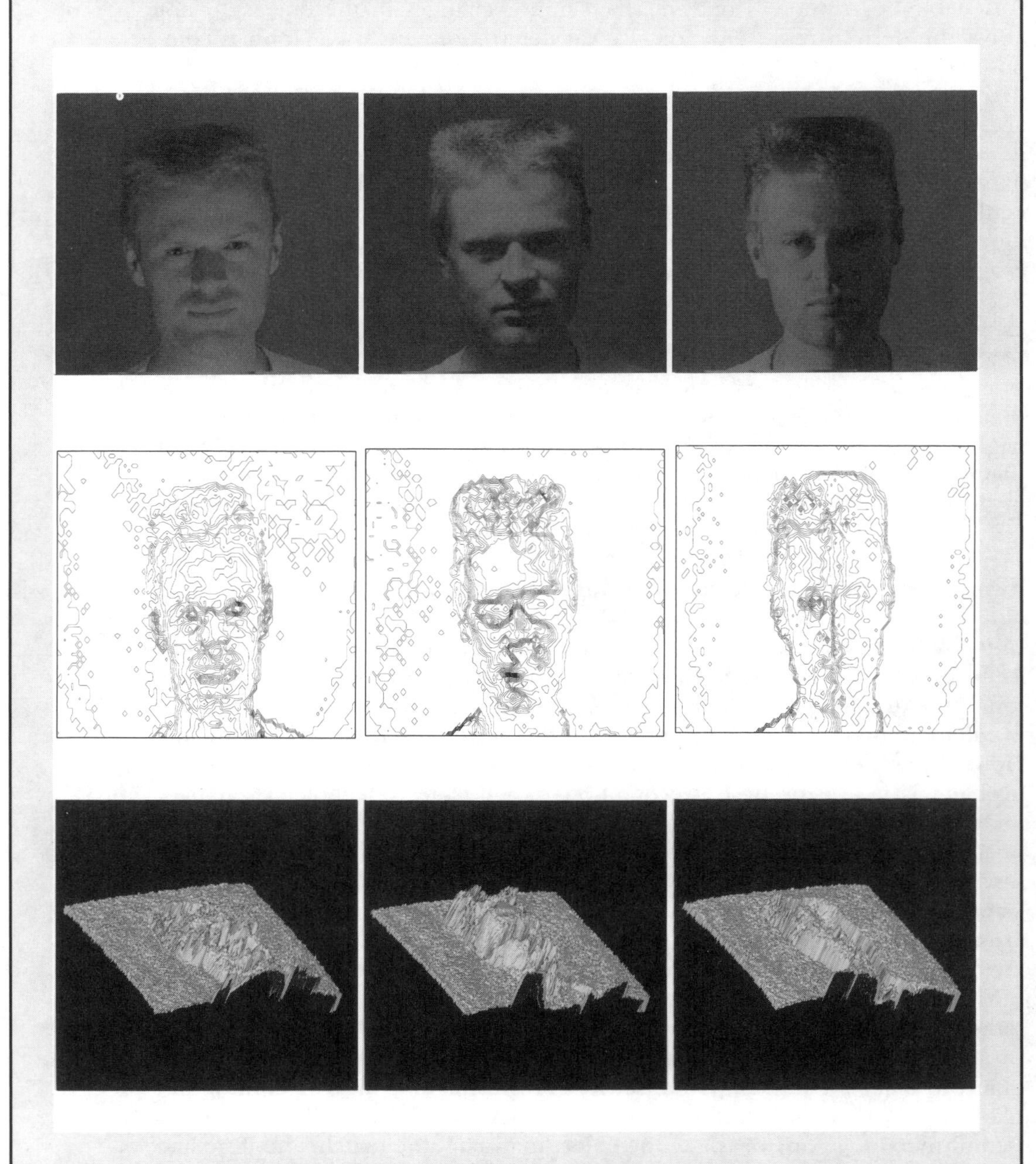

4.3 Top: Three pictures of the same individual photographed under different lighting conditions; middle: Contour maps generated by treating the brightnesses in the image as defining a landscape; bottom: the brightness landscape of each of the images is rendered by shading.

Visual perception can be thought of as a process of extracting information about a visual scene from images. Faced with a picture of a scene it is difficult to see where the problem lies. This is because we are used to seeing and interpreting images, and our visual systems have evolved special mechanisms to deal with this kind of input. However we can represent the number eight in a variety of ways, for example, VIII, 8, or 1000 (binary) and we can represent the information content of images by other means than by the brightness of picture elements or pixels. In computers, images are encoded as arrays of integer values. There are many ways of visualizing the array. If we represent the array in terms of image brightness we will call that representation an image or a picture. We could also consider the array as a landscape with values indicating the local altitude or height of a surface. The surface could be rendered as a wire frame drawing or a contour map, by joining up points of equal altitude, or the surface could even be presented as a hologram. Alternatively the information could be represented in a more obscure way, as an array of characters. None of these manipulations alter the information content of the array.

notion of the template which can incorporate these sorts of transformations. We will consider what needs to be done to build more effective template systems later in the chapter.

It is difficult to imagine that we have stored in our heads templates for all possible objects. Could we have a template for the faces of all the people we know or for each of the possible faces we might meet on our daily journey to the office or to the lecture hall? How do we encode the faces of people we do not know? One way of getting round this difficulty is to define some elemental parts which we can fit together in many ways to make more complex objects. The building blocks are generally called features or **primitives**, although the term feature is also used for distinctive and diagnostic attributes of an object, for example, the dent or scratch on your car that would allow you to identify it as yours in a line up of identical models. For faces people are sufficiently clear on which are the most important features that have been given names, the nose, the hair, the brow, the mouth, and so on. Penry (1971) the inventer of Photofit developed a system which allowed eye-witnesses to build up a face out of these elementary building blocks and it is possible that the visual system could reverse this task. It could decompose a new face into a finite set of separate features. This process of breaking the whole down into elemental parts provides an efficient coding system which is perhaps best seen in symbolic systems, like the alphabet or a number system. It would be wildly inefficient, as well as illogical, to have a different symbol for each integer from negative to positive infinity. However, the advantages of part decompositions generally come from the prospect of having a relatively small finite set of symbols. To apply this approach effectively to objects we need to be able to define a set of primitives that can fairly represent the variation seen in the target object. To return to the question of face recognition, at the very least, two of the faces we are interested in representing need to possess the same nose, say, to derive any efficiencies in encoding.

Although we could encode objects as a set of features and nothing else we would not be able to distinguish between two objects with the same set of features but different organization. Clearly it is not simply the presence of the features that is important but also their spatial relationships from one to another. We need a **structural description** to connect together the elemental parts and specify the interrelations. The feature/structural description framework for object recognition generates a number of more specific questions, which we now address:

- What are the primitives, the elemental parts of the description?
- What relationships should be encoded?
- How does recognition and identification proceed once we have the description?

SAQ 4.2 When a man shaves his moustache or beard the effect can be startling or go entirely unnoticed depending on the individual concerned. Consider how the strategies outlined above would cope with the deletion of features.

In thinking about object perception it is useful to split the problem of the perception and visual analysis of objects into two separate problems. The first is the problem of *object constancy* which we considered above. How can we recognize an object as the same object under a variety of transformations which affect the image of the object? The second is this: How does the object description interact with stored information about objects, a process that allows us to draw inferences about the object or recover details about the object from memory? In the scenario when Mom picks up the phone she knows that it breaks up into a handset and a body. She would be tremendously surprised if the body of the phone stayed fixed to the handset. But there is no way that she could know this, for certain, by just looking at the phone. Both of these problems will affect the kinds of representations we might wish to have if we were attempting to design an object recognition system and it is to the problem of representations for object recognition that we now turn.

Intermediate Representations

The color of an object can be an important clue to its identity, but that is not an issue that is developed here in any detail and therefore, for our purposes, we can think of the input to the visual system as a two-dimensional array of brightnessess. As we noted earlier, it is important to realize that this is an abstract data set that can be visualized in a variety of ways: as an image, by encoding each value in its natural form, or alternatively, as a landscape, as is shown in box 4.1. The problems facing the visual system in object recognition are usually best thought of with reference to the landscape representation since it is easy to become seduced by the apparent ease with which the human eye and brain can interpret images.

Cartoons

Most theories of object recognition require some intermediate representation between the image and the generation of the final representation of the object which we take to be recognition. The fact that we can readily recognize objects and people from silhouettes, line drawings, and cartoons has led to suggestions that the first stage in object recognition should be the extraction of edges or object boundaries, leading to a sparse representation of the input image in which only points which are correlated with significant brightness changes in the image are represented. In both Marr and Nishihara's system and Biederman's (1987) approach, 3D shape descriptions are derived directly from a representation of image contours.

However we should be wary of over emphasizing object boundaries at the expense of the internal features of objects. It is not necessarily the case that because we can recognize shapes from line drawings there must be a similar stage in normal visual processing. While we may readily classify a line drawing as a car rather than an airplane, it would be much more difficult to recognize the line drawing as being one's own car rather than a similar model. Silhouettes which are recognizable as particular objects are highly constrained. Only leftward or rightward facing heads are likely to provide sufficient information for recognition. There is a need to understand the perceptual processing of both line drawings and pictures. We cannot substitute an understanding of one for an understanding of the other.

The alternative to taking a 2D representation of object boundaries as the starting point for object recognition is to attempt to recover and represent geometric properties of the surface of the object. In common language we usually describe shape by analogy, for example, cylindrical, banana shaped, bulbous or furrowed, but to describe the shape of a banana as "banana shaped" is unhelpful. We need to find a way of describing shape which does not draw on our knowledge of the shape of objects. To be at all precise about what we mean by the shape of a surface we have to adopt the language of **surface geometry**. In order to represent surfaces in a flexible manner we have to characterize how surfaces vary. Essentially surfaces vary in their location in space, in their orientation and in their curvature. Methods of specifying these sources of variation and calculating the relevant values is the objective of surface geometry.

Surfaces Primitives

Marr and Nishihara's 2½D sketch

Marr and Nishihara (1978) referred to a description of the surface geometry of objects as a 2½D sketch. All depth cues can in principle contribute to the representation. The *2½D* sketch is a representation of surface distance and orientation from a particular viewpoint in which information from all depth cues are combined. Surface position can be represented in an image in which the value at each point represents the distance from the eye to the surface. Although a description of surface distance completely determines the shape of the surface, higher-order properties like orientation and curvature are left implicit and therefore inaccessible.

Surface orientation requires two parameters. The local orientation of surfaces is typically encoded in terms of the **surface normal**, an arrow of unit length pointing directly away from the surface (figure 4.4). The surface normal can be specified for each point of the surface so long as the distance to the surface varies smoothly. We need two parameters to specify the surface normal, surface slant, and surface tilt. Surface slant can be defined informally as the angle that a particular surface normal makes with the line of sight, and surface tilt refers to the direction in which the surface is slanted, which can be specified as an orientation in the image plane.

Koenderink's local shape index

We could also represent how the surface normal is changing, at all points on the surface. This measure would encode information about the local curvature of surfaces. At any point on a surface we can find a direction in which surface is maximally curved and a direction in which the surface is minimally curved, unless the surface is a sphere, in which case curvature is the same in all directions. The directions of maximum and minimum curvature are always at

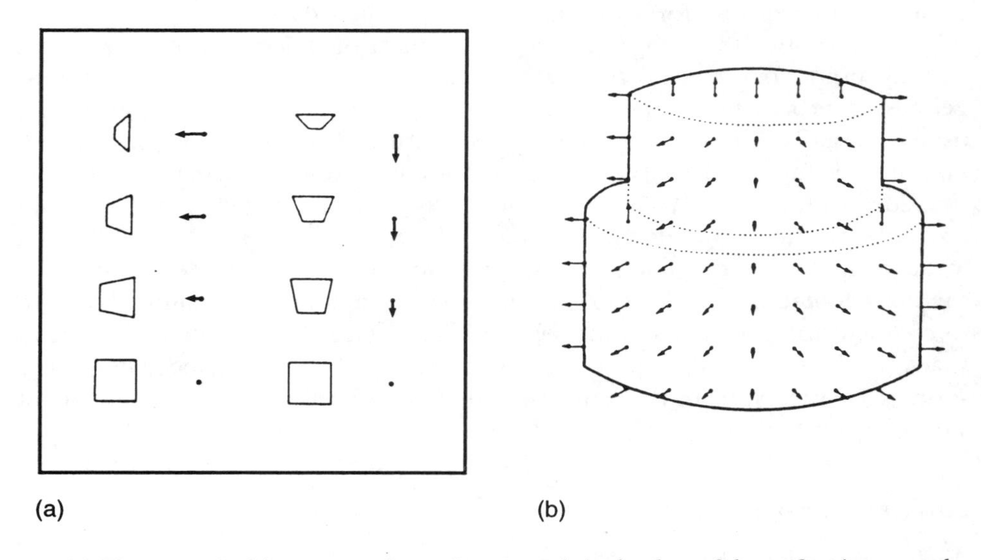

4.4 (a) The arrows in (a) represent the surface normal. As the slant of the surface increases the normal projects a longer line in the image plane. The tilt is a direction in the image plane which indicates the direction of slant; (b) A pair of cylinders with the surface normals indicated.

right angles to each other. These are called the principal curvatures. Their directions are called principal directions and their values characterize the **local shape** of the surface patch. When the principal curvatures are both positive then the surface patch is convex, when they are both negative the patch is concave. When one is positive and the other negative the surface is saddle shaped and when one of the principal curvatures is zero the surface is cylindrical. A description in terms of curvature is a rich source of information about the shape of a surface patch, but a curvature description does have a disadvantage since it varies with scale. The curvature of a sphere is the inverse of its radius. A soccer ball is less curved than a golf ball yet we would wish to say both are spherical.

Koenderink (1990) has offered an elegant way of describing local variation in shape, called the local shape index. Koenderink describes shape in a way that has the advantage of separating out changes in shape from changes in scale. At any point on a surface we can derive two numbers, the values of the principle curvatures. We can then use those numbers as x and y Cartesian coordinates to plot a point on a graph (figure 4.5). The position on the graph

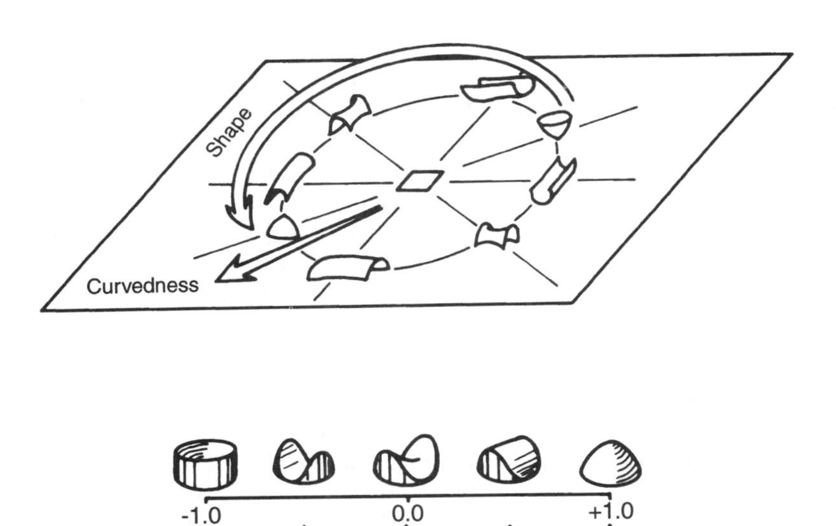

4.5 Top: The values of the principal curvatures completely determine the local shape of a surface at a point on the surface. We can indicate the local shape as a point on a 2D plot in which the axes represent the values of the two principal curvatures. Since local shape is the relative value of the principal curvatures all points along a line which make a specific angle with respect to the axes have the same local shape. The scale of the surface or 'curvedness' is encoded as the distance from the origin; bottom: Here the angular measure has been scaled between –1 and +1.

indicates which one of the many possible surface shapes can be found around that particular point on the surface. A change in scale, produced by magnifying or shrinking the surface, results in change in the value of both principle curvatures and a movement of the point towards or away from the origin of the graph. The scale of the surface, called curvedness, is given by the distance from the origin – the further from the origin the more curved the surface patch. Our intuitive notions about the shape of a surface is captured by the relative values of the principle curvatures. Local shape encodes relative curvature, the ratio of the two principal curvatures, and is therefore insensitive to changes in the absolute values of the principle curvatures. Local shape varies around the origin from convex spherical through cylindrical and saddle-shaped to concave spherical (figure 4.5). In fact local shape can be thought of as the angular coordinate in a polar coordinate system imposed on a graph in which the Cartesian axes represent the values of the two principal curvatures. We are still at the level of local surface properties but we can extend our classification of surface variation in regions by including the embedding of one surface type in another to produce ridges, furrows, and dimples. The value of Koenderink's formulation will be seen later when we consider **invariant** properties of shape representations.

Transformations between representations

The surface-based approach to the representation of 3D scenes relies heavily on the techniques and concepts of surface geometry. If the surface can be represented as a **range map**, an image containing the distances to the object, then procedures can be used to calculate the surface normal, principal curvatures, local shape, and other properties of the surface. As we noted earlier these properties are implicit in the range map but they can be made explicit by applying the appropriate computational procedures. We can imagine an **image processing pipeline** in which local higher-order properties of a surface are derived from a symbolic representation of depth or distance, that is an internal representation of depth rather than a direct measure of depth. Indeed, the whole idea of an intermediate representation is that it lies between one kind of description of an image and another. We can ask the question whether this occurs in human visual processing.

The symbolic pipeline idea implies a degree of sequential dependence in image processing. Our ability to judge changes in slant should depend upon our ability to judge depth differences, and our ability to judge changes in curvature should depend upon our ability to judge changes in surface orientation. However, Johnston and Passmore (1994) have shown that under certain illumination conditions, curvature discrimination thresholds for surfaces defined by shading and disparity are a factor of ten lower than would be predicted on the basis of slant discrimination thresholds for the same surface patches. We can conclude from this that surface slant and curvature are encoded separately and directly from the retinal images. Both representations could be considered as intermediate. When we look at a curved object we may be able to tell that one part of the object is further away than another part, but that information is not fed forward in order to compute curvature. The perception of curvature requires a separate system.

One advantage of the surface-based approach is that we only need include in the representation that which can in principle be known by processing the retinal images. The attribution of **volumetric primitives**, which we consider in the next section, goes beyond the information given and therefore involves making, possibly erroneous, inferences about the contents of the scene.

Volumetric Primitives

We have found it convenient to give a number of very simple objects names like sphere, pyramid, ellipsoid, and cylinder. We can specify these objects by indicating the type of object and giving values for various parameters of the object. For example, for the cylinder we only need two parameters, the length of the straight axis and the radius. Thus with only two pieces of information we can specify all possible cylinders. The addition of two more parameters specifying the orientation for the main axis allows us to specify all cylinders and their orientations in 3D space. In terms of information content this is a huge saving on a range map. Also the information is held in an explicit form, which is specific to the object. We identify the type of object, specify the sources of variation, and fill in the values for the parameters. Binford (1971) described a volumetric object called the **generalized cone**. This object is the solid swept out by a **connected curve** which is drawn along an axis in 3D space (figure 4.6). The size of the cross section is allowed to change as the curve moves along the axis. It would also be possible to specify the curvature of the axis if required. Both Marr and Beiderman used generalized cones as the primitives of their object description system. Beiderman (1987) argued for a limited number of primitives called geons which include generalized cones,

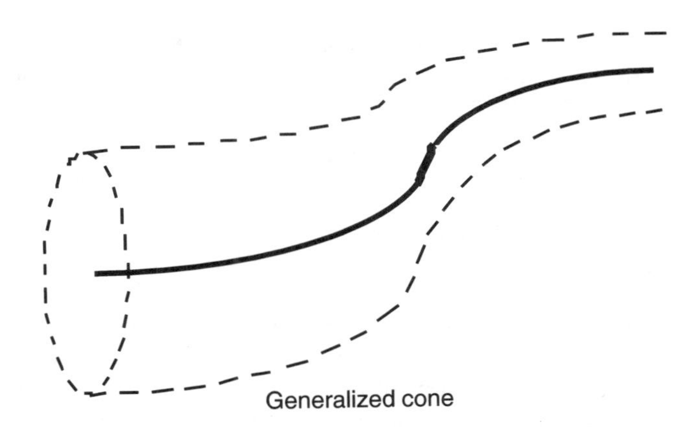

4.6 A volumetric primitive generated by sweeping a connected curve along an axis.

wedges and cubes, and there have been other suggestions along similar lines (Pentland, 1986).

The problem with volumetric primitives is that they cannot capture the sources of variation seen in general objects. By a general object we mean a smooth bounded volume that can take an arbitrary shape like a potato or an arbitrarily moulded lump of clay. The drive, thus far, in thinking about how we describe objects is directed towards finding the most abstract means of describing an object, which is most resistant to transformations of the object and which is most efficient in that it requires the fewest pieces of information. This strategy begins to unravel when we consider general objects. For potatoes there is no opportunity to reduce the description further, to provide a specification of a volumetric form plus parameters. We can of course recognize the object as a potato and with sufficient experience of any particular vegetable we would surely recognize it again. There may be no better way to encode the object than at the level of the local description, in terms of the local shape and local orientation.

This leads us to a puzzle. If the best we can do in the case of general objects like potatoes or lumps of clay is to provide a description in terms of local orientation and local shape why do we need a more extensive global representation for a simple object like a sphere? Because it is possible to form a very simple description for regular objects does that mean the visual system is obliged to do so? Here a thought experiment can contribute to the argument. Suppose we take an arbitrary lump of clay and start to role it into a ball. Does our perception of the shape of the clay alter in any fundamental way at the transition when we form a perfect sphere? We will return to these puzzles later in the chapter. Having considered the question of the representation of shape up to the level of local shape descriptions and volumetric primitives, we can now consider the problem of how we recognize objects as remaining the same under various transformations.

SAQ 4.3 In how many ways can we represent the shape of a bottle? Does the we way represent the shape determine how we are able to pick the bottle up?

Object Constancy

Invariants and Transformations

There are two general strategies used to attack the problem of object constancy: the invariants approach and the transformation approach. These two strategies can be illustrated with a simple example. Suppose you were asked to say whether two arbitrary triangles were the same or not. You might try to solve this problem by measuring the angles and the lengths of the lines and then noting down how the lines are connected. If both triangles gave rise to the same description you could say they were identical. Alternatively you could rotate and translate one of the triangles to exactly cover the other. The first example captures the essence of the invariants strategy and the second the transformation approach. They are not, of course, mutually exclusive and a robust system might employ both approaches.

Gibson was a vociferous advocate of the invariants approach. In Gibson's (1979) view, the visual system had to separate the properties of the retinal image due to changes in the viewpoint and the illumination from the properties of the retinal image, which were dependent upon the object and,

Devil's Advocate Box 4.2 Volumetric primitives

One is tempted to claim that for general objects like potatoes, lumps of clay, or three-dimensional amoebae there is no simpler description than the object itself. However, an astute reader might wish to take issue with this. One counterargument might be that the shape of an object depends upon some notion of the magnification or scale at which we view the object. An object has surface features down to the microscopic level and therefore any description we form of the object must involve some degree of abstraction. Of course, this is a valid argument, any representation of an object involves a notion of scale. Real objects are simply physical entities. Ideas of scale do not apply unless we attempt to measure or describe the object and, therefore, it makes no sense to talk about an object being represented by itself. Perhaps it would be better to revise the initial statement to say that (a) general objects must be represented at the finest level of detail available to the eye/brain and (b) there is no adequate representation of the object in terms of other simpler objects.

However this might provoke the counterclaim that we could form a good representation of a potato from volumetric primitives, like spheres and ellipsoids collapsed together to form a volume which had the same form as the general object, and that this would require less pieces of information. This is not too difficult to deal with because we already think of the object as being represented by local descriptors. A local shape description can be thought of as equivalent to patches of spheres, ellipsoids, and saddles sewn together to form the surface and, therefore, the volumetric primitives do not add anything further to the surface-based approach in which volume is left implicit. In order to accommodate the detail of a general object, the simpler volumetric descriptors end up being equivalent to the local shape descriptors that we started with.

But surely, to continue the argument, we could rub down the high points and fill in the hollows to generate a description which would capture the global shape of the object. Surely this would be a simpler description. Indeed it would be a description of a simpler object, but it would not be a simpler description of the original general object. The description would not capture the detail that we want to represent. The arguments are important because they place limits on the value of progressive abstraction of spatial information. We do not need any stage of abstraction beyond the local description of surfaces in order to see objects.

therefore, *must be invariant*, because we assume the object to be invariant. Clearly, the retinal image itself is not invariant under a change in viewpoint so there are no invariants in the retinal image per se. However the influence of the object on the retinal image persists under a change in viewpoint. It should therefore be possible to derive a measure from the retinal images which characterizes some property of the object and this measure should be invariant if the object is invariant. This is what Gibson meant by the detection of higher-order invariants in the optic array. In the invariants approach the aim is to form a description of the object which does not change radically with changes in viewpoint, rotations of the object, or other changes. We could then attempt to match the invariant description to some description in memory in order to identify the object.

The transformation approach aims to undo the transformation produced by object rotations and image formation in order to allow a simpler match between the image and an internal model. This is often the approach taken in robot vision systems. One can either proceed by attempting some image transformation or, as is more usual, if the computer system already has a model of the widget it wishes to pick up, the system can rotate, scale and render the widget in order to build a template that can be used for an image-based identification. Ullman's (1989) approach involves the alignment of pictorial descriptions and can be thought of a sophisticated development of the template technique since the endpoint is a pictorial descriptive or image-based match. The problem of object recognition is seen as involving a normalisation or registration stage followed by a search through a set of internal models. Normalization involves first segmentation of the object and translation to some canonical location, the object may then be scaled and rotated before being matched to a model store. Rotation of the object can involve rotation in a 3D space. The degree and direction of rotation could be calculated in one step by matching a set of critical points on the object and model. With sufficient numbers of corresponding points Ullman demonstrated that the correct 3D transformation to align the critical points could be derived and therefore it should be possible to normalize the input pattern. Thus he attempts to get round the problems of object constancy by a normalization stage. With the normalization stage completed we only have to worry about the stage in which we match to memory. The match need not be on the basis of the raw image. Ullman's reference to a pictorial *descriptive* code requires the description to be localized in an image but it can be a qualitative description asserting the presence of an image feature like an edge or something more complex like a ragged edge. We will come back to transformations, but first let us look in more detail at the invariants strategy.

SAQ 4.4 Can you think of any problems in trying to implement the transformation approach in a biological visual system?

Invariant Image Properties

We have seen that the invariants approach aims to describe objects in a way which does not depend upon viewpoint. We would get off to a good start if our recognition system started by encoding image features which were themselves invariant with respect to changes in illumination and viewpoint. Any subsequent processing would then be illumination or viewpoint independent. This is the approach taken by Beiderman (1987). Beiderman, like Marr and Nishihara (1978) proposes that object recognition starts with a process of edge extraction. As we have seen, this reduces the problems resulting from changes in illumination. In Beiderman's model, non-accidental properties of edges are taken as evidence for particular elemental 3D volumetric structures in the object called geons (Witkin and Tenenbaum, 1983). The non-accidental properties include collinearity, parallelism, symmetry, and co-termination. These relationships are unlikely to occur in the 2D image if they are not also present in the 3D object. A straight line in 3D space will remain straight whatever the viewpoint and lines which meet in 3D space will also meet in the image. Thus the viewpoint has only a limited effect on whether the correct geon is recovered from the image.

Intermediate Representations

In the surface-based approach to object recognition we could also attempt to represent an object using measures which are invariant with respect to the viewpoint. Marr's 2½D sketch involves an image-based description of surface distance and surface orientation. Since distance and orientation are dependent upon the viewpoint this description would change as the viewpoint changes or when the object moves. Ideally, to follow the invariants approach we would wish to represent properties of the object surface since properties of the surface depend upon the object, not the viewpoint. Surface curvature is a property of the object not the viewing conditions and therefore, if we could recover curvature, the use of this measure would bring with it certain invariant properties. However, curvature depends upon size and we would wish to recognize a model horse as easily as a real horse. A representation in terms of Koenderink's local shape index would be scale independent as well as view independent since what is characterized by local shape is the relative values of the two principal curvatures at any point on a surface. Changes in scale or size do not affect relative curvature. If we were able to reduce an object in size the local shape description would remain unchanged, although its spatial representation in the image would shrink.

Reference Frames

Marr and Nishihara's solution to the problem of invariance with respect to viewpoint was to describe the object using a coordinate system centered on the object itself – an object-centered rather than a viewer-centered representation. Although this would provide an effective way of introducing an description which would be independent of the orientation of the object, the problem facing this approach is how to define a procedure for allocating the principle axis for an object which will ensure that the same axis is arrived at irrespective of the orientation of the object. A general rule would be to attribute the principle axis to the long axis of the largest object component part. However, a horse presents a very different profile when walking toward you than it does when it walks across your path. In the first case the axis would be placed through the head and shoulders because of the **foreshortening** of the body. In the second case the axis would run through the body of the horse and would stay parallel to the ground. If we start with different principle axes we will inevitably end with different descriptions for the same object and this would violate the uniqueness constraint.

SAQ 4.5 How would you represent the shape of a horse using Marr and Nishihara's scheme? How different would your representation be if you were asked to represent a giraffe?

Canonical Views

The aim of a view invariant representation is to allow object recognition irrespective of viewpoint. Any effect of pose, the orientation of the object in 3D space, would be attributed to some problem in deriving an adequate representation. However there is evidence that viewpoint makes a difference. The critical question to ask here is how easy is it to recognize a known object from an entirely novel viewpoint? Bulthoff and Edelman (1992) investigated the recognition of novel three-dimensional objects that could either look like twisted paper clips, called wires, or rubber potatoes with some parts of the surface drawn out into rounded conic protrusions, called amoebae. Subjects saw a movie sequence in which the objects oscillated (+/– 15 degs) about a particular axis. Two positions were chosen; the reference position at 0 deg and an additional position at +75 degs rotation. Recognition was tested using a forced choice procedure in which subjects had to indicate whether a static image was from the target object or not. The target view could either (a) lie between the set of views presented (**interpolation** condition), (b) lie outside the set of views presented, but result from rotation about the same axis (extrapolation condition), or (c) result from rotation about an orthogonal axis (ortho-meridian condition). Figure 4.7 illustrates the three conditions.

They found that errors, the number of targets missed, were highest in the ortho-meridian condition, lowest in the interpolation condition and intermediate in the extrapolation condition. Errors increased as a function of the

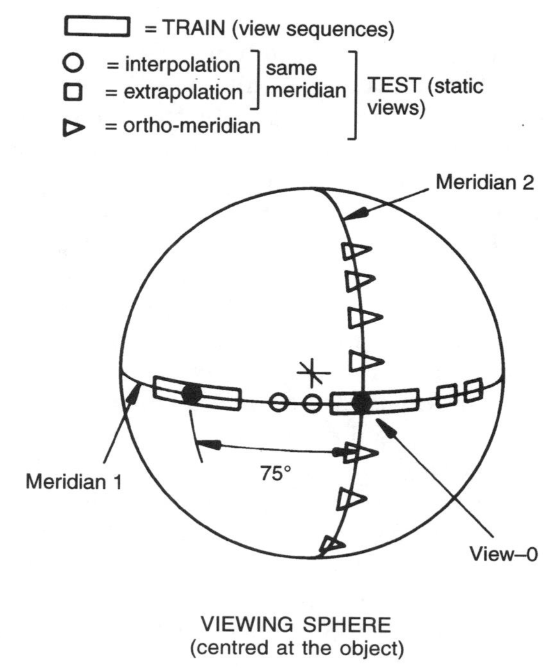

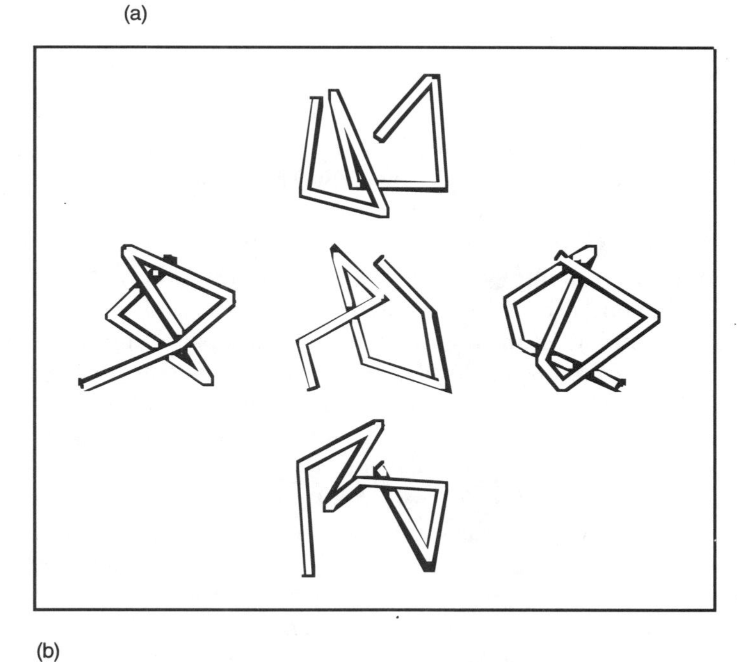

4.7 (a) The viewing sphere indicates all possible views of an object. The black dots indicate the average positions of the views in the training set. Test views could either lie between the training views (interpolation condition) or outside but along the horizonal axis (extrapolation condition). Additional test views were drawn from along the orthogonal meridian (ortho-meridian); (b) A number of views of a test object.

degree of rotation away from the reference view. A straightforward interpretation of this finding is that recognition depends upon being able to interpolate between learned views of an object. Bülthoff and Edelman conclude that we need not involve a 3D representation at all, we can simply interpolate on the basis of the 2D images generated by objects. However, they do not discuss the effects of changing the illumination of the objects, which might be expected to severely impair any strategy based on interpolation between 2D images of particular views. Nevertheless, the data are difficult to accommodate within an object recognition model which emphasizes view-invariant encoding or alignment of three-dimensional models.

To summarize, although it is possible to recognize objects from a variety of views, not all views of an object are equally recognizable. An invariant-based theory or a transformation-based theory would need to be able to account for these effects. A view interpolation model naturally predicts effects of viewpoint or **pose**, but current approaches are image based and would therefore be very sensitive to changes in illumination. Also, a view interpolation approach can only account for identification on the basis of how objects appear: it has no knowledge of objects or transformations since objects are represented as a collection of 2D images. If we move our favorite armchair to another part of the room and cover it with new material how are we to recognize it?

Part Decomposition

Suppose that we had a system which solved the problems of object constancy by circumvention, for example by delivering a suitably transformed normalized version of the object to be recognized, we could proceed by matching this image to stored representations in memory. However, we would need a representation in memory for each object we were likely to see. This would be a very inefficient use of resources. How can we find a match if part of the object is missing? What if the object has moving parts that articulate or change position? Problems of this kind are attacked by decomposing the object into parts. Notice that part decomposition does not help with the problem of object constancy since the difficulties outlined earlier apply to parts as well as to whole objects.

Part decomposition is the essential first stage in forming a structural description of an object. Marr and Nishihara advocated the use of generalized cones as primitive elements and a hierarchical distributed coordinate system for the specification of part relations in which minor components are specified in relation to major parts. Fingers are specified relative to hands, hands in relation to arms, and the arm is specified in relation to the body (figure 4.8). This allows part relationships to be specified in a flexible way that provides for both a general description covering a whole class of objects and a detailed description which can be used for discrimination between objects. Once a description has been generated recognition proceeds by comparing the de-

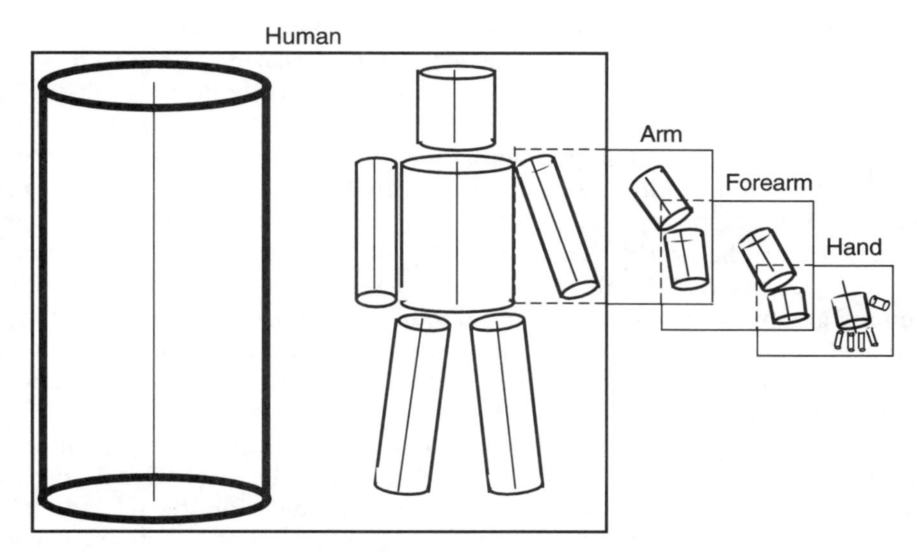

4.8 A distributed coordinate system for the specification of articulated objects.

scription against stored models. The models stored in memory embody constraints on possible component part relations, for example, for a man there are physical constraints on the articulation of the leg so that the leg can flex backwards at the knee but not forwards. The aim of the process of interaction with the model store is not a match, as such, but the formation of a description, which is more precise than that available in the model store. It is a matter of choosing the most appropriate model and specifying the parameters of the model given the constraints that would normally apply. Marr and Nishihara's answers to the questions posed in the introduction are that the primitives are volumetric, that relations between the primitives should be encoded in a distributed fashion, and that recognition proceeds by generating precise structural descriptions on the basis of stored models, which embody constraints on the organization of an object and the visual input.

Decomposition Strategies

Both Marr and Beiderman address the problem of part decompositions from the perspective of the geometry of the object. In particular, part boundaries are tied to concavities in the image contours which relate to the boundaries of objects. Whereas it is often the case that concavities coincide with part boundaries this is not always true. Light bulbs, bells, plastic shampoo bottles, and ornate table legs have concavities that do not correspond to component parts. Babushka dolls, shampoo bottles, thermos flasks, and telephones have components which do not segment into parts at concavities. At a single glance, without a chance to investigate, it is not clear whether a telephone segments into a handset and a body.

It is quite reasonable to assume the organization of the visual system reflects the constraints on images and image formation that are produced by the nature and characteristics of the physical world in which it has evolved. The laws of perspective, the geometry of motion parallax, the results of occlusion, and the effects of changing illumination, to take a few examples, are so ubiquitous that it would be strange if the visual system had not evolved visual mechanisms which exploited these constraints in the recovery of general properties of the world, like the layout of surfaces in the environment. However, in the case of object recognition it is not clear that there are any general geometrical rules that determine how objects, often recent man-made artifacts, should be decomposed into parts. Part decomposition can not be achieved a priori since it relies on one's experience with objects and in particular of how objects change (Johnston, 1992).

If we ask the question, "How many parts make a teapot?" to a designer of children's plastic toys the answer would probably be two. This is because manufacturers find it convenient to produce moulds for the left and right sides of the teapot and then join the two together. The rest of us are likely to say four: the bowl, the spout, the handle, and the lid (figure 4.9). The reason we do not decompose the teapot into left and right sides is because these are always mirror images. The shapes of the left and right sides always co-vary and, therefore, there is no reason to encode them independently. On the other hand, spouts, bowl lids, and handles can vary quite independently of each other with only minor constraints placed by one component on another resulting from functional adequacy or, as happens quite often, variation in fashion and style (Johnston, 1992). If object part boundaries coincide with geometric indicators, like concavities in bounding contours, this may well be entirely coincidental.

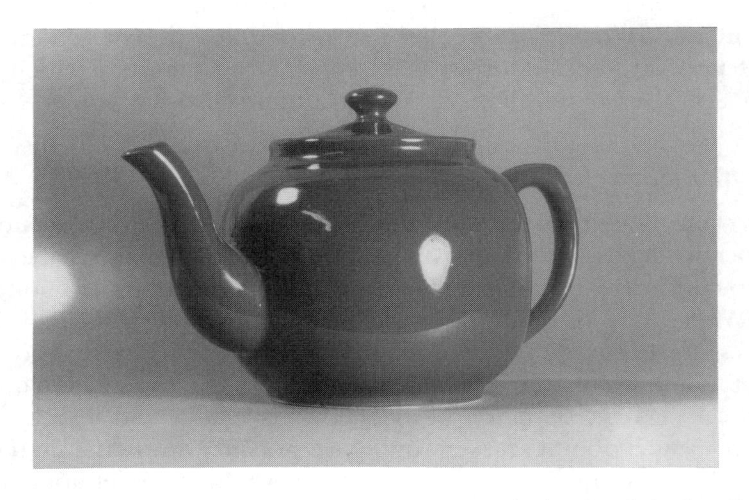

4.9 A teapot has four identifiable parts, the lid, the spout, the bowl and the handle.

Faces

It may be useful in evaluating the various approaches to surface and object perception outlined above to consider theories of object recognition in the context of a particular kind of object, a face. Faces form a particularly interesting class of objects since they have what appears to be almost infinite variety, yet faces are readily distinguishable from other objects. They undergo non-rigid transformations, changes of expression, as well as rigid transformations, for example, rotations, and they also act as channels of communication, controlling conversation and indicating subtle changes of mood. There is also evidence that new-born infants are particularly interested in faces.

Inverting Faces

The clearest evidence that faces are not represented in an object centered or view-invariant way is that faces are more difficult to recognize when they are presented upside down (Yin, 1969). This is also true of other objects which are usually seen in a particular orientation, like houses, but the effect of inversion on recognition is more profound for faces than for other objects. It is likely that this effect reflects the fact that we are more used to seeing upright faces than upside-down faces. The strength of the effects of inversion depends upon our experience with a particular object class. Diamond and Carey (1986) investigated performance in recognizing individuals drawn from a sample of pedigree dogs and found that inversion posed more difficulties for expert judges of the breed than for novices. One might expect if face recognition is influenced by expertise and familiarity with an object class that individuals might find it easier to recognize faces from their own ethnic group. Indeed this has been shown to be the case. People may be less sensitive to the sources of variation of other-race faces than to the sources of variation in own-race faces (Shepherd, 1981). Surprisingly, the inversion effect is greater for other-race faces (Valentine and Bruce, 1986a). Expertise of itself does not in a simple and straightforward way dictate a bigger inversion effect. An explanation which gets round the apparent contradiction is to say that it is more difficult to analyze a face, in relation to relevant dimensions of variation, when the face is inverted. Experts encode objects in an elaborate way and find it difficult to recover the same elaborate code when presented with inverted images. Other-race faces are encoded less efficiently in any case and the recovery of the appropriate description is further disrupted by inversion.

Photographic Negatives

People are remarkably bad at recognizing people from photographic negatives. Why might this be? The manipulation does not have a radical affect on the image. Negation simply inverts the sign of the contrast. Light areas are

Box 4.2 Face recognition in infancy

Faces might seem just like very complicated objects. However, there are a number of facts about faces that make them different from other objects. One of them is that if faces are turned upside down they are particularly difficult to process compared with faces the right way up. Recognizing or discriminating pictures of houses, for example, does not suffer anything like as much when the pictures are inverted. A second fact is that a few people with brain damage cannot recognize faces – even their own immediate family – while they have no trouble with any other kind of object or picture.

Another interesting thing about faces is that they seem to be responded to at birth! Babies do not have to learn that faces are important, they are born with that knowledge. One kind of experiment involves showing the baby a drawing and seeing

4.10 (a) The arrangement of experimenter and infant in the replication of the Goren et al. study. Drawing by Priscilla Barrett; (b) Mean head and eye turning for the real, scrambled and blank face stimuli. The face is followed significantly further than the other stimuli.

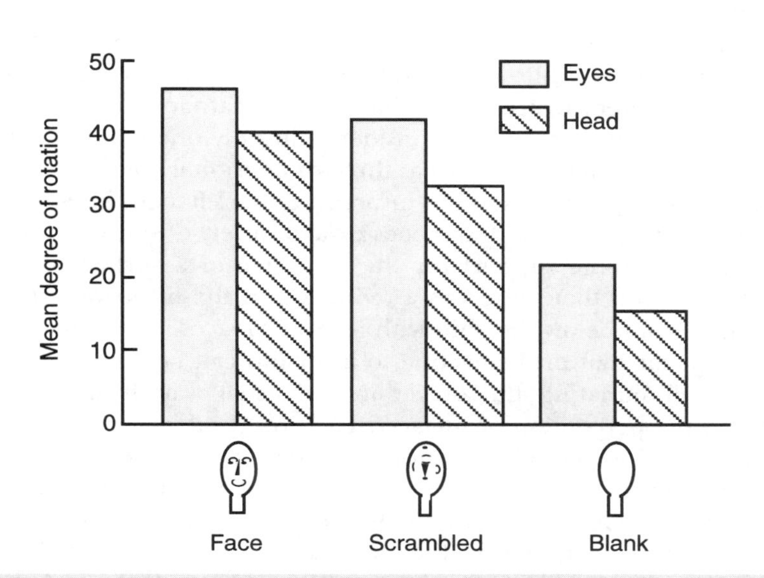

how far they would turn their heads to follow it. They followed the schematic face, A, further than the control stimulus B. Johnson and Morton (1991) did this experiment with infants who were about 30 minutes old. In another experiment with similar results the infants were only 7 minutes old, on average! It should not be thought that human faces are specified completely in the infant brain. After all, the infant only needs to pick out a face from other stimuli that might appear. Human society guarantees that this face will be human. When the infant turns towards it then learning mechanisms can take over.

Learning does seem to be very fast with respect to faces. At 3 or 4 days of age, infants prefer to look at their mother than at another female face. And this is under conditions where the infant cannot smell the mother and where there are no vocal cues. What is learned seems to be related to the form of the hair rather than any detail of the facial features, since if the two women wear head-scarves, the baby loses the preference for mother. It is only at about two months that the features are enough to discriminate mother.

represented as dark areas and vice versa. Of most interest is the amount of information that remains invariant. The boundaries of the face; the shape of the hairline; the outline, size, and position of the features; virtually all spatial measures of the face remain unchanged. The fact that changing the sign of contrast in cartoons and line drawings makes so little difference to our ability to recognize depicted faces underlines the point made earlier that the perception of line drawings and pictures have to be considered separately.

One effect of negation is on the recovery of surface shape from shading and shadows. Shadows are by definition dark and in the case of shading dark regions of a surface indicate that the region is at an angle to the prevailing illumination. Difficulties in recognizing faces in negatives point to problems in the recovery of the three-dimensional structure of the face, although other explanations are possible: the brightness of the hair and the pupils are also dramatically altered. Bruce (1988) has argued that faces have been treated as two-dimensional patterns often considered in only one or two poses and that we should be more aware of faces as three-dimensional objects. The advantages of a surface-based code is that it is invariant under changes in illumination and it allows the integration of depth cues from a variety of sources. Johnston et al. (1992) provide some support for the existence of a surface-based code for faces. They found that there was a reduction of the inversion effect for faces illuminated from below the chin with respect to faces illuminated from above. It would appear that problems due to inversion can be compensated to some degree by illuminating the object in the usual way, from above, thereby enhancing the perception of surface structure. If subjects were using pictorial cues one would expect illumination direction and inversion to have additive effects. Further evidence for the importance of internal surface cues comes from an experiment using faces which were defined solely by shading. Bruce et al. (1993) found that altering the curvature of the surface at the top of the nose biased subjects judgments about the sex of faces. If the surface was more protruding then the face was more likely judged to be male than female.

Features or Configurations?

In an earlier section we considered the idea that it might be useful to break up an object into component parts or features in order to reduce the amount of stored templates one would need to recognize an object or to form the basis of a structural description. A structural description provides information about the relations between the primitives of the description. It is usual to consider the configuration of the face as specifying relative positions of the features. We will return to the question of how best to think about configuration in a later section. Here we consider whether it is best to think about the perception of faces on a feature by feature basis or in terms of a global pattern. The photofit approach mentioned earlier provides a prototypical feature- based approach to constructing and encoding faces. Sergent (1984) asked subjects to decide whether pairs of faces were the same or different. The faces were photofit constructions in which two types of chin line, eyes/eyebrow and inner space (the distance between the nose and the mouth) were factorially combined. Sergent found that if the faces differed on more than one dimension reaction times to indicate that the faces were different could be faster than when the single most salient feature was the sole difference between the faces. This suggests subjects were responding on the basis of the configuration of features rather than single features. The effect did not occur when the faces were inverted.

Young, Hellawell, and Hay (1987) demonstrated that stimuli constructed by aligning the top and bottom halves of two different faces look quite different from the individuals whose faces were combined to form the composite.

Composites were formed from famous faces. They compared recognition times for reporting the identity of part of the composite face when the parts were spatially aligned against conditions in which the component parts were shifted relative to each other. Subjects were quicker to identify the component faces when the components were shifted, showing that the component faces had been drawn into a new configuration (figure 4.11). Again this effect was reduced when the faces were inverted. Taken together these experiments provide good evidence that subjects are sensitive to the global properties of the face and that it is difficult to recover configural information from inverted faces.

Caricatures and Prototypes

Caricatures provide a view on how information in memory might be organized with respect to objects and with respect to faces in particular. A good caricature of a politician or celebrity appears to capture the essence of that

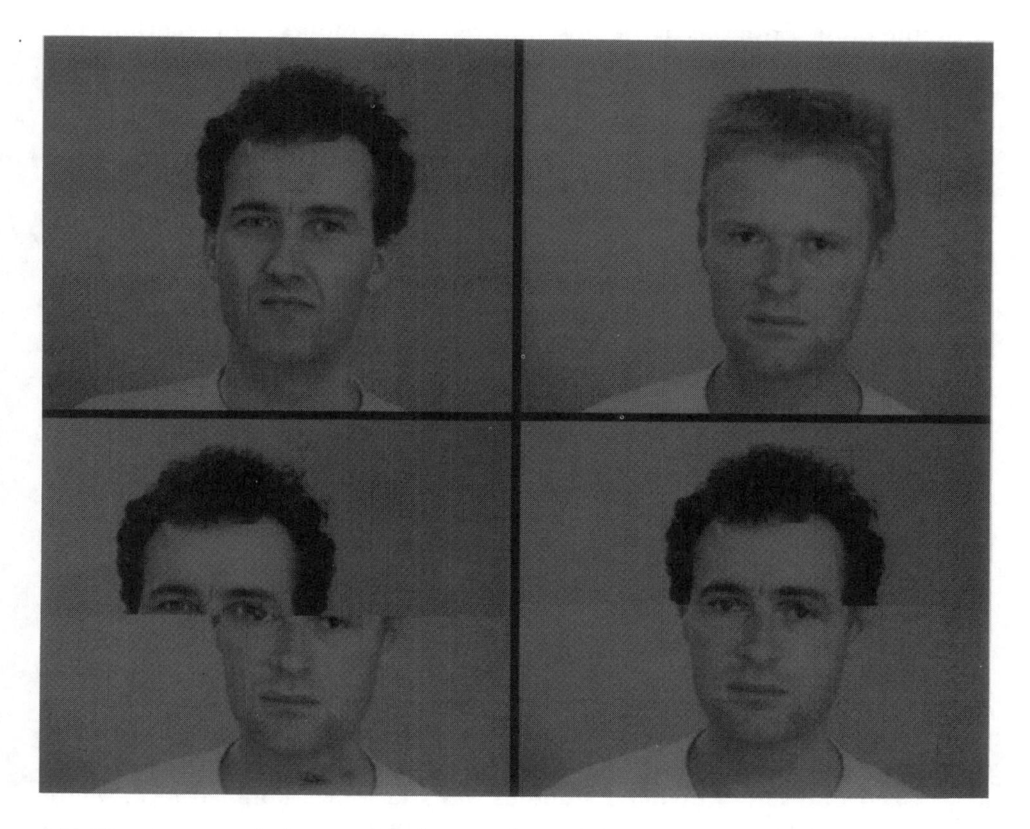

4.11 The picture on the bottom right is an amalgam of the two faces above. The faces appear to merge into a new configuration. It is easier to detect the component faces when they are spatially displaced or when the whole figure is inverted.

individual. A caricature involves exaggeration and, therefore, a move away from an accurate rendering of an individual, but nevertheless the distortion only seems to enhance the strength of the signal. Indeed, caricatures have been considered superfidelity stimuli, which may be more easily recognized than the true image.

The caricaturist's art is to take those aspects of an individual's face that are distinctive and particular to that individual and to exaggerate those elements. In order to operationalize this process it is necessary first to generate an average or prototypical face with which to compare any particular individual. The idea of generating a facial prototype has a long history. Sir Francis Galton (1822–1911) the Victorian scientist and pioneer of the measurement of human individual differences believed that by averaging together the faces of known criminals one could eventually arrive at a prototypical criminal face: an early version of offender prototyping. One difficulty that has to be overcome before averaging over faces is to decide how the individual faces are to be scaled and aligned. Brennan (1985) used line drawings of faces. Control points were designated which marked critical points on the face, allowing the encoding of the positions of the lips, the eyes, the contours of the hairline and the chin. The positions of the list of points can be specified for any particular face and an average formed over the ensemble. Each individual face is scaled with respect to the inter-pupil distance. We can think of the average as a prototype for that particular sample of faces. To generate a caricature the coordinates for the control points of an individual face are compared with the average and the list of displacement **vectors**, which takes each control point from the prototype onto the particular face, are increased in length by an overall scale factor. Figure 4.12 shows the effects of this procedure on a line drawing of Jack Nicolson. Benson and Perrett (1994) have demonstrated that when the amount of distortion is optimized for particular faces, subjects recognize caricatures more quickly and with less errors than is found for accurate line drawings.

SAQ 4.6 Does the fact that we can produce caricatures mean we must have a representation of a prototypical face?

Evidence for the idea that we might encode faces as deviations from an underlying norm come from recognition studies using distinctive and typical faces. Valentine and Bruce (1986c) found that distinctive faces are recognized more quickly than typical faces and that this effect does not depend upon any differences in the familiarity of the faces. On the other hand, when subjects were asked to decide whether a stimulus was a face or a jumbled face, where the sections including the eyes, nose, or mouth were randomly positioned, subjects reaction times were quicker for typical faces than distinctive faces. The closer a face is to the average the easier it is to classify it as a face and the more difficult it is to identify it because of interference or competition from other typical faces (Valentine and Bruce, 1986b).

Constraints on Faces

Marr (1982) advocated the understanding of the visual mechanism at a number of levels. The computational theory level requires the specification of the goals of the computation and the principles by which the goals might be achieved. At this level of analysis it is necessary to examine how real world constraints might help us to formulate strategies for achieving these goals. The study of visual perception is to a large extent the study of how we recover and represent the sources or causes of the variation we see in images. Thus an understanding of the factors which cause faces to appear the way they do is key. Bruce (1988) points out that faces are constrained by the growth processes that form the skull and the musculature which controls the actions of our face and our expressions. The main point here is that the growth of the skull involves a global transformation and changes are correlated rather than independent in nature.

Whereas the basic shape of the face depends largely on the structure of the skull, changes in the configuration are caused by the action of the musculature

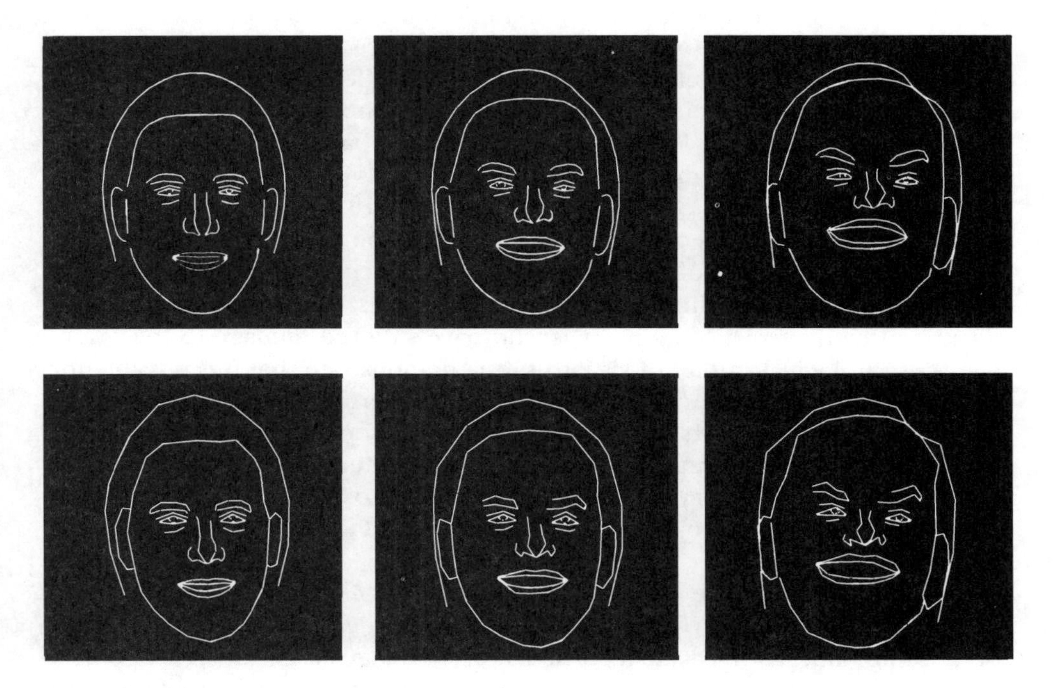

4.12 Line drawing caricatures. Left to right: anticaricature, veridical, caricature line drawings of actor Jack Nicholson. In the top row the control points have been joined with curved lines and in the bottom row with straight lines. From Philip J. Benson (1994), Visual processing of facial distinctiveness, *Perception* 23, 75–93. Reprinted with permission of Pion Limited, London.

acting on the jaw and the resistive elastic properties of the connective tissue and skin. Ekman and Friesen (1978) identify six basic expressions: happiness, anger, fear, surprise, disgust/contempt, and sadness. Each of these are achieved by a pattern of muscle activity. Thus changes of expression require correlated changes in muscle groups. The dynamic patterns one finds are drawn from a subset of the possible configurations of muscular activity that could be generated and, therefore, we can say that the shape of the face is constrained by the mechanisms controlling coordinated action. Often facial gestures are very characteristic of the individual, and it is the art of the mimic to abstract the facial gestures of a familiar politician or celebrity and reconstitute those gestures within their own face. The important point here is that the codes which are used to describe faces should reflect these kinds of constraints.

Characterizing the Global Shape of Objects and Faces

In previous sections we have considered the problem of characterizing the local shape of surfaces and objects, but this falls short of a theory of the **global shape** of objects. A theory of global shape can make use of the fact that global changes due to unitary causes have multiple but correlated effects at a local level. For example, when one is about to sneeze there are preparatory changes throughout the face which have local correlated effects on the shape of the face. The pattern of local effects generalizes over individuals and is uniquely related to a single cause. The goal of a theory of global shape is to capture the effects of global transformations and the constraints on those transformation as applied to a particular class of objects (Johnston, 1992). In an earlier section we saw that we do not need a theory of global shape in order to explain how we see objects, a theory of local surface shape is adequate, but we do need to explain our understanding of the nature of objects. This distinction can be illustrated by the work of individuals who have suffered damage to the brain. Neuropsychological studies of visual agnosia demonstrate that it is possible for individuals who have a lesion in the brain to see perfectly well, but be unable to name objects or tell that a drawing of a composite animal, a rabbit with a fishes tail, say, is not a real species, or recognize their wife (Humphreys and Riddoch, 1987). A theory of global shape needs to account for our understanding of the spatial properties of objects, but what would constitute a theory of global shape?

The first point to make is that *image-based accounts are insufficient.* Theories of object perception have in the main been concerned with how we identify objects and allocate objects to a particular class. Image-based theories of object classification or identification do not, of course, offer a theory of how we characterize the shape of objects. The advantage of a theory of object shape is that it allows the prospect of representing and reasoning about three-dimensional objects and how they transform. Without a theory of object shape the visual system is limited to reasoning about relationships between images of objects, that is associative relationships based on what objects look like. An associative account would have it that we recognize profile and full-face images as views of the same face simply because we have seen them sequentially. There need be no representation of information which is specific to 3D faces. The information is basically associative because causal processes are not represented, they are left implicit, and, if required, they would have to be inferred.

It is useful at this point to review the distinction between local and global shape and reiterate the point made earlier that *local intrinsic images are sufficient to account for immediate 3D vision.* The platonic solids are so regular that we have given them names like cube and ellipsoid. All ellipsoids can be specified in terms of the global form, ellipsoid, and two parameters giving the lengths of the major axes. A simpler or more efficient description is difficult to contemplate. However, potatoes are bumpy ovoid shapes that have no general description that would, for example, allow us to identify a particular exemplar. There is no simpler description which can carry the shape information than that provided by a intrinsic image containing local descriptors, like local surface orientation or the shape and curvedness parameters of Koenderink's local shape index. We have no difficulty in seeing or identifying individual potatoes and, therefore, we can safely conclude that a higher-order global description is not necessary for the appreciation and identification of objects. If this is the case, there is no need to strive for the most general description possible even when a simpler description, like cylindrical or ellipsoidal, is available. Simple objects are seen using the same mechanisms as more general objects such as lumps of modeling clay. In Bülthoff and Edelman's experiments, subjects have been presented with both kinetic depth and stereoscopic cues. They can clearly see the three-dimensional structure of the amoebae and so what subjects appear to be learning in their experiment is how the local shape descriptions or the appearance of the amoebae changes under transformation.

Computer graphics requires the specification of a 3D object or scene, the pigmentation of the object, the light sources and reflective surfaces in the environment, the image plane, and the viewpoint. Given the information and necessary physical models, it is now possible to generate excellent images of 3D scenes. In computer graphics we represent the effects in the image of causes in the environment. Typically the visual system is faced with the inverse problem of recovery of causes from effects. Surface perception requires the recovery of the geometry of the scene irrespective of the illumination and viewpoint. In object perception *the goal is to extract and represent the causal factors which underlie the natural variation that we see in objects.* **Inverse problems** can only be solved by embodying natural constraints on the transformations of objects like those considered above. When a general object changes its shape it changes its identity unless the transformation is sufficiently simple and global in its effect, like a simple stretching along one dimension. In the case of

Devil's Advocate Box 4.3 The shape of biological objects

4.13 Drawing of a tree.

In the discussion of the shape of an object we usually have in mind general smooth objects like potatoes or household artifacts that would be found in the kitchen and which are referred to in the scenario. However, there are numerous objects which have a very different mode of construction and as a result a very different shape. Trees are a good example of this kind of object. It would be hard to imagine extending Marr and Nishihara's structural description approach to trees, not because their theory could not be scaled up to cope with complex objects, it is just that it is beyond the bounds of imagination to think the human mind would encode the position and relationship of every branch, twig, and leaf on a tree. In computer graphics (Foley, van Dam, Feiner, and Hughes, 1990) growth processes can be simulated by starting with a pattern say ABBAA and then replacing all the As with B{B}A{B} and all the Bs with another pattern AA to give a new sequence, at which point the process can be repeated. The brackets indicate a branching point. This is called an L-grammar, a style of describing structure devised by Lindenmayer. Here a global impression is built up from the repetition of microstructure. When we recognize a particular species of tree we may recognize the microstructure or we may recognize the shape of the bounding volume of the tree.

complex objects, like faces, a simple cause can have many effects but the effects are likely to be correlated, for example, preparing for a sneeze or smiling can have a number of results in terms of the surface structure of the face. A theory of global shape needs to address those aspects of the shape of an object that can be influenced by a parameter the scope of which includes the whole object.

If the main role of the representation of objects is to encode the way that objects transform, then it makes sense to aim to *represent the object in terms of the major sources of variation.* These dimensions should reflect the regular and lawful transformations of the object. The consequences of this approach are that the representations have to be learned since we have no a priori knowledge of how objects vary, and that they must be specific to a particular object class, and it is for these reasons that we find dependencies on the visual diet (Johnston, 1992). The representation can be envisaged as a generalization of the "form" plus parameter means of encoding global shape. For example, the face has a particular abstract form which could be parameterized in terms of its major sources of variation. The use of the term form is to distinguish the representation from an encoding based on distance from a prototype. We might think of a prototypical ellipsoid which is average over a set of exemplars, but the form of the ellipsoid is consistent for all exemplars. Recognition does not involve extracting the invariant information over exemplars but extracting the parameters of the form.

SAQ 4.7 In a computer program the value of a parameter can be local to a procedure or global. A global parameter has a value that applies throughout the programme. Can this distinction between local and global parameters be combined with Marr and Nishihara's proposal of a modular, distributed coordinate system?

We can ask the question what constitutes an object part and also what criteria govern the point at which the most elemental part is assigned. In this framework decomposition into parts occurs *to the extent that parts can vary independently of the whole.* Taking the example of the teapot: the handle, the bowl, the spout, and the lid can all vary independently. If there is some correspondence between them, we say we have a particular style. In this way decomposition is based on natural variation and therefore part decompositions are learned: they are not based upon geometry. Dissociation occurs when the transformations of one class of objects differ from the transformations of another. Thinking about part decomposition in this way leads naturally to a hierarchical decomposition like that considered by Marr and Nishihara. When the arm moves, the hand, forearm, and upper arm can all move together as a unit and therefore one would expect them to be grouped at the level of the arm. However, they can also move independently and therefore also need to be considered as independent sub-units. The distinction is analogous to the idea of the scope of a parameter in functional computer programming. A geometric approach has difficulties with this dynamic interactive aspect of part decomposition. Sometimes transformations appropriate to one object can be grafted onto another. This facility has been used to great effect by advertisers and animators providing us with dancing chocolate bars, cars with faces, and flirtatious sauce bottles.

Summing-up

We started the chapter with a number of questions to which we can now return. The problem of object perception has been characterized as: (a) how do we recognize an object as the same object under various transformations and (b) how do we characterize the shape of the object in relation to the transformations an object can undergo? A number of approaches to the problem of object constancy have been considered, including invariant- and transformation-based techniques, and we have also considered the kinds of representations that might link the image and the final scheme for describing objects. The emphasis has been on representations and strategies rather than mechanisms because we consider that it is more important, given our current state of knowledge, to explore the structures that are necessary for the development of specific theories about particular perceptual processes.

Further Reading

For an excellent review of work on face perception read *Recognising Faces* (1988) by V. Bruce. *To See or Not to See* (1987) by G.W. Humphreys, and M.J. Riddoch provides a very readable account of disorders of object recognition. Evaluating object recognition theories of computer graphics psychophysics (1993) by H.H. Bülthoff and S. Edelman which appears in T.A. Poggio and D.A. Glaser (eds), *Exploring Brain Function: models in neuroscience*, provides a clear and readable account of computational approaches in object recognition including more detailed discussion of the evidence for the view interpolation approach to object recognition.

Exercises

1 How easy is it to recognize people from silhouettes? Find some newspaper cuttings of celebrities or well-known public figures and make two photocopies. Fill in one copy with black ink to produce a silhouette. Ask a friend whether they can recognize the individual? How easy is it to recognize the person on the basis of the silhouette? Check using the other photocopy that they do recognize the celebrity when more information is available.

2 Find a black and white picture of a landscape that you can draw on. Make some copies on a photocopier and pass those around a group. Explain what is meant by a surface normal and then ask each person in the group to draw surface normals directly on the picture. Collect the pictures together and compare the efforts of each member of the group. Is each description the same? Where do the discrepancies lie?
3 Does viewpoint matter? Next time you are out with your camera take some pictures of objects from strange angles or close up. Are the objects easily recognizable? If not, why not?
4 What is the minimum amount of information you need for object recognition? Make some animals out of pipe cleaners. Get other members of the group to guess what animal you have made. How good are they at this? What information are they using?

CHAPTER

5 Producing and Perceiving Speech

Outline

In this chapter we give a brief overview of the essential background defining terms such as phoneme. We go on to develop these for understanding the acoustic structure of speech. The acoustic background is then used as the principal basis for understanding theories of speech production and perception. We outline selected theories of speech production, namely locus theory, Henke's look-ahead theory, and Jordan's artificial neural network account and show how these relate to theories of word production. The perceptual theories we cover are motor theory, natural sensitivities, and Elman and Zipser's neural network theory. We also mention how phoneme-based theories relate to theories of spoken word recognition. In the final section we cover the general requirements when explaining both production and perception in a satisfactory theory. This partly involves reassessing theories, such as motor theory as well as considering other theories which have as their main focus the explanation of this relationship.

Learning Objectives

After reading this chapter you should be able to:

- understand some of the basics of articulatory phonetics, such as what is meant by a phoneme
- understand how air in the lungs is transmitted through the vocal tract and produces speech
- interpret the information on a spectrogram
- explain what is meant by coarticulation and recognize the characteristics of different types of coarticulation
- evaluate theories of speech production
- evaluate theories of speech perception
- evaluate theories about the relationship between speech production and perception

Key Terms

- categorical perception
- coarticulation
- manner of production
- phonemes: place of articulation
- spectrogram

Introduction

The scenario (chapter 1) involves people in verbal interaction. There are two distinct aspects of what they are doing when engaged in this process: they are producing speech and recognizing and responding to other people's speech. How cognitive mechanisms allow participants to achieve these goals is the focus of this chapter.

One of the most important aspects of speech production is how, once a speaker has decided to produce a word, this is brought into effect. A widespread assumption is that speakers organize their speech production processes by loading a buffer with the elemental parts of sounds (phonemes) and issuing them to the articulators. However, as we will show phonemes vary considerably even when the speaker speaks the same phonemes. We will outline the evidence for this, as well as accounts about how to explain it.

When the perception of speech is examined, we will only have space to deal with a rather limited range of things that the listeners in the scenario are doing. These will concern some of the basic findings which pertain to perception of phonemes and the different theoretical approaches that have been taken to explain them. Though we will mainly focus on perception of phonemes, we will indicate how phoneme perception fits into an illustrative model of word recognition.

Basic Background on Production and Perception of Speech

The notion of a **phoneme** is basic to the understanding of speech perception and production:

> The easiest way to understand the nature of phonemes is to consider a group of words like 'heed', 'hid', 'head', and 'had'. We instinctively regard such words as being made up of an initial, a middle and a final element. In our four examples, the initial and final elements are identical, but the middle elements are different. It is the difference in this middle element that distinguishes the four words . . . Such distinguishing sounds are called phonemes. (Denes and Pinson, 1973, p.14).

Here we are mainly going to concentrate on how phoneme strings are produced and perceived. First, some general remarks are given justifying why it is necessary to consider phonemes. All the models of speech production discussed here propose that word production proceeds by the output of a phoneme sequence from a sequentially-organized buffer that contains phonemes. The models of speech perception attempt to explain how phonemes are retrieved from the acoustic pattern that is heard. What is going to be presented, then, are some of the basic ways by which the production and perception of phoneme units takes place. Even with models that include higher-order constraints you are likely to have to understand phoneme output or input in order to evaluate most if not all the models, since either phoneme production or perception (as appropriate) are usually components of these models.

In order to understand the important characteristics of phonemes, we first need to know what vocal organs, or articulators, are involved in producing them. The articulators we will be concerned with are shown in simplified form in figure 5.1.

Three technical terms are used in connection with describing phonemes: voicing, **place of articulation**, and **manner of production**. Plosive stop consonants and vowels will feature extensively in our survey of theories of speech perception and production. For this reason, we will illustrate how the three terms apply to these two classes of phonemes. In this chapter, phonemes will be indicated by symbols between //.

Manner refers to the magnitude of the maximum constriction in the vocal tract brought about by moving the articulators along the base of the vocal tract towards those articulators at the roof. This characteristic distinguishes phoneme classes, such as the plosives from the vowels: In the case of plosive stop consonants, an articulator along the base of the vocal tract is brought into

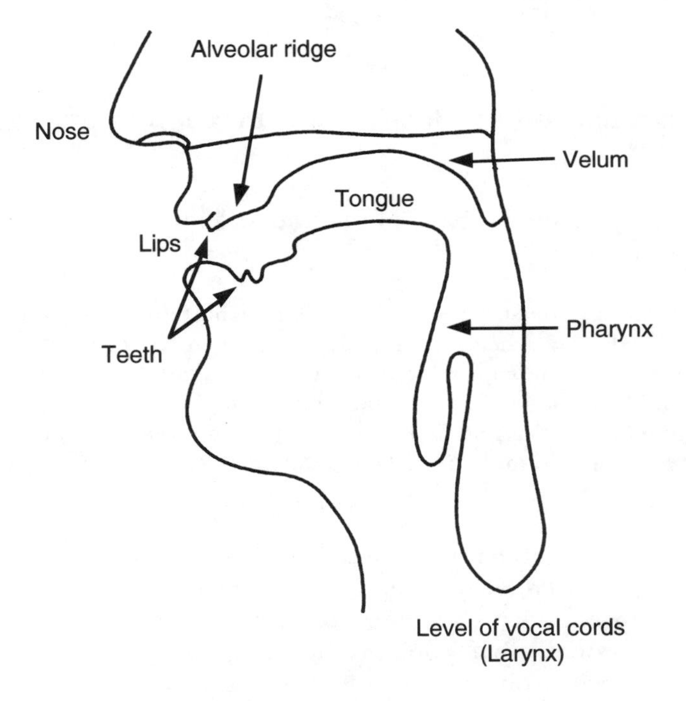

5.1 Schematic cross-section of the vocal tract indicating some of the important articulators referred to in the text.

contact with part of the roof. In vowels, the base and roof are not in contact at any point along their length at any time that they are spoken.

Though manner indicates how close together the articulators come, it does not indicate at what point in the vocal tract the articulators come closest: the place of articulation specifies that characteristic. The lips, tongue, and velum (shown in figure 5.1) are the principal articulators that can deform the vocal tract to produce constriction at different places along the vocal tract. The different plosives used in English are produced either by closure at the lips (labial), contact between the tongue and the gum ridge (alveolar), or with the tongue against the soft palate (velar).

Positioning the articulators to produce a plosive at a particular place, as just described, does not cause any sound. The source of energy that leads these and other phonemes to be heard is air expelled from the lungs which passes into the vocal tract through two muscular flaps called the vocal cords (figure 5.1). The modes of operation of the vocal cords in controlling the release of the air from the lungs involves the third technical term we mentioned above, voicing. The different ways this is brought about lead to, what is referred to as, different **sources of excitation**. The vocal cords may be adjusted either to bring them together so that the pressure of the airstream causes the cords to vibrate (when a **voiced source of excitation** is said to occur) or to keep them apart so that they do not vibrate as the air passes through (a **voiceless source of excitation**). Both voiced and voiceless plosive stop consonants occur in English, and both occur for each of the three places of articulation of the plosives. The characterization of the six plosive stop consonants of English in terms of place of closure and voicing are summarized in table 5.1.

SAQ 5.1 Try saying words that include the plosive stop consonants to yourself. Be sure to check whether the articulators at the base and roof of the mouth are coming into contact and whether the position corresponds with the description given in the text. Place your fingers on your **larynx** (see figure 5.1). When saying voiced plosives can you feel a tickle (the vocal folds vibrating) but none when saying a voiceless plosive? Say "bad boy." Concentrate on the /d/ at the end of bad and the /b/ at the beginning of boy. Say them slowly and feel the tongue contacting the alveolar ridge and then the lips coming together. What closures do you make when you say these sounds rapidly? Get someone else to say these words slowly and rapidly. Do you hear a /d/ in bad? If not, what implications do you think this might have for speech perception? Could a child playing a game called "batball" or being told that he is a "bad boy" tell the difference between bat and bad when the words are spoken rapidly?

Vowels, our second manner class, are characterized by voiced excitation when spoken normally. During the voicing of a vowel the vocal tract is open along its length, although at some point the base and roof of the mouth are closer together than elsewhere. Vowels can either be produced with a constant point

Table 5.1 The six plosive stop consonants of English. Voicing is indicated along the top and place of articulation on the left-hand side.

	Voiced	Voiceless
Place		
Labial	/b/	/p/
Alveolar	/d/	/t/
Velar	/g/	/k/

of maximum constriction or the point of the constriction can change during the course of the vowel. For our survey of speech production and perception, we only need to consider vowels where the constriction is at the same position throughout. The main way a speaker determines what vowel is produced is by locating the tongue at different positions in the vocal tract. Two factors are used in describing tongue position in vowel production: (1) The point where the tongue and roof of the mouth come closest together (e.g., near the lip end or further back), and (2) how close the tongue is to the roof of the mouth (degree of constriction). In articulatory phonetics, these are called front-back position and tongue height, respectively. Some examples of vowels with a single constriction are summarized in table 5.2.

SAQ 5.2 Say the following words noting where your tongue is located using the front/back and height descriptors employed in table 5.2. Some cases in which the place shifts during the vowel have been included. Can you tell which these are? Heat, hay, hoe, bit, bye, gate, father (first vowel), now, fur. Get a friend with a different regional accent to say these words. Do you think this speaker's accent affects where his or her tongue is positioned? Does this affect the vowel? If the speaker uses different vowels when asked to read the same word, consider what implications this would have for theories of speech perception.

So far, then, we have an idea about what the articulators are doing when two classes of phonemes are uttered. The final step we need to cover is the translation of the articulatory configurations into sound. The importance of this step is that, on the one hand, the acoustic waveform is the end-product of speech production and, on the other, it is the starting point of speech perception (the sound a listener receives). We will focus on the acoustic properties of voiced plosive–vowel syllables again, for the reason that they provide essential information for appraising the theories of speech production and perception that we will consider later. We will see how voicing, place, and manner are all helpful in understanding what sound issues from the vocal tract during the production of these syllables and what a listener has available

Table 5.2 Some of the vowels of English. Front/back position is indicated along the top and tongue height on the left-hand side. As well as the phoneme symbol, a word containing the vowel is also indicated.

		Front/back position.		
		Front	Central	Back
	Hi	/i/ heed	/u/ who'd	
		/I/ hid	/ə/ heard	/U/ could
Tongue height				
	Med	/ε/ head	/ʌ cud	/ɔ/ hawed
	Lo	/ae/ had		/a/ hard

to determine which particular voiced plosive–vowel syllable has been produced.

First, since we are only considering the voiced plosives and since the vowels too are voiced, voicing occurs on each of the phonemes in the syllables being considered. Recall that during voiced sounds, the air issuing from the lungs is modulated by the opening and closing of the vocal cords. The cords open slowly (airflow is allowed) and close quickly (airflow is stopped). Thus, during any opening and closing cycle, an asymmetric pulse of air enters the vocal tract. The pulsing flow repeats continuously throughout the time that voicing is applied, so the pulse shape repeats relatively regularly. The flow into the vocal tract is depicted in figure 5.2. The x axis is time and the y axis is a measure of the amount of energy (the higher the line is at any point in time, the more energy enters the vocal tract). If we were able to listen to this pulse, it would sound like a buzz, not like speech. Later we will see what the vocal tract does to this acoustic energy to give it its distinctive speech qualities.

Besides representing the energy entering the vocal tract in terms of energy and time as in figure 5.2, the energy in this source of excitation can also be represented in terms of its frequency content. You will notice that the waveshape that repeats in figure 5.2 has a section labeled "P." P is the period

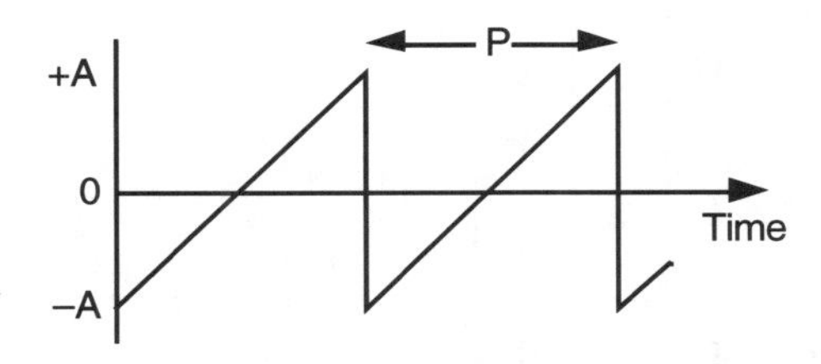

5.2 Amplitude-time changes of a sawtooth waveform. Time is on the x axis and amplitude on the y axis. Amplitude fluctuates between +/– A. The period of the sawtooth is indicated (P).

of time of the repeating waveshape. Periods can be converted into frequencies by the simple formula:

Frequency in the units Hertz = 1/ Period in seconds.

Thus, if the sawtooth in figure 5.2 took 0.01 seconds to repeat, the frequency of repetition would be 100 Hertz.

An important theorem due to Joseph Fourier, tells us that complex periodic waveshapes, like our sawtooth waveform, can be decomposed into simple sinusoidal components (smoothly changing up and down movements which are also periodic but, unlike the sawtooth, are elemental frequency constituents). Fourier's theorem also tells us that the constituent frequencies of a complex periodic waveform have an exact relationship with the original waveform: they can only be an integer number of times the frequency of that of the original waveshape (1 x it, 2 x it etc.). Thus, in the 100 Hertz example, only frequencies of 100 Hertz, 200 Hertz and so on can be components of this waveshape. The amplitude of the sinusoidal components that make up a periodic signal (such as the sawtooth waveform) can be represented by an amplitude spectrum, which is a graph of frequency (x axis) versus amplitude (y axis). When the signal contains a sinusoidal component at a particular frequency, a line is drawn whose amplitude is given by the y axis. The amplitude spectrum of a sawtooth waveform is shown in figure 5.3.

So far, we have described the properties of the source signal that enters the vocal tract in the frequency domain. Next, the alteration that the source undergoes as it passes through the vocal tract has to be ascertained. To keep things simple, we will consider vowels which have a stable vocal tract position throughout. From what has been said earlier, you know this class of sounds has

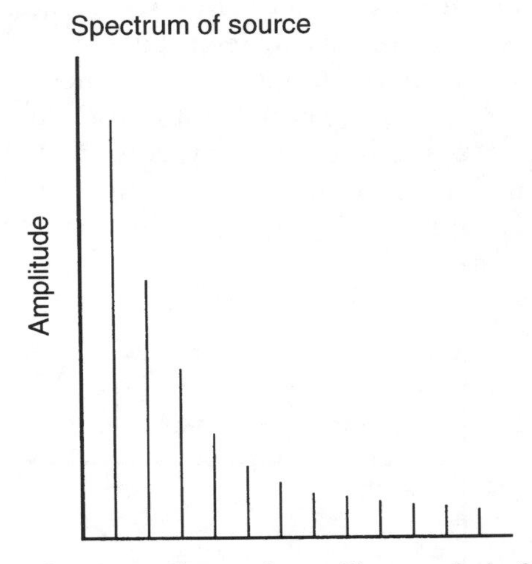

5.3 Amplitude spectrum of a sawtooth waveform. The x axis is frequency and the y axis amplitude. The components present are indicated by a line whose position on the x axis indicates its frequency and whose height indicates its amplitude.

a particular manner (tract open) and different vowels are produced with constriction at different places in the vocal tract (if necessary go back to SAQ 5.2).

To figure out the alteration that the source undergoes as it passes through the vocal tract, we first need to represent the vocal tract operation in the frequency domain as we did with the source of excitation. The transmission of certain frequencies through a vocal tract with a given shape is better than that of others. Thus the vocal tract is said to **filter** the frequencies. How well the vocal tract transmits a range of frequencies can be established directly by applying sinusoidal vibration at a particular frequency and known amplitude at the bottom of the vocal tract. The output amplitude at this same sinusoidal frequency is then measured at the lips. This procedure is then repeated over the range of freqencies of interest (see Fujimura and Lindqvist, 1970, for full details about how this is done or Rosen and Howell, 1991, for a more elementary description). Graphs representing how the amplitudes of different frequencies are affected by that vocal tract shape are shown in figure 5.4 (called the amplitude responses of the vocal tract shapes); in this figure, various vocal tract shapes correspond to the vowels indicated in the insets. The main things to note are the frequency regions where energy is passed best (the peaks in the graph), and that the position in frequency of these regions depends on what the vocal tract shape is.

Those areas where frequencies are transmitted well are usually broad and may extend over several sinusoidal components of the source energy. The regions in which good transmission occurs are called the formants (they are the resonant frequencies of the cavities of the vocal tract). The formants are numbered from the lowest frequency up, F1 refers to the first (lowest) frequency formant, F2 to the second lowest frequency, and so on. The position in frequency of the formants depends on the shape of the vocal tract. Since the

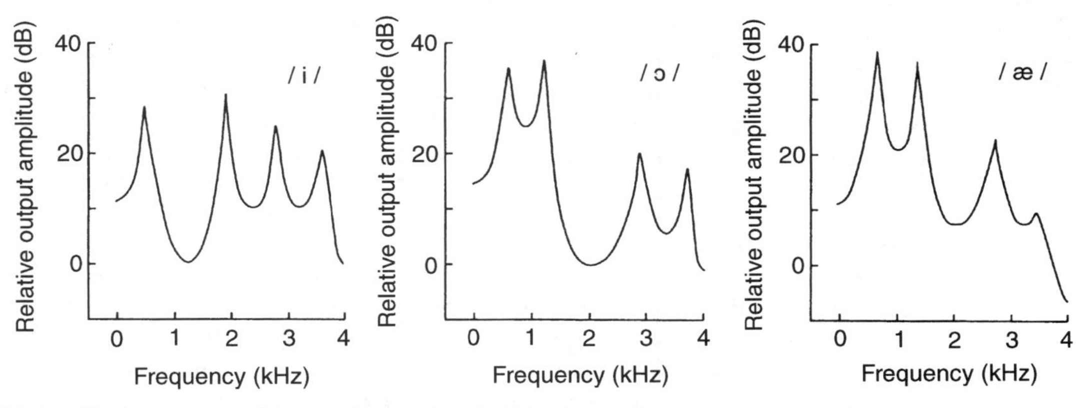

5.4 Amplitude response of the vocal tract when held in the position appropriate for three different vowels. The vowels are indicated by the symbol at the top (/i/ /ɔ/ and /ae/). Frequency is along the x axis and amplitude along the y axis.

position in frequency of the formants is determined by what vowel is being spoken, the formants are the important determinants used by a listener to establish the identity of the vowel (and, for the same reason, for identifying other phonemes).

Up to now, the source and vocal tract filter have been considered separately. The description of the filtering action just given is the output of the vocal tract when a range of sinusoids of the same amplitude are applied as input. However, the consideration of the properties of the source given earlier indicates that the energy that enters the vocal tract (the sawtooth waveshape) has only certain frequency components and these are not all at the same amplitude. To establish what the vocal tract output would be when a sawtooth source enters, the outputs like those in figure 5.3 need to be processed so that only the sinusoidal components that are present are selected, and allowance must be made for the variation in their amplitude (because the level of the components in the source vary in amplitude). Once this step has been achieved, we will have a representation that reflects what the combined action of the source and vocal tract filter is like (i.e., like the vocal tract output a real speaker would produce).

The procedure to combine the amplitude spectrum of the source and the amplitude response of the vocal tract is to multiply the amplitude response of the vocal tract by the amplitude spectrum of the source at corresponding frequencies. When this multiplication is performed, only those frequencies that are present in the input occur in the output because the source signal is zero at all but these frequencies. Multiplying the amplitude spectrum of the source by the amplitude response of the system also takes account of any variation in the amplitude of the components of the source signal. The steps taken to derive the amplitude spectrum of the vocal output are summarized diagrammatically in figure 5.5.

SAQ 5.3 Perform numerical checks of the processes summarized in figure 5.5 to show the effects of passing a source of excitation that has components that decrease in amplitude as frequency increases and to show what happens when there is no energy in the source at certain frequencies through an amplitude response of your choice. If you need help, consult Rosen and Howell (1991).

One final important aspect of speech acoustics is the representation of the way the frequencies in a sound change over time. The formant frequencies of the vocal tract change relatively slowly over time and follow the movements of the articulators. The **spectrogram** is a plot of the frequencies in the signal (vertical axis) represented over time (horizontal axis). The amplitude at a particular frequency and time is represented on a grey-scale (the darker the point, the higher the amplitude). The type of spectrograms encountered in this chapter are all broad-band spectrograms which are particularly suited to

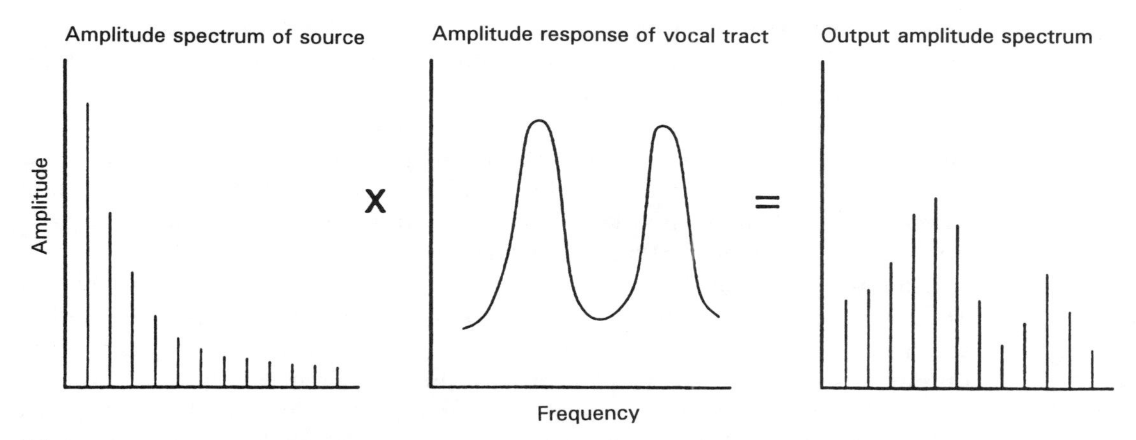

5.5 Summary diagram indicating the steps involved in establishing how to determine the output amplitude spectrum of speech when the amplitude spectrum of the input and the amplitude response of the vocal tract are known. The amplitude spectrum of a sawtooth source is shown in the far left, the amplitude response for the vocal tract this is passed through next to it and the spectral output is on the right. The figure indicates that the steps involved in obtaining spectral output are to multiply the source by the amplitude response at corresponding frequencies.

revealing the formant movements over time (see Rosen and Howell, 1991, for other types).

As we saw earlier, the formants are a direct result of the vocal tract shape. As the vocal tract changes in shape (for instance going from bilabial closure at the start of a /b/, to the shape for the vowel that follows), the formants move over time. Thus, this representation is also revealing as an indication about what speech movements the speaker is making. In addition to the use of spectrograms in understanding speech production, the human auditory system monitors the same properties of sound that are depicted in a spectrogram (i.e., frequency, amplitude, and how they change over time). Thus, the information in the spectrographic representation has been considered to be a clue about what information the speaker has available for making perceptual decisions about speech. We will see later how spectrograms have been used to investigate some of the fundamental problems in speech production and perception.

Production Accounts

Now, using the ideas we have been considering, we can turn to the problem of how, when a speaker intends to communicate a message, the brain controls the muscles that move the appropriate articulators to produce the required speech sounds. The basic problem is that though a speaker might be required to produce a particular speech sound (either voluntarily or because an experimenter asks him or her to do so), we cannot find a direct relationship between that sound when it is spoken in different verbal contexts and the resulting action at the muscular, articulatory, or acoustic levels. How might this variation arise? Variation on phonemes due to contextual variation, whether in articulator position, muscle commands, or acoustic properties, can be considered as instances of coarticulation. It has been defined as " . . . the influence of one speech segment upon another; that is, the influence of a phonetic context upon a given segment" (Daniloff and Hammarberg, 1973, p. 239).

Later we assess locus theory. This is an example of a theory in which a phonemic string represents the specification for the message which is issued to the articulators. It specifically attempts to explain why variation in *acoustic* output arises for the same phonemic input. Problems in this early theory spawned a number of accounts that attempt to get round some of its deficiencies. Following on from that, we consider Henke's look-ahead theory which attempts to explain how variation in muscular commands, articulatory position, and acoustic output occur when a phoneme is realized in a different context. This theory too has problems but introduces the important concept that the realization of a phoneme at the articulatory, muscular, and acoustic levels is not a result of physical constraints (as in locus theory) but can arise at the input specification to these processes. Some of the concepts employed in look-ahead theory are also employed in the next theory considered, Jordan's (1986) recurrent neural network account.

Production Theories of Sequencing Coarticulation – Locus Theory

As just stated, the central concern of production theories is to explain how variation at the acoustic, muscular, or articulatory level occurs for a particular speech sound when it appears in different contexts. Though all workers have accepted that the acoustic energy associated with a particular phoneme varies with the context in which it occurs, early workers denied that the muscle commands for a phoneme varied with the context the phoneme occurred in. They claimed that the variation could arise simply from the mechanical constraints that are necessary when going from the shape appropriate for one phoneme to another.

The locus theory is an early example of a theory which explains the variation between a consonant and the vowel that follows on the basis of such mechanical effects (Delattre, Liberman and Cooper, 1955). The theory was originally developed to explain the acoustic variation observed with each of the plosives when spoken in different vowel contexts. The main assumptions in the theory are that fixed commands are issued whenever a particular phoneme is produced, and that variation arises because the speaker has to move his or her articulators from the position specified for one phoneme to that of the next, which will vary with what the adjacent contextual phonemes are.

Let us illustrate this with a particular example, the voiced alveolar plosive, /d/ before various different vowels. The excitation is applied continuously (i.e., with no break) across the plosive and vowel. So, in this case, voicing starts at the /d/ and continues until the end of the vowel. Next, we need to consider the articulatory movements and the effect this has on the sounds that emerge from the tract. As already seen the articulatory gesture for /d/ starts with closure in the vocal tract at the alveolar ridge irrespective of the following vowel. At closure no sound emerges. As the tongue is released it describes a trajectory towards the vowel configuration. As the degree of opening in the vocal tract increases a sound is emitted. The tongue at this point is partly towards the vowel configuration. Thus, the formant frequencies are not quite at the values they would have when the closure is released, since the vocal tract has changed in shape. The formants will move in the direction appropriate for the following vowel as the vocal tract is now nearer that shape. As time goes on, the tract shape and, therefore, the formants get progressively nearer to those of the vowel until the vowel itself is reached.

The consequences of the movement from plosive closure to vowel on the first and second formants is illustrated in the stylized spectrogram shown in figure 5.6. Vowels have been used in these spectrograms which have the same first formant frequency. For the second formant, the point where the dashed lines start represents the frequency value which would be produced if sounds could be transmitted when the vocal tract is closed (the "locus" of the stop consonant). This locus frequency is approximately constant for any particular speaker whatever vowel the /d/ precedes. The solid lines represent the actual formant transitions observed before the different vowels. Where the formants finish and the transitions that are made is determined by what the final vowel configuration is. Note that the direction the formant transitions move is determined by *both* the vowel and the stop consonant.

SAQ 5.4 Describe how the formant movements shown in figure 5.6 arise and how the formant transitions in the first part are affected by the vowel context.

This explanation of why the formant transitions for the /d/ phoneme vary with vowel context does not assume that different muscular commands produce the various transitions appropriate for the /d/, indeed, we started the description of locus theory by assuming that closure at the alveolar ridge would always occur for the /d/. Since this closure could be achieved by the same muscles contracting whatever the vowel context, the muscle commands for the /d/ phoneme may be identical whatever the vowel context. Similarly, the muscular instruction specification for the vowel may also be fixed. The only reason the transitions differ in the theory is through the mechanical requirement of the vocal tract having to go through different intermediate states when responding first of all to the fixed commands for a /d/ and then the fixed commands for the following vowel. A point that we will return to later, as can be seen from figure 5.6, is that information about the vowel is available before the steady state is achieved and available for the consonant from release of the stop consonant to the steady state formants associated with the vowel alone.

Coarticulation explained by locus theory has been termed phoneme sequencing (Howell and Harvey, 1983). As we have seen, the case can be made that phoneme sequencing is largely, if not completely, accounted for by the mechanical requirements on the articulators in the way described by locus theory. However, a problem with the locus theory is that it does not offer a complete explanation of all coarticulatory phenomena, as all instances of coarticulation cannot be explained in terms of these constraints.

Coarticulation in the syllable /stru/ is an example of coarticulation which is difficult to reconcile with the assumption that the same commands are issued to the articulators whatever context the phoneme occurs in (Daniloff and

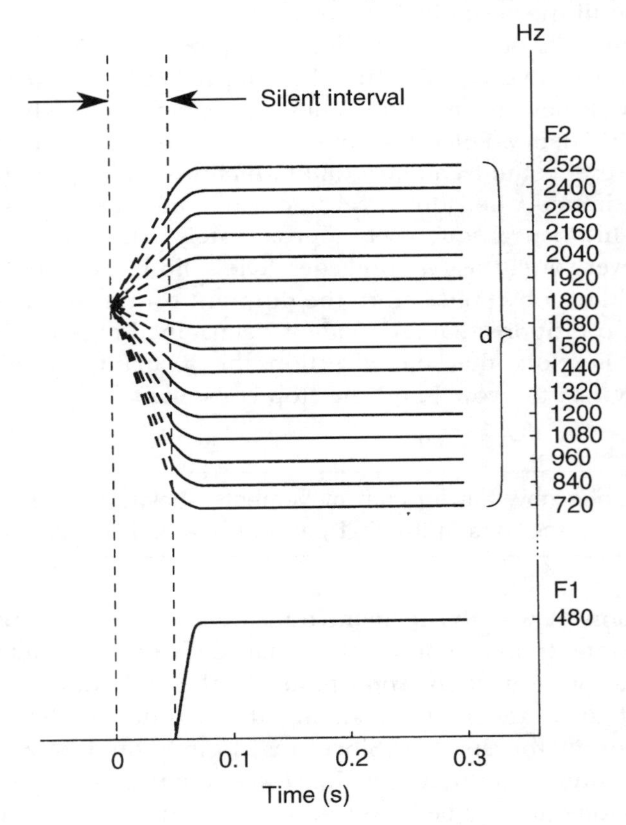

5.6 Formant transitions producing /d/ before the steady states on the right. Time is plotted on the x axis and frequency on the y axis. The solid dark bands indicate the actual formant transitions that would be measured. The transition at the bottom is associated with the first formant (labelled 'F1') and the top transitions are those of the second formant (labelled 'F2'). The dotted lines extrapolate the transitions back to the point at which the gesture commences (time 0). Since there is a silent interval before sound onset, no sound is heard (in this example) until some 0.05 sec after the gesture starts.

Moll, 1968). In this syllable, the lips round at the phoneme /s/ and remain rounded in anticipation of the vowel, which does not occur until three phonemes later. This lip rounding affects the acoustic output too. The differences in the phoneme /s/ in the syllables /si/ and /su/ can be heard on close listening. The /s/ in /su/ is produced with lip rounding as in /stru/ while that in /si/ is not. In the /stru/ example, coarticulation occurs over three phonemes and has been observed over as many as seven phonemes. It seems unreasonable to argue that mechanical effects like those inherent in locus theory could operate over time intervals of this length, so the account is not complete.

A second example to illustrate the same point concerns speech spoken at different rates: in these cases speakers restructure their articulations in a complex way. If speakers issue the same commands whatever the rate of the speech, then the articulators would never reach the target positions at fast rates (referred to as target undershoot). Undershoot of the tongue occurs for vowels (Lindblom, 1963) and undershoot of the velum for consonants (Kent et al., 1974). However, speakers change the velocity of the lips (Gay and Hirose, 1973; Gay et al., 1974) and mandible (Abbs, 1973) to minimize such undershoot. Thus speakers appear to alter the muscular commands to the articulators at fast rates so as to achieve the target. This restructuring of the commands implies that speakers do not use fixed muscle commands for a particular phoneme as required by locus theory.

Production Theories of Coarticulation – Henke's Look-ahead Theory

One potential way in which the latter class of coarticulatory, as opposed to sequencing, effects could be explained has been offered by Henke (1966). We can illustrate his theory by considering the word "bond." This word includes an /n/ sound. This consonant is produced by dropping down the flap of skin called the velum (see figure 5.1) which allows air to flow through the nasal cavities. For this reason it is one example of the manner class "nasal." You will have noticed that the /b/ and /d/ of this word are plosive stop consonants, so at some point airflow has to be completely stopped. Thus, these consonants cannot have the nasal cavities open otherwise it would not be possible for the air to stop (in other words, though the lips might be closed for the /b/, air would escape via the velum into the nasal cavities and out into the environment). Thus, for the /n/, it is obligatory that the nasal cavities be open whilst for the /b/ and /d/ the opposite is the case. The only phoneme left is the /ɒ̧/ vowel. For English vowels, it does not matter whether the velum is open or closed, the vowel will be interepreted as /ɒ̧/ whether it is spoken with the velum in either of these positions. Henke would refer to the situation that occurs on the /ɒ̧/ vowel as a "don't care" nasalization condition since it does not matter whether the nasalization happens or not. We can summarize the state of affairs with respect to nasalization on the phonemes of "bond," then, as follows:

	Phoneme			
	b	o	n	d
Nasalization	Not nasal	Don't care	Nasal	Not nasal

Henke argued for a "look-ahead mechanism." The tenet behind the theory is that a speaker will start to position an articulator appropriately as early as possible. So by "looking ahead" we can see that the /n/ will need to be nasalized. Is it possible to start the nasalization on the first phoneme? No, because an open nasal passage is inconsistent with production of a plosive stop consonant. Can it start on the /ɒ/ vowel. Yes, because nasalization on English vowels does not affect what vowel is identified. Since this is the earliest point that nasalization can start, Henke's theory tells us that this is the point at which nasalization will commence. So far we have considered where we can start nasalization, in anticipation of when it is required. The same mechanism could be applied to determine how far nasalization will perseverate. It has to occur on the /n/ since this is a nasal consonant. Can it continue from the /n/ onto the /d/? No, for the same reason that nasalization cannot occur on the initial /b/.

SAQ 5.5 Henke's look-ahead mechanism was described in the text for nasalization of English vowels. Languages like French have nasal vowels ("mot" means something different to "mon" even though these are both CV syllables with the same consonant and vowel but with the vowel in "mon" nasalized). First convince yourself that the vowels have the same tongue positions in these words. Next, consider what Henke's look-ahead mechanism would predict about perseveration of the nasalization onto the vowel in "mot"; see Howell (1981, 1983) for some experimental tests of coarticulation of nasalization between nasal consonants and vowels. Work out for yourself what Henke's theory would predict for other articulatory features (for instance frication in Daniloff and Moll's (1968) /stru/ example discussed earlier).

Henke's theory is specifically about production of coarticulation and he would need to propose, if coarticulation is primarily there to help production, what the effects it might have on perception: production might be made easier by "looking ahead" but this would be bought at the expense of making perceptual processing more difficult. Thus, people would have to identify a vowel phoneme as an /ɒ̃/ in our earlier example when it is and is not contaminated by nasalization. One possible way round this problem can be ruled out. It might be considered that the basic units of perception are not phonemes, but syllables or words. The force of this argument would be that coarticulation within the phonemes of a word would not matter if the word was the unit of perception. You would have one pattern to recognize for the word "bond," not separate patterns for the constituent phonemes of the word. Consequently,

Devil's Advocate Box 5.1

A child learning to speak his or her native language will hear words from the caregiver. Though this might suggest that children start learning words, some authors have proposed that this is not a very efficient procedure for acquiring extended vocabularies; in principle, you only need to learn how to produce 40 or so phonemes to speak all the words in a language. Consider these arguments and decide whether they are exclusive alternatives. If they are, which do you think is favored on balance?

it would not matter that an /ɒ̃/ phoneme in one word differed from that in another. However, this solution to the problem does not work since coarticulation occurs between words (Howell, 1983), not just between phonemes. Therefore, contextual variation between words faces just the same problem as that between phonemes.

Production Theories of Coarticulation – Jordan's Artificial Neural Network Theory

Chapter 2 described various modeling approaches that have been taken in cognitive science. One of the approaches that is outlined there has been developed with speech production and perception in mind, namely, recurrent networks. As stated in chapter 2, simple feed-forward networks are not very good for mapping inputs and outputs that change over time, and time variation is a crucial property when considering how speech is produced and perceived. Here we will consider the application of recurrent networks to the problem of explaining coarticulation (Jordan, 1986).

First, let us specify some of the implementational details of Jordan's recurrent neural network (you might need to reconsult chapter 2). The input to the network, for predicting coarticulation, is a plan of the phonemes to be produced. The output will be some feature of articulation. So, for instance, in the example considered later in this section, we will be concerned with the spread of lip rounding between the phonemes in the plan (i.e., coarticulation of lip rounding). Thus, the output activation is supposed to indicate whether the lips are rounded (say high activity) or spread (low activity). The mapping of inputs to outputs is via a layer of hidden units. The hidden units are necessary, as stated in chapter 2, so as to get round limitations in the computational power of the networks which would otherwise apply. Since the networks are of the recurrent type, some of the output is fed back to the input. Once the networks have been set up, training data are presented in which the desired output for given phonemes is known (rounded, not rounded, don't care); the weights on the connections are adjusted during training until the desired output occurs. The assessment, after training has been completed, will involve ascertaining whether the network exhibits properties that corrrespond with observed data on coarticulation.

Now let us go through the operation of this network explaining how it accounts for coarticulation. The recurrent connections in the model associate a static plan with a serially-ordered output sequence. The first element of the phoneme input string is taken and fed through the network where it appears as output. The output element (here the first phoneme) is fed back to the input layer where an association is generated with the next element in the sequence. This then serves as input which generates a further output and so the sequence repeats. The feedback from the output to input (the recurrent connections) allows the network's hidden units to remember its own previous output so that behaviour can be shaped by previous responses.

When the first element is presented as input, it is not influenced by forthcoming elements as none have appeared. Consequently coarticulation does not appear in the network as described so far even though data indicates that this happens: in the sequence /stru/ discussed earlier, for example, lip rounding for the /u/ is observed as early as the initial /s/ in the sequence. To circumvent this problem, the training regime is altered by preactivating elements to anticipate later occuring elements in the sequence. Lip rounding is then produced on the /s/ because the state of the input layer at the onset of the sequence is similar to its state during the production of the /u/. More generally, similar inputs tend to lead to similar outputs unless training data act to prevent it. So, if during training, lip rounding is not disallowed on the /s/, (i.e., error is generated when rounding is produced on this element), then it will tend to appear on this phoneme.

Relationship of Phoneme Production to Higher-Order Units

As stated earlier, most models of speech production have a step translating lexical items into a string of phonemes. Thus, what we have been through in the earlier part of this chapter is relevant to these models. To take one recent illustration: Levelt (1989) has outlined a complete theory (or blueprint as he calls it) of speaking. This involves specifying a phonetic plan and realizing that plan in articulation. Added components of his model which have not been covered here are stages involving initial conceptualization of the message (which is explicitly designated preverbal and, so, is not relevant to a chapter specifically to do with speech) and formulation. One other aspect of Levelt's (1989) perspective is the emphasis on perceptual monitoring of speech which, again has not been covered here. This process starts for Levelt with the overt speech, which is processed by the auditory mechanism into a phonemic string. That topic will be the focus of our next section.

Perception Accounts

The task facing a listener is to follow the speech message. In the face of the variable signals that are delivered to the listener by the speaker, this is not an easy matter. We discuss some of the theories that sought to explain how this is achieved.

Motor Theory Explanation of Perception

All the theories considered up to now have been specifically concerned with production. One early theory set as its goal a much more ambitious enterprise – to explain how production and perception are linked. This theory was motor theory. The specific concerns it attempted to explain were the puzzling aspects associated with perception.

One of these puzzling aspects was the perception of a particular phoneme in the face of different cues in different contexts. In the discussion of locus theory in the production section of this chapter, the basic phenomenon of sequencing was described and it is clear that the formant transitions for a plosive stop consonant depend on the vowel that follows. In particular, the formant transitions for a plosive at a particular place vary with the vowel context (though only place of production of plosives was discussed, similar arguments apply to perception of other phoneme **contrasts**). If perception of place in plosives depends on the formant transitions, the mapping between the acoustic structure and phoneme percept is not straightforward as there appear to be no invariant formant transitions that could be used to identify these phonemes directly.

A potential way out of this dilemma is that speakers produce other acoustic cues that identify the place of the plosives. One possibility is a brief burst of noise which occurs as the plosive is released. Spectrographic analyses have shown that the frequency of the noise burst varies with place of production of the plosive. Consequently, this acoustic cue might offer a straightforward indication about the plosive's place of production (i.e., all the listener might have to do is to determine the frequency at which the noise burst is centered and he or she would know the place of the plosive).

This explanation of perception of place in plosives would require that the burst cue is invariantly related to place whilst the formant transitions are not, and that the bursts should always occur so that it can fulfill this role whenever a /d/ occurs. With respect to the first issue, the noise burst shows different frequencies before different vowels because some vocal tract opening is required before it can issue from the vocal tract. For example, a noise burst at 1440 Hertz is heard as /p/ before /i/ but as /k/ before /a/ (Liberman, Delattre, and Cooper, 1952). With respect to the question of whether the noise burst is an ubiquitous cue, Dorman, Studdert-Kennedy, and Raphael (1977) have shown that the release burst does not always occur when a stop consonant is produced. Thus, the problem of acoustic variation of the cues for place in plosives cannot be solved by adopting this cue rather than the formant transitions. Consequently, the motor theorists consider that the problem of how perception might take place in the light of this complex mapping still needs to be faced.

The evidence indicates that a direct mapping between single acoustic cues and phoneme percept is untenable: both burst frequencies and formant transitions exhibit considerable contextual variability so no acoustic property can be identified which indicates what phoneme is being spoken. Though it is possible that the information specifiying, for instance, the place of articulation of a plosive is provided by the weighting between two or more acoustic cues (such as burst frequency and formant transitions), or that hitherto unidentified acoustic cues mediate this, the authors of motor theory took a different line: rather than an auditory process, they proposed that perception takes place by interpreting the acoustic input in terms of articulation (motor theory, Liberman et al., 1967).

The central tenet of motor theory which arose out of these studies, is that though the information that carries information about a phoneme varies with the other phonemes it is spoken with, the particular phoneme is the consequence of a fixed articulatory intent whatever the context. Thus, if the listener had access to the supposed articulatory invariance for this phoneme, this could be used to interpret which phoneme was intended from the variable acoustic cues. Work conducted at the inception of this theory seemed to support the view that there is a simple invariant relationship between a particular phoneme and the muscle commands that gave rise to this phoneme whatever context the phoneme occurs in. Thus, though invariance is absent in the acoustic signal, it seemed that it was present in the myomotor commands (see the section on locus theory for contrary evidence). In sum, early versions of motor theory proposed that perception took place by interpreting the acoustic cues in terms of the muscular commands which, in turn, indicated what phoneme had been said whatever its context. Though this general theoretical framework has implications for both perception and production, its main impact has been in the area of perception.

Studies on the perception of synthetic speech continua were conducted to test predictions of motor theory. The locus theory discussed earlier provided some of the information that is necessary to understand how such synthetic speech continua are constructed. Here we will run through how a place of articulation continuum (i.e., a continuum varying between perception of a plosive from say bilabial place to alveolar before a particular vowel) can be produced, before looking at how results on the perception of this continuum have been interpreted in terms of motor theory.

A typical continuum going from bilabial to alveolar places has F1 with an upward transitional movement during the first 50 or so ms, which then stays at a fixed frequency (at which point the "speaker" has reached the vowel). F2 transitions, before the steady state F2s of the vowel are shown in figure 5.7, start below the F2 steady states for /b/ and above for /d/. The transitions for /b/ move up and for the /d/ down over the 50 ms transitional region. The different F2 starting value that cues place can be interpolated by specifying different F2 starting frequencies (see figure 5.7). If a listener is asked to label a randomization of sounds drawn from this continuum as /b/ or /d/, he or she classifies sounds with low F2 onset frequencies as /b/s and sounds with high onset frequencies as /d/s. One important aspect of perception on this type of continuum which was noted was the abrupt way that the phoneme labels given changed across the continuum from one category to another. These categorization curves are supposed to be sharper than those which occur with nonspeech (Liberman et al., 1967). The point on the categorization function where each category is heard equally often is called the phoneme boundary (in speech) or the category boundary (in nonspeech).

Discrimination of items along the place continuum has been examined using an ABX discrimination task. In this task, each trial consists of a triplet of sounds presented in sequence (designated A, B, and, X respectively). A and B stimuli are always different and always have the same physical difference – for example, adjacent sounds on an evenly spaced continuum. The third (X) sound is the same as either the first (A) or the second (B) sound, and the subject has to indicate which he thinks the third sound is like. Performance

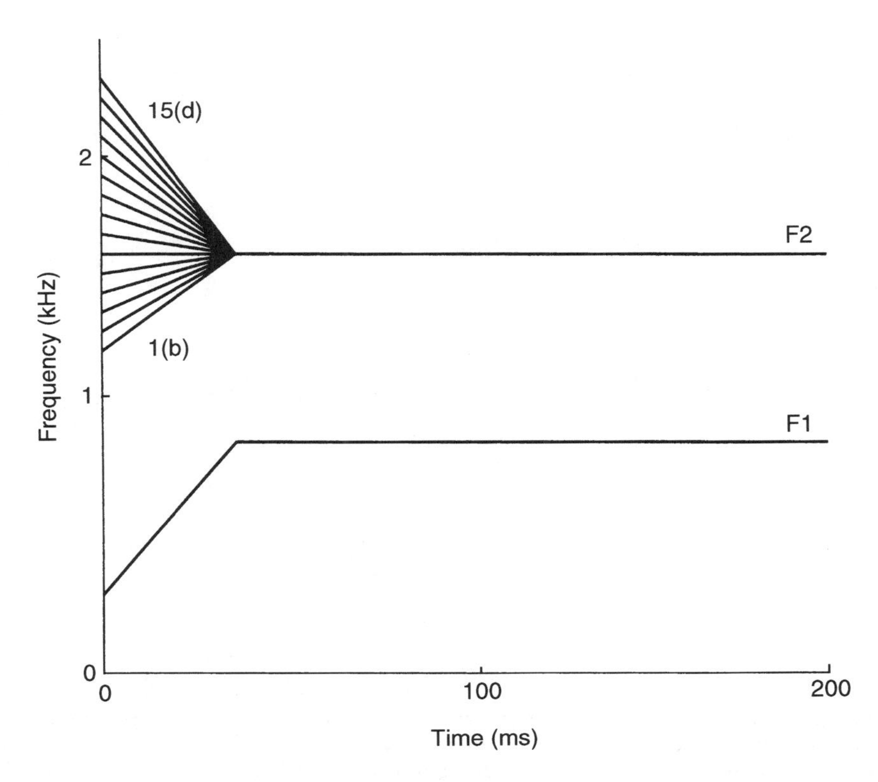

5.7 Stylized spectrograms indicating construction of a plosive place of articulation continuumn. F1 is the same for all stimuli. Fifteen different stimuli are indicated (1 and 15 labeled) indicating the formant transitions before the same F2 steady state. Stimulus 1 has a low rising F2 whilst stimulus 15 has a high falling F2.

with two sounds drawn from the same category is almost at chance but performance with stimuli straddling the phoneme boundary is considerably better. In other words, there is a discrimination peak at the phoneme boundary. Nonspeech continua did not seem to show such discrimination peaks (see Miller, 1956, for a contemporary review of identification and discrimination experiments with nonspeech). This type of perception originally observed in certain speech continua was termed **categorical perception**.

The three main characteristics of categorically perceived continua are summarized in figure 5.8. These are a sharp identification function, chance discrimination within a phoneme category, and a peak in discrimination performance at the phoneme boundary. Originally categorical perception was interpreted in terms of listeners employing phonemic labels during perception. Discrimination was poor for two stimuli in the same phoneme class because then the listener would give the first two sounds of an ABX triplet the same phonemic label and would have to guess which the third sound was most like. If the A and B sounds were labeled differently (as happens when a pair

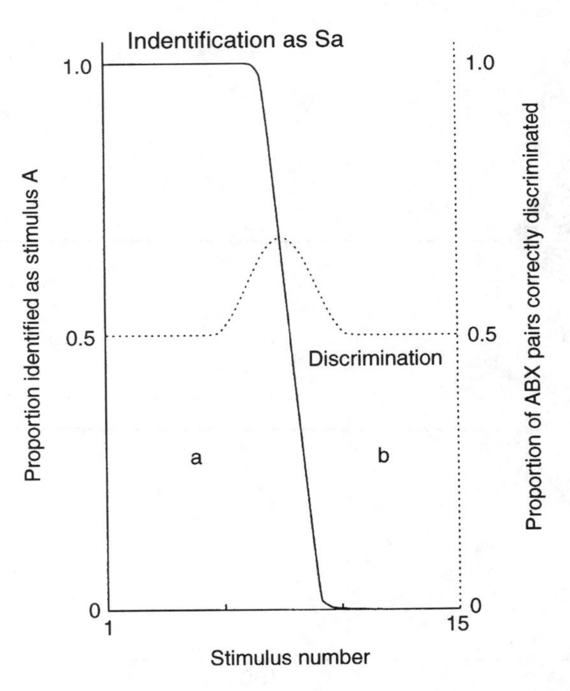

5.8 Summary diagram indicating the important characteristics of categorical perception. Two sets of results are plotted: identification and discrimination. The abscissa represents the stimulus number (as shown in fig. 5.7). For the identification responses, a binary response is required and the solid line indicates the proportion of stimuli identified as stimulus A. For the discrimination test, an ABX discrimination is required (described in the text). The ordinate for discrimination performance accuracy for the X stimulus and performance across pairs at different points along the continuum is represented by the dashed line.

straddle the phoneme boundary) the listener only has to ascertain what the label of the third sound is, to perform the discrimination correctly (Liberman et al., 1967). Thus all the features of categorical perception are predicted by this account. Since the use of phonemic labels do not apply to nonspeech sounds, categorical perception was considered specific to speech.

Here we will now continue our assessment by critically examining categorical perception as one piece of evidence for the position that mapping acoustic information onto phoneme percepts has to involve a specialized speech processor (in motor theory, a mechanism that involves reference to the articulatory muscles).

It is important to note that the explanation offered by motor theorists invokes a specific relationship between identification and discrimination. A second point to note is that no such relationship between identification and discrimination was thought to occur for any nonspeech continuum which would not be interpreted as phonemes. Also, unlike the case in speech, where discrimination seemed to be limited by identification, discrimination in nonspeech was normally thought to be far better than identification.

The sharpness differences between the identification functions of speech and nonspeech continua cannot be quantified for a variety of reasons (see Rosen and Howell, 1981). Thus, it is questionable whether these data alone indicate differences between the way speech and nonspeech are perceived.

Similarly, though the unusual discrimination performance was linked to the phonemic labels in the early experiments, this aspect of categorical perception was thought to be specific to speech. Views changed in the middle 1970s when it was reported that categorical perception occurs with certain nonspeech continua more complex than those used in the classic psychophysical studies (Miller, Wier, Pastore, Kelly, and Dooling, 1976). At around the same time, it was reported that chinchillas showed similar identification (Kuhl and Miller, 1975) and discrimination functions (Kuhl, 1981) to humans on a /ba-pa/ speech continuum. The former finding indicates that categorical perception is not peculiar to speech and the latter finding suggests that even nonhumans (which would not need a specialized mechanism for the perception of phonemes) may show categorical perception.

SAQ 5.6 Explain what is meant by categorical perception. Describe how perception of nonspeech sounds differs from categorical perception of speech.

Natural Sensitivities Account of Perception

The findings of categorical perception for nonhuman animals and for nonspeech continua led to an alternative interpretation of categorical perception which presupposes that, despite motor theorist's claims, there are invariant auditory properties that mediate phoneme perception (Stevens, 1981). Thus, from this theoretical perspective, perception of speech would be similar to perception of nonspeech sounds (not involve different mechanisms as proposed in motor theory). Specifically, humans and possibly mammals in general, appear to have sensitivities to some dimensions that auditory signals, including speech, vary on, or, to put it another way, the auditory system does not respond to equal physical differences equivalently. Changes at some points on physical continua are barely noticeable, while other equally large differences at different points on the same continuum are clearly distinct.

Stevens (1981) has argued that speech has evolved to take advantage of these perceptual discontinuities to distinguish one phoneme from another. Peaks in the discrimination function of nonspeech continua are important for the clue that they may give to where discontinuities in the auditory system exist, which may be used to advantage in the discrimination of phonemic classes. Such discontinuities might also serve to divide signals ranged along such continua into discrete categories and explain the "sharp" identification functions on categorically perceived continua. This account is one version of what is called a "natural-sensitivities" account. Though such an account could explain identification performance on categorically perceived continua, it is the discrimination results which are most important because they truly reflect perceptual discontinuities, while, as mentioned above, sharp identification functions alone do not definitely establish these discontinuities.

The evidence that has been taken as support for a natural-sensitivities account of speech perception is difficult, if not impossible, to account for by a theory such as motor theory that holds categorical perception to be a feature of a special speech-processing mechanism in the human brain. Those theorists who continue to maintain that speech is special have offered other evidence to show that it is necessary to propose a specialized speech-processing mechanism.

Natural-sensitivities theory has not finally won the day however. There are reasons to doubt some of the results that form the basis of natural-sensitivities theory. For example, Cutting and Rosner (1974) found that nonspeech sounds varying in rise time were identified as sounding like a plucked or bowed string, and discriminated in the way that is held to be evidence for categorical perception. Rosen and Howell (1981) failed to find the mid-continuum peak in the discrimination function that Cutting and Rosner did. Instead subjects were best at discriminating sounds at the "pluck" end of the continuum and became gradually worse as stimuli were drawn from nearer the "bow" end. Thus, this particular instance of categorical perception of a nonspeech continuum is still in doubt.

Elman and Zipser's Artificial Neural Network Account of Perception

As was the case with speech production, attempts have been made to get artificial neural networks to learn the structure of speech sound categories (Elman and Zipser, 1988). Elman and Zipser's (1988) network is similar in its architecture to Jordan's (1986) network. The main difference is, as explained in chapter 2, that the recurrent feedback that gives the network its memory is from the hidden units to the input rather than from the output to the input as in Jordan's model. In one of their studies, Elman and Zipser applied spectra of digitized real speech waveforms as inputs to their networks. The outputs that they required were the phoneme labels (see below). They arranged for the input patterns to span a range of different plosive–vowel contexts: the speech sounds were the three voiced plosives in three vowel contexts (/a/, /i/, and /u/). Thus, it would be expected that their input tokens included different acoustic inputs due to sequencing variation. Elman and Zipser (1988) were able to train three different networks to learn all the plosives, all the vowels, or all nine plosive-vowel syllables perfectly. Thus, although there is acoustic variability between, for instance, the plosive stop consonants, the networks appear to be able to locate context-independent information for the recognition of these phonemes.

Elman and Zipser have attempted to explain one specific thing about speech perception, namely the mapping of variable cues in real speech sounds onto phoneme categories. Their networks appear to achieve this task successfully. However, it should be clear that they will need to address other topics such as those addressed by motor theory to provide a complete theory of speech perception.

SAQ 5.7 It has been stated in the text that Elman and Zipser's neural network approach needs to explain a variety of phenomena that have been highlighted earlier. Collect together what you think these are. Try and think of crucial tests which would allow an assessment of whether an adequate theory will involve a neural network approach or a theory such as motor theory.

Relationship of Phoneme Perception to Higher-Order Units

In the final section of the production section, we considered the role of production of phonemes in the context of models which look at higher-order units that are more meaningful (Levelt, 1989). In particular, the relationship of what we had been considering to words was covered. Besides the production of phonemes, we also saw that Levelt's (1989) model involved a process of phoneme perception too, as he lays stress on monitoring (i.e., listening to) speech for controlling later speech. We can see, then, many of the issues discussed in connection with phoneme perception have direct relevance for this theory of speaking.

Other theoretical attempts have been concerned with specifying the details of the interaction between phonemic information extracted directly from the phonemes (so-called bottom-up information) and knowledge that we have as a result of knowing the words of a particular language (top-down information). The relationship between bottom-up and top-down processing has been captured in various ways. For example Marslen-Wilson and Welsh (1978) have outlined one such version. One factor that informs their theory is that the beginning phonemes of words (principally consonant strings) restricts the number of candidate word possibilities dramatically. Thus, a perceptual mechanism working just on word onsets would do a reasonable job of word perception without identifying all the phonemes. In Marslen-Wilson and Welsh's (1978) theory, words are removed from a candidate pool as phonemes are successively processed. Thus, their model offers some indication of how top-down processing may modify the need for full processing of phoneme events. One feature of the model is that, although it is described as a theory of word recognition, it is based on data from both perception (mispronunciation detection) and production (shadowing). It may be the case that these data sources introduce (production) or extract (perception) into/from the speech signal, cues that are specific to both production and perception. In our summing-up, we will consider the relationship between perception and production.

Summing-up

It has been seen that the produced speech message varies due to a variety of influences. Moreover, when considering speech perception, the variation that occurs in a speech segment requires complex processing by the auditory system. Theoretical accounts have been considered along the way, which attempt to offer solutions to some or all of these problems. For instance, Henke's (1966) theory attempted to give an account of coarticulation phenomena. Stevens' (1981) theory presents an explanation of categorical perception in particular. Each of these theories offers an account in one particular domain, Henke (1966) in production, Stevens (1981) in perception.

The premise behind these theories is that there are anatomically-distinct structures responsible for speech production and perception respectively. Though it is a question of empirical investigation whether these theories account for phenomena in their own particular domain, on their own they will fail to offer a complete account of all the phenomena that have been discussed (i.e., across production and perception domains). One solution to this problem is to develop separate theories for each domain.

Stevens has taken this latter approach. He has attempted to examine how phoneme categories come to be distinguished in speech production: in his quantal theory (Stevens, 1972), he attempts to show what are the distinctive properties of vowels that occur universally in the languages of the world. He finds the universal vowels (the so-called point vowels) have stable acoustic properties. Thus, for this particular sub-set of vowels there may be distinctive acoustic properties which forge the link between speech production and perception.

Stevens' approach in developing independent accounts of the distinctive acoustic properties of produced speech sounds and his attempts at seeking corresponding perceptual discontinuities (e.g., in the case of pluck-bow and the related affricate–fricative sounds) constitutes one possible approach to explaining phenomena in speech perception and production. The critical feature is that the independent theories relate to each other at higher cognitive levels.

An alternative approach to explaining the problems of perception and production has been to propose that the domains are inherently linked. One particular version of such a theory has been discussed – motor theory. To recap, it was considered that muscle commands are invariantly related to the properties of the phonemes. The acoustic signal was considered by such theorists as too complex for the phonemes to be perceived by analysis of acoustic attributes of phoneme categories (as in Stevens' natural-sensitivities theory). Rather, the solution proposed was that the perceptual mechanism deciphers the speech by interpreting the acoustic signal in terms of the motor commands responsible for them. Two main drawbacks to this theory are, first, that the assumed simple relationship between motor commands and phoneme categories does not apply. Second, the theory is based upon the view that speech is processed by a different mechanism to other sounds as it involves reference to phonemic events. Such a position is untenable in the light, for instance, of the evidence showing that chinchillas show categorical processing of speech continua like humans.

On the other hand, there are two distinct advantages of motor theory. First, it attempts to offer a general theory both in the sense that it explains most phenomena in each domain as well as between domains. The second advantage is that it posits one mechanism, responsible both for the production and perception of speech (albeit at the cost of proposing a different mechanism for speech and nonspeech sounds).

The general approach behind motor theory continues to influence certain avenues of research. Separate perceptual and production theories have emanated from the same laboratory that led to motor theory (Fowler, Rubin, Remez, and Turvey, 1980, for production and Fowler, 1986, for perception): Action theory, as is Jordan's (1986) network account of speech control, is a general theory of action. Action theory applied to speech attempts to explain how variable muscular commands could arise when it is necessary to produce the same phoneme in different contexts (a failure already highlighted in connection with motor theory). The perceptual theory concerns direct perception, which again is a general theory of perception which has its origins in the work of J. J. Gibson in visual perception. The latter theory circumvents, then, another inherent limitation in motor theory – that concerning the special processor required for speech sounds. The emphasis in this theory on how motor factors are involved in perception has some parallels with motor involvement in perceptual processes that have been highlighted in connection with motor theory. Also, Gibson's emphasis on abstract relational invariants and rejection of a feature-analytic approach to visual perception has parallels with the basic assumption behind the motor theory of speech perception.

It is noteworthy that these advances in theorizing are achieved by moving the explanatory basis away from peripheral levels of explanation (the motor commands) to higher cognitive levels. This makes the distinction between these theories that can link perceptual and production processes and autonomous theories, less clear.

Before this chapter ends, mention needs to be made of another linked theory that also offers a general account of many of the phenomena discussed in connection with speech production and perception. Lindblom (1990) has proposed hyper-hypo (H&H) theory. This theory focuses on how the speech production mechanism adapts its performance dynamically in answer to changing perceptual demands. There are two main arguments of H&H theory: (1) Speech motor control is "purpose-driven" and exhibits a degree of plasticity. When such constraints are not operative, speech production defaults to a low-cost mode. This is termed "economy" of speech motor control. (2) Speech perception involves the discrimination of lexical items, and lexical access is therefore a function of the distinctiveness rather than invariance of acoustic stimuli.

Speakers are assumed to tune dynamically their production to their assessment of the perceptual demands of their audience arising from considerations like the need to speak clearly. A crucial point in the theory is that speakers do not need to ensure that their productions are realized using acoustic invariants, only that they posses sufficient contrast (i.e., discriminative power sufficient for lexical access). Speakers are also assumed to develop a feel for the survival value of phonetic forms. Intra-speaker phonemic variability can then be seen as arising from systematic variations along a hyper- to hypo-speech dimension.

H&H theory was developed to explain intra-speaker phonemic variation. In the theory this variation exists to serve production, in that production is the result of both different phonetic context (i.e., coarticulation) and variations in the precision of speech motor control (i.e., variation in the degree to which coarticulation effects are tolerated). It specifies, like motor theory, an inherent link between speech production and perception. It is likely to be a theory that generates much experimental testing in the near future.

Further Reading

Ladefoged (1975) provides an elementary introduction to phonetics. Rosen and Howell (1991) give a non-mathematical treatment of the background in acoustics. Harnad (1987) is an edited collection that provides an extensive treatment of categorical perception and the research into this topic. The issue of the *Journal of Phonetics* which includes Fowler's (1986) paper, also has commentaries by theoreticians including many of those discussed here. Lindblom (1990) also discusses the relationship between his ideas and many of the theories discussed here.

Exercises

1 How does the source-filter theory of speech production aid understanding of speech production?
2 Compare and contrast alternative theories of coarticulation (including sequencing).
3 What are the empirical findings concerning categorical perception and in what alternative ways have they been explained?
4 Outline and assess views about the relationship between speech perception and production.

CHAPTER

6 How Many Routes in Reading?

Outline

The processes involved in reading are complex. We will be concerned with just a portion of these processes, the ones involved in dealing with words. Words may be recognized as wholes, or letter by letter. We will investigate the way in which this might be done, the way the recognition systems develop during acquisition and what happens when these systems are damaged. More than this, we will be concerned with the relation between experiment and theory in putting together an information processing account of mental function and describing the relation between data from normal subjects and that from brain-damaged patients.

Learning Objectives

After reading this chapter you should be able to:

- comprehend the scope of the concept of a lexicon
- grasp the relationship between semantics and phonology in reading
- appreciate the difficulties with designing rules for converting letter strings into sound
- distinguish word identification from lexical decision
- differentiate between deep and surface dyslexia
- understand the stages involved in learning to read

Key Terms

- dyslexia
- grapheme–phoneme conversion
- lexical decision
- logogen
- metalinguistic tasks
- phoneme awareness
- priming

Introduction

Reading, we know, is the process of converting print to meaning and to sound. The full story of reading is very complex and would require consideration of a lot of things that have to do with language in general, memory and problem solving as well as with reading in particular. As you read this page, you are accessing linguistic processes that have to do with the structure of sentences and with the relation between sentences. You are calling on your general knowledge and on your experience in order to try to understand what is written. Mention of *linguistic processes* might have reminded you of English lessons at school. And when you are reading a novel, you might well recall whole chunks of your own life in order to understand what is happening on the page. Our objective here will be much more modest. We will focus on single words. We agree that you cannot get a whole lot of meaning out of single words, but we can get enough to make do. And we can certainly get enough sound.

The Nature of the Problem

Do we have to go through sound to get to meaning? Take the word:

elephant.

When you read it to yourself you will probably hear an "inner voice" saying the word. In fact, when presented with a single word it is rather difficult to stop that inner voice from operating. For some people it is obvious that you convert the print to sound and then listen to the inner sound in order to understand what the word means. For other people, the inner voice is an optional extra. In fact, the debate between the two positions has gone on for over 100 years.

But, before we go any further, we had better get inside the head. The process of reading aloud could be seen in the form of figure 6.1. The print is out there in the world; and when we read a word out loud the sound is out there. What we need to do is work out what is going on in the brain between the two.

The first thing we do is adopt the convention that as cognitive scientists we are only interested in explaining what is happening inside the head. Thus, the social causes of illiteracy may be of interest to us as citizens, but, as cognitive scientists, we have nothing to say about that topic. Once inside the head we note that the brain is made up of neurons. However, it would not help us particularly to look at the neurons involved during reading. To start with there are too many of them, but, more important, we want to describe what the brain is doing not how it is doing it (see also chapter 1 on levels). What we have to do, then, is to start thinking about the kinds of representation and kinds of process necessary to do the job we want to do.

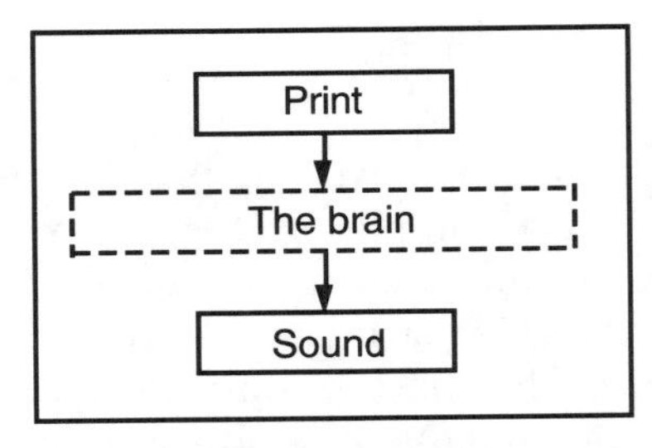

6.1 The basic information processing model for reading aloud. Our job is to characterize the processes inside the dotted box.

We will begin with an input **lexicon** and an output lexicon, as in figure 6. 2. (see devil's advocate box 6.1 for a different use of the term lexicon). The input lexicon is the set of representations of lexical items, specified by their visual appearance only. We can assume for the moment that the same visual input lexicon is used for upper case and lower case words. The output lexicon specifies the phonological form of lexical items – the way in which they are spoken. The output lexical entry for <elephant> would be accessed if we read or heard the word, saw the picture or were asked to give an eight letter word which meant a large animal with a trunk. In all these cases we have to produce the same output.

With respect to the determination of the meaning of the written word, there are a couple of possibilities, diagrammed in figures 6.3 and 6.4. In both cases we have called the process concerned with meaning, "semantics," and we will not be too concerned in this chapter as to how it functions. What concerns us is its relationship with the other processes. In both cases we have an arrow from semantics to the output lexicon, to indicate that we can generate the name of the animal from a definition. The two options have to do with access to semantics from the input lexicon. In one case the access is direct; in the

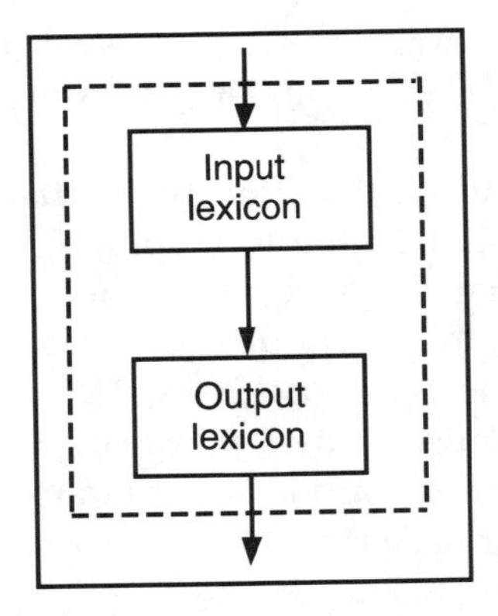

6.2 The first stage in modeling reading: to separate input and output systems.

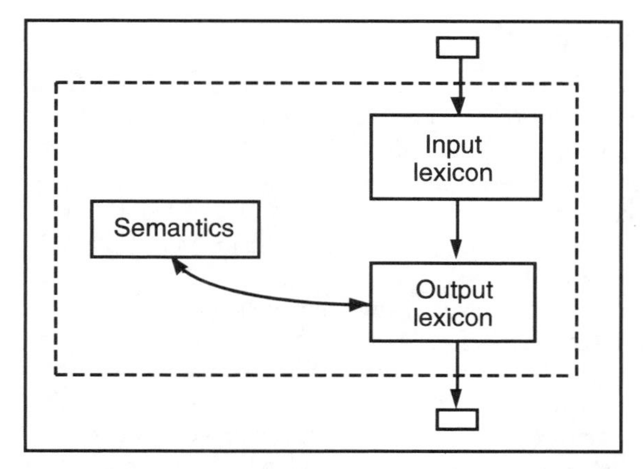

6.3 A model which claims that the meaning of a printed word can only be determined by saying it to oneself.

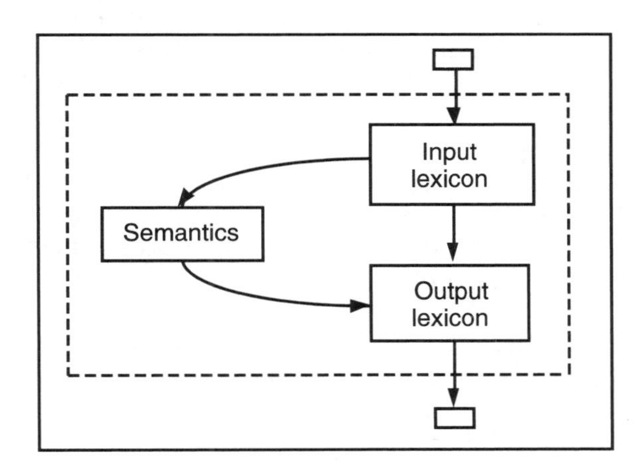

6.4 A model which allows meaning to be determined directly from print.

other case we have to obtain the phonological form – the "inner voice" – before we can discover the meaning.

We have a number of reasons for preferring the first option. One reason has to do with design principles – a matter of aesthetics, one might say. It would seem more efficient to have direct access to meaning from print. For that reason the direct route option would be the default – the one that would be chosen to produce a system that would work. Fortunately, there are other kinds of evidence which indicate that the principle of efficient design could be the correct one in this case:

Devil's Advocate Box 6.1 The lexicon

In the expanded logogen model there are at least five modules which contain information relevant to words: auditory input lexicon; visual input lexicon; phonological output lexicon; graphemic output lexicon; and the semantic system. An alternative approach to the idea of a lexicon is taken by linguists, as you will see in the next chapter. It is more convenient for linguists to conceive of the lexicon as a single entity. A lexical entry, in this framework, includes a specification of phonological, semantic, and syntactic features all together.

Which is correct? Are there many lexicons or is there only one? It is a matter of viewpoint. The two approaches in this book are compatible because they are doing different tasks. It would, however, be possible to advance a psychological theory with only a single lexicon. But, from the arguments in this chapter, it would not be able to account for crucial data. Note that it is a separate empirical question whether or not the neuronal activity relating to lexical processes is located in a small area of the brain or widely distributed. A wide distribution of activity would still be compatible with a functional unit.

1 It is possible for us to understand the meaning of words we cannot pronounce. A good example of this is "The Russian Novel Phenomenon." When some of us read a Russian novel, we rarely try to pronounce the names of the different characters out loud or even to oneself. Nevertheless, we never have any difficulty in recognizing them and knowing who they are, every time they are mentioned. In our scenario, the card from Granny mentioned the Polish town of Szczecin. Mother would not need to learn how to pronounce this word, or even try to pronounce it in order to be able to write the word and recognize it on the postcard next summer.

SAQ 6.1 Think about experiments you might do to explore the "Russian Novel Phenomenon."

2 There are a number of experiments showing the facilitation (also called "priming") of the processing of words by stimuli which are themselves masked – that is, are presented under conditions where the subjects are unaware that the **priming** stimulus has been presented. Now, we assume that the phonological code produced by the output lexicon will be available to consciousness. Since we are unaware of the masked prime, it follows that there could not have been any phonological output and thus that the output lexicon could not have been involved. Since the prime has a semantic relationship to the stimulus word, the input lexicon must have been activated, and so it must have a direct access to semantics (of course, the assumption that the phonological processes are always available to consciousness may be wrong, but it is the simplest one to make).

3 Certain brain-damaged patients, called *deep dyslexics,* exhibit particular errors in reading single words, producing as a response a word which is semantically related to the stimulus word such as saying "thunder" when presented with the word ***storm***. This indicates access to semantic information independently of phonology. We will discuss these and other dyslexic patients later in the chapter.

What assumption has been made about the cognitive architecture without justification? The assertion has been made that the input lexicon and the output lexicon are separate from each other. Are there other options? Well, one option is that there is only one lexicon. Within this lexicon, there would be, in effect, a specification of the visual (and acoustic), phonological, and semantic features for each item. This is the kind of lexical specification found in some linguistic theories. In fact, this was the first form of a model for word recognition called the **logogen model** (Morton, 1969), as shown in figure 6.5 (see also box 6.1). The central idea was that whenever a word was available as a response, the same element was involved. So, if we read the word *elephant* to ourselves, hear the word spoken or see the animal, the same logogen would be active. The initial form of this model has nothing between the logogen system and the outside. However, we need to be able to represent the state of a word being available as a response without it actually being spoken. The way chosen to do that was to interpose a response buffer between the lexicon and the actual response. This addition is shown in figure 6.6. The response buffer was also meant to be the store in which information was held in short-term memory paradigms.

The next stage in the reasoning requires us to look at priming effects in word identification.

Priming Effects

It is a fact about the human processing system that, almost without exception, if a particular stimulus has just been processed, it is easier to process it again if it occurs in the immediate future. This is what is known as *identity priming.*

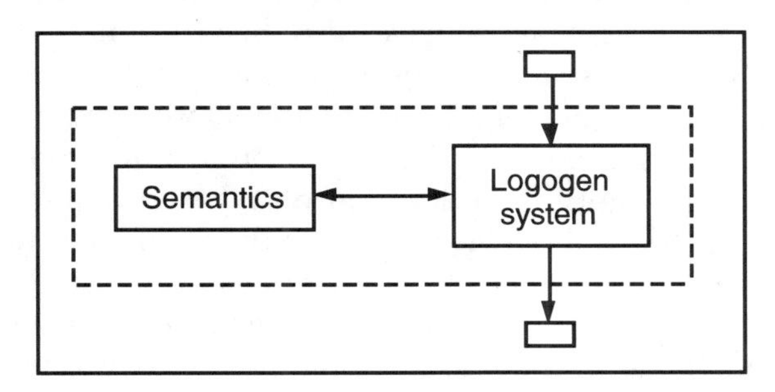

6.5 The first form of the logogen model, in which input and output lexicons were not separated.

Box 6.1 Logogens and lexicons

You may be wondering how to recognize a logogen when you see one. The original idea was not just that there were separate units for individual words (or morphemes), in addition, the form of the mechanism of operation was specified. In fact it was a parallel activation model, full of units which accumulated information from the input. We can suppose that the input system analyzes the input – say, an individual word – and passes on any information it determines. This information would include the length of the word, features of individual letters – such as the presence of ascending letters like t and l or descending letters such as g and y – as well as the identity of letters and something about their position in the word. All of this is sent on to the set of logogen units in parallel. Each unit has a threshold, and when a particular unit has collected more than the threshold amount of information, then it fires, and sends information about the identity of the stimulus to more central processes. This process, like other visual processes, would be fast and would work even though the stimulus was incomplete. It could also lead to errors. Where continuous prose was being read, these errors could be corrected by the central processes which dealt with the meanings of sentences as a whole.

A contrasting mechanism would be the way a spell-check works on your computer. For that program, the word "cat" is made up of *c*, then *a* and finally *t*. All three are required, in that order. It is not possible for the spell-check to be confused about the order of the letters and report back that it was not sure whether the word was *cat* or *act*. Furthermore, the way the computer codes information does not allow any question about the nature of the individual letters. The spell-check never reports that the word might be either *cat* or *cot*. But these are exactly the kind of things that happen when you are trying to read something that is blurred or is in small print or has flashed in front of your eyes.

Thus, having read the word *elephant* in the previous paragraph, if you were now to see a blurred image of the word *elephant*, you would be much more likely to see it than you would a blurred image of the word *crocodile*. This form of priming is called *perceptual priming*. This phenomenon cannot be accounted for simply on the basis of output effects. Thus, if you were asked right now to produce the name of an animal, most of you will come up with "elephant" or "crocodile," because these items have also been primed in your output lexicons, in your semantic systems or in the connections between the two. But if you were asked for the name of an African animal with a very long neck, and a little time later were presented with the word *giraffe*, you would not be facilitated in your recognition of the word. What this means is that the perceptual priming effect takes place in the input lexicon and that this lexicon is not affected when you produce the name in response to a definition (or, in fact, in response to a picture of the item or after hearing it spoken).

Now we would assume that the same semantic system was used irrespective of the kind of input. We would also want to assume that the same output lexicon is used irrespective of the kind of input. The results of the priming

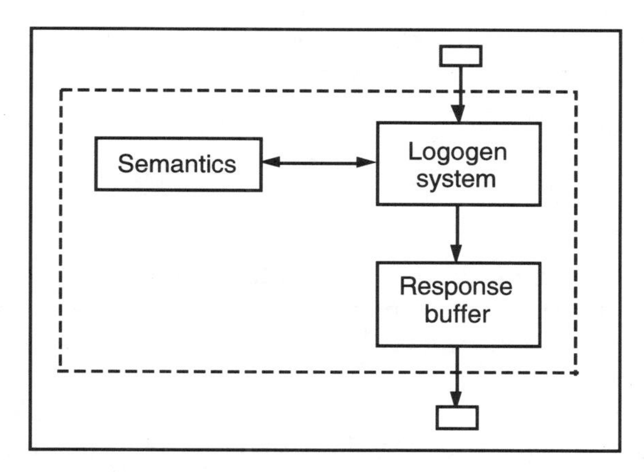

6.6 A modification of the model in fig. 6.5 which allows us to say a word to ourselves.

studies, then, mean that the facilitation effect must be taking place in the visual input lexicon, rather than in the semantics or in the output lexicon, since, of the three, it is only the visual input lexicon which is unaffected by picture naming. If we do not separate the two lexicons, we cannot account for the data. The model in figure 6.6 cannot be the correct one; that in figure 6.4, with the addition of the response buffer, is to be preferred.

Priming studies can tell us something further about the visual input lexicon. In one experiment, subjects were shown a mixture of handwritten and printed words. The handwritten words were scrawled in such a way that individual letters were not easy to distinguish, although the words themselves were clear. Half an hour later they were given very brief presentations of the words, but all in printed versions. The facilitation (compared with words which had not been presented in the first part of the experiment) was about equally great for those words which had been seen in a handwritten form compared with those that had been seen in a printed form (Clarke and Morton, 1983). In combination with the other experiments, this result indicates that the level at which the priming takes place is specific to the analysis of visually presented words, but is more abstract than the form of the stimulus itself.

Reading without Words

In our scenario, when Mom tries to read Granny's letter she gets into trouble at one point because she cannot make out exactly what the letters are in one of the words Grandma has written. "nolinay" is the best she can do as a first attempt. The problem was to do with identifying the letters, not with turning the letters into sound. This is the opposite to the problem with "Szczecin," where the letters were clear but the pronunciation was not.

These events lead us to the fact that while reading words can be seen as being based on the existence of the words themselves, there is also the possibility of reading strings of letters which do not constitute words. These can be simple or complicated, such as:

zib
plaft
grodgernaught

What is also the case for the above is that all English native speakers will agree on how these strings are to be pronounced. The problem with *Szczecin* is that it contains sequences of letters that do not occur in English.

SAQ 6.2 Before proceeding further, think about what information you would include so that a device could translate from strings of letters to strings of phonemes.

The most obvious method of producing sound from print is to have a set of rules which encompass the mapping from one to the other. These could be represented as follows.

z → /z/ (where the bold, italicized letter is understood as the printed form and the letter in slashes "//" is a phonetic symbol)
i → /I/
b → /b/

SAQ 6.3 There are some obvious problems with such a simple system. If you have not thought of them already, think about them now.

The most obvious problem is that in English, as in other scripts, there are a number of sounds that are represented by more than one letter:

sh →/ʃ/, as in "shun," vs. ***s*** → /s/, as in "sun"
th → /θ/, as in "thin," vs. ***t*** → /t/, as in "tin".

These double letters are called digraphs. They do not present us with any problem other than the necessity of the digraph rule dominating the single letter rule. How that is done would depend on the method of implementation. If we were literally using rules which were applied serially, then the digraph rule would be applied before the single letter rules. If we had a parallel system for implementing the mapping, then the digraph relationships would have to inhibit the single letter ones. Note another complication, these rules only work within a single syllable. Words such as **glasshouse**, which contain ***sh***, have to be split into components before the grapheme–phoneme conversion takes place.

SAQ 6.4 Note that the digraphs are not limited to consonants. How would you assign the letters of the words *queen* or *sleigh* to sounds?

A more difficult problem is that a particular letter or digraph can represent more than one sound. For example, we find the digraph ***th*** in two forms, as in "this" or in "thin." The letter ***g*** can be as in "gin" or "gun." And with the vowel sounds, there are a number of options, some of which are regular. In particular, we have the rule whereby a final ***e*** in a monosyllable has the effect of lengthening the vowel sound. Thus, compare the pairs:

hat	hate
win	wine
dot	dote
cut	cute

In all cases the vowel is longer with the final *e*, but it is not simply a question of drawing the vowel out in time. What is formed is a double vowel sound, or diphthong, with two completely different sounds. If we were writing a conventional linguistic rule we would express such context sensitive constraints in the form:

a → /əI/ | _C*e*

This means pronounce ***a*** as /eI/ where it is followed by any consonant followed by an ***e***. One would write the rule in this general way to capture the various forms like ***babe***, ***mace***, ***made***, ***page***, ***male***, ***mane***, and so on. If you were going to write a series of rules to be carried out serially, these context sensitive rules would have to precede the simple rules which give the short vowels.

We are not finished yet, however, for there are exceptions to be taken into account. Try saying the following items out loud:

mave
cade
tane
bafe
have
fape

With a bit of luck you will have pronounced both ***mave*** and ***have*** to rhyme with ***rave*** (if it didn't work with you, try it on some unsuspecting passer-by). So why is ***have*** pronounced as if it were ***hav***? There is no answer to that question other than, because! and there are lots of other examples of words like ***have*** which cannot be pronounced using the regular relationships between graphemes and phonemes. Another example is ***pint***, and the ***yacht*** which Granny is going on is an even more extreme example. All of this complicates the way we think about the translation from letters to sound.

The natural step is to think of reading words as in some way different from reading nonwords. We will see later that the same distinction is made in learning to read. There is some dispute as to the sense in which the two are different, but these turn out to be mainly at the level of implementation (see chapter 1, "Strategies for Cognitive Science"). The issues will be explored later. So far as our model of the reading processes is concerned, we have to add a means of identifying letters. This is shown in figure 6.7, where we show the model, with letter identification as a separate process. We will use this model as a framework for the rest of the discussions. Note that we will sometimes refer to the conversion to speech via the input lexicons as the *lexical route*, and the grapheme–phoneme route as the *non-lexical route*. Some alternative positions are mentioned in devil's advocate box 6.2.

Lexical Decision

One further task that we should discuss is one in which an individual is presented with a string of letters and is asked to press one of two buttons

Devil's Advocate Box 6.2 Words and nonwords

There are two kinds of alternative approach to explaining the way we process nonwords. The first involves saying that nonwords are read by reference to words. These are called analogy theories (see Henderson, 1982). The idea is that the nonword string would be processed by the lexicon, but none of the lexical entries would be able to match the string completely. What would happen is that all the partial matches to the letter string would produce their phonological equivalent, and some kind of summation of these outputs would determine the result. For example, if the stimulus string were *baze*, then *bed*, *bad*, *bade*, *maze*, *haze*, and so on, would all be partially activated, and the response "baze" would result from the summation of all these. In this case there would be no conflict in the production of the response, given that the part-to-part mappings for each partially activated item were known. If the stimulus string were *bave*, on the other hand, there would be conflict between the phonology produced by the partial match to *rave*, *cave*, and so on, and that to *have*. The resolution of the conflict in the latter case would mean that it would take longer to read *baze* than *bave* – which has been found.

The second alternative view is that found in distributed connectionist models (see figure 6.8). These do not represent lexical items directly and have no means of telling words from nonwords. The fact that strings of letters making up words, such as *haze*, affect semantic outputs as well as phonological ones, whereas strings like *baze* only affect the phonological outputs, is seen as a consequence of the learning history of the networks.

depending on whether the string makes up a word or not. This is called the **lexical decision task**, and has proved very productive in exploring various aspects of word processing. You should be wary of the confusion in much of the literature which is brought about by reference to the results of experiments on "word recognition," which do not specify the nature of the experimental technique used. The word identification tasks referred to in an earlier section used a technique where the stimulus word is presented for a brief period of time. The limit to performance is brought about by a lack of stimulus information. In the lexical decision task the stimulus is usually presented clearly and the limitation in performance is the speed of processing. The two tasks appear to depend on different processes. The identification task is not subject to cross modal effects and is taken to reflect the operation of the visual input lexicon. The lexical decision task seems to depend on a wider variety of process, and there are some experiments which suggest that the visual input lexicon might not be the crucial factor in the decision making. For example, cross-modal priming is found in lexical decision experiments. The cross modal priming in this task leads us to conclude that the lexical decision task relies on more central information. Word recognition in the lexical decision task, then, might involve there being some semantic information available. In addition, certain variables operate in different ways in the two tasks. Most striking of

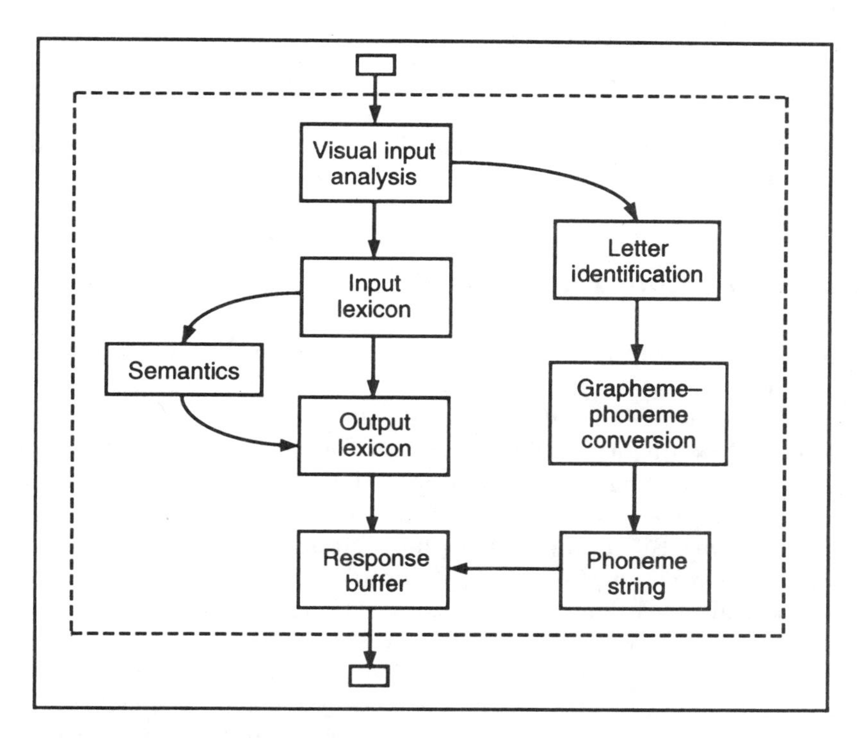

6.7 The addition to the model which will allow the identification of letters and their translation into phonemes.

these is word frequency which is a very potent variable in lexical decision but unreliable in its influence on word identification.

One thing the lexical decision task tells us is that the various routes in our model operate in parallel during the performance of the task. Thus, regular words, such as ***save*** are responded to faster than irregular words like ***have*** (when word frequency and other variables are controlled for). It is supposed that this is because the regular words are recognized via the grapheme–phoneme route as well as by the lexical route (if you tried to recognize ***have*** through the grapheme–phoneme system it would not sound like a real word). In addition, nonwords which sound like real words (such as ***playt***) take much longer to respond "no" to than control strings (such as ***clayt***). This finding requires that the model be complicated further by the addition of a route from the grapheme–phoneme system back into part of the system where words are represented. Other experiments indicate this as well. For example, Lukatela and Turvey (1994) recently showed that if you are asked to name a sequence of words as quickly as possible, that is if you have related words, like ***toad*** and ***frog***, next to each other, then naming is facilitated. They also showed that the sequence ***tode***, which sounds the same as ***toad*** (and is called a pseudohomophone) facilitates the subsequent naming of frog almost as much as the word ***toad*** itself. In addition, the fact that you can readily understand a sentence written in nonsense such as

y kann ewe reed theez werdz?

indicates that there is a ready access to the semantic parts of the system from the phonology part.

Simulation of Word Processing

The main conflict with respect to the modeling of reading is whether there is both a lexical and a non-lexical route. Until recently, challenges to the dual route position have been difficult to evaluate since they were never explicitly formulated. Recently, a parallel distributed processing (PDP) connectionist model of word reading has been proposed by Seidenberg and McClelland (1989) – see chapter 2 for background on such models. This is explicit in that it takes the form of a computer program that accepts a letter string as input and produces a form of phonological output (see figure 6.8). The input units in this model are fairly complex and can be thought of as each being sensitive to 1000 randomly chosen triples of letters. These include the word boundary character which we represent as #. Thus the word cat, presented to this system, would activate any unit that includes #ca, cat or at# in its specification. The phonological output units are also counterintuitive in their specification.

The performance of the Seidenberg and McClelland (S&M) model has

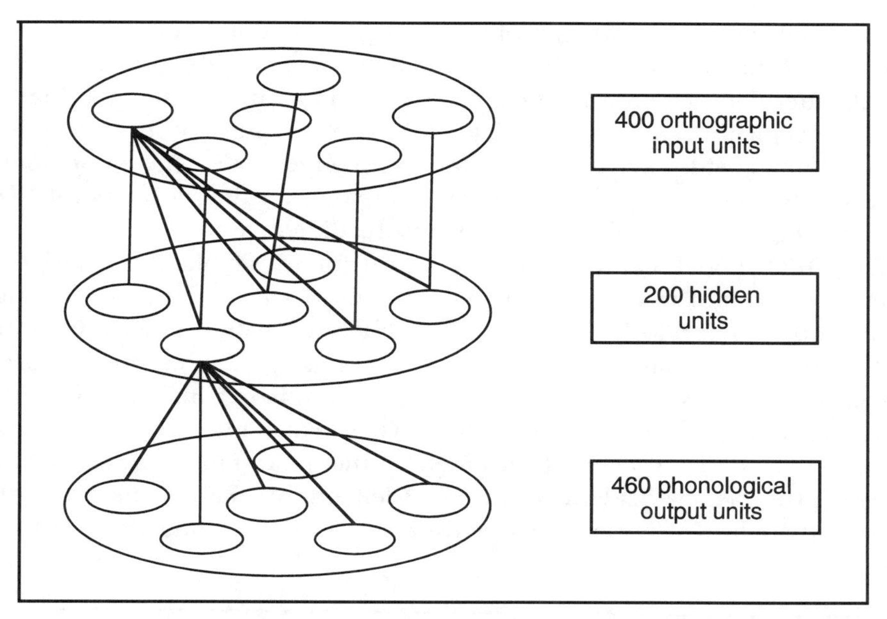

6.8 Outline of the Seidenberg–McClelland model of reading aloud. All units in any layer are connected to all units in the next layer, and all the hidden units have backwards operating connections to all input units.

been discussed in some detail by Coltheart et al. (1993; see also Coltheart and Rastle, 1994). It reads exception words much as human subjects would, making errors on low frequency ones. When reading nonwords, the S&M model only got about 60 percent correct where human subjects were correct 90 percent of the time (under speeded naming conditions). A connectionist system with the same architecture but which uses rather more plausible conventions for input and output has been proposed by Plaut and colleagues (in press). This performs much more like human subjects with nonwords.

A completely different approach has been used by Coltheart and his colleagues who developed a dual route cascade (DRC) algorithm. On the crucial nonword test, this model got 98 percent of the items correct. The model, shown in figure 6.9, is a fully computerized version of the one put forward in figure 6.7 with two particular features in its functioning. The first feature is that there is both excitatory and inhibitory feedback at each stage in the lexical route. If a partial analysis at one level gives rise to a single interpretation at the next level, the partial analysis will be boosted. This means that later stages will effectively clean up the earlier stages. The second feature is that the model operates in a cascade fashion. This means that rather than the lexical units having a threshold, all activation is passed on from level to level. Decisions are, in effect, made by the system as a whole rather than by any one process. The letter identification and word recognition components of the

DRC model are a generalization of the interactive activation model of McClelland and Rumelhart (1981). This means that the letter identification system has 26 letter detector units for each possible position in input letter strings. All letter detector units are connected to all word recognition units in the input lexicon either with excitatory or inhibitory connections. The phonological lexical system and the phoneme system are based on that developed by Dell (1986), which also uses the principles of spreading activation.

The DRC model has proved very successful in simulating a variety of phenomena found in experiments on visual word and nonword recognition. We can note, to start with, that it gives us a difference between exception words and regular words. This effect occurs when processing of an exception word at the phonemic stage is still in progress when conflicting information arrives from the non-lexical processes. What is of particular interest to us here are those findings that specifically distinguish the DRC model from the PDP models proposed by Seidenberg and McClelland and Plaut. The first difference is that the DRC model works from left to right along the word as it

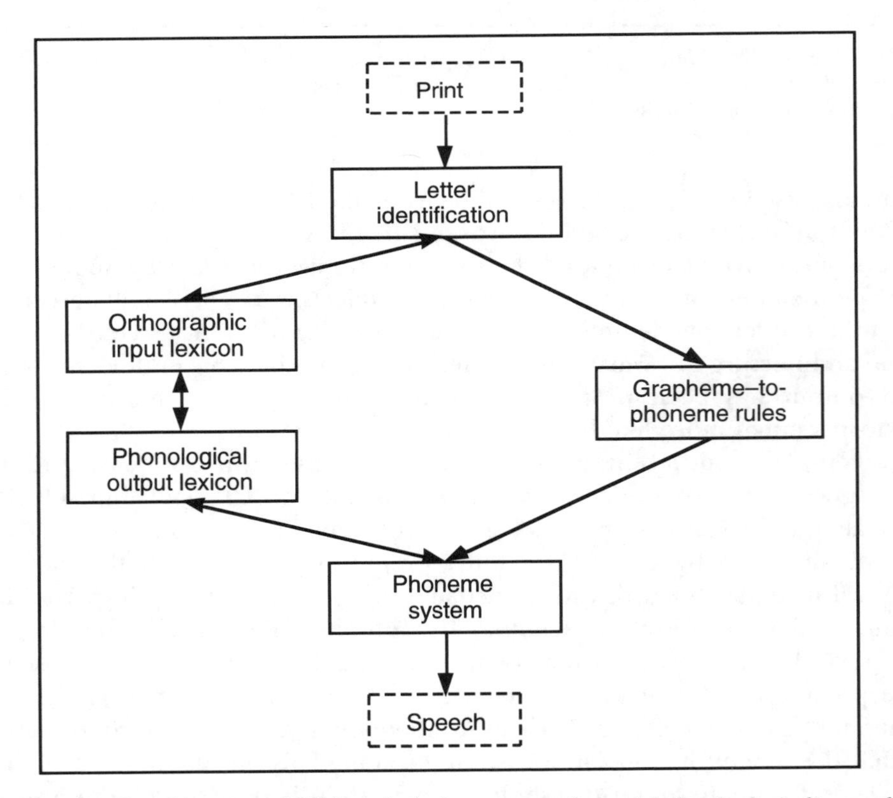

6.9 The overall architecture of Coltheart's dual-route cascade model. On the lexical route there is excitatory feedback and at all levels there are inhibitory connections to enable one candidate to win.

converts from a graphemic representation to a phonemic one. This would be the usual (though not necessary) way for a rule governed system to operate. On the other hand, it is an essential feature of the PDP models mentioned above that there is no temporal aspect to the processing, since activation of the phonemes occurs entirely in parallel. Suppose, then, we take a set of words which have the source of their exception pronunciations at different positions in the words. Examples are *chaos* (with the irregular correspondence for the first phoneme), *deadly* (second phoneme), *crooked* (third phoneme) *famine* (fourth phoneme) and *debris* (fifth phoneme). The later the irregular correspondence, the less should be the interference effect, since the more likely it would be that the lexical route would have finished its processing before the conflicting information arrived. Coltheart and Rastle (1994) carried out the test of this hypothesis using a total of 96 target words. The size of the irregularity effect was a clear function of the position of the irregular correspondence. Such a finding constitutes a clear challenge to the *parallel* component of PDP models.

A second challenge to the network models we have described involves priming with what are called *pseudohomophones.* Homophones are words which sound alike, such as *phrase* and *frays.* Pseudohomophones are nonwords which sound like real words, such as *phrays* or *brane* or *brayn.* In the model shown in figure 6.8, the overlap between pairs of homophones or between pseudohomophones and their word equivalents must primarily be in the phonological units, with a much smaller involvement of the affected hidden units. Since the network is in principle distributed, there is nothing that knows that a particular stimulus is a word. The phonological units which are active when the network has just read *phrase,* will be the same as those active following *frays* or *phrays.* All priming effects should be symmetrical, then. In the model in figure 6.7, there would be priming in either of the lexicons. This should mean that the nonword would not be primed by the words and that the words should prime each other but not be primed by the nonword. In the model in figure 6.9, nonwords will prime the homophonic words because of the feedback from the phoneme system to the phonological output lexicon. Again, however, nonwords should not be primed by words. The experimental findings were that words were primed both by homophonic words and by the pseudohomophones. This would not be predicted by the model in figure 6.7. However, the nonwords were primed neither by other nonwords or by words. Only the model in figure 6.9, of those we have described, could predict this result. In addition, no model hoping to make do with only a lexicon and no non-lexical route would be able to account for this result.

On the basis of these and other data, we are left with two routes for converting print into **phonology** as well as with a need for representations of words. At the moment there is no network capable of satisfying the constraints imposed by these data, though this is not to say that such a network would never be possible.

The Cognitive Neuropsychology of Reading

It is an essential characteristic of models of human performance that they can give an account not only of correct normal performance but also of errors made by normals as well as the performance of patients. Models of the kind developed in this chapter have been used to great effect over the last 20 years in helping us to understand the variety of deficit found in reading tasks following brain damage. Prior to 1966, reading deficits were described in a very crude way, and there was little liaison between neuropsychologists who looked at patients, and experimental psychologists who studied normal performance. This was changed by John Marshall and Freda Newcombe. The first kind of patient described by these workers were known as **deep dyslexics** (Marshall and Newcombe, 1973). The most salient characteristic of these patients is the occurrence of semantic errors when they are trying to read out individual words without time pressure, stimulus degradation, or context. Examples of these errors are:

close → "shut"
thermos → "flask"
projector → "camera"
negative → "minus"

We might note that the patients also produce semantic errors in picture naming, such as responding "needle" when shown the picture of a thimble.

SAQ 6.5 Semantic errors seem like a paradox within some theories of psychology. How is it possible, people asked, that these patients recognized the words (otherwise they could not produce semantic errors) and yet did not recognize the words – since they read them incorrectly. How has a cognitive model resolved this apparent difficulty?

In addition to the semantic errors, patients with deep dyslexia often make visual errors such as:

idiot → "idol"
hassock → "hammock"
pivot → "pilot"
saucer → "sausage"

In a few cases these forms of error are combined, with the visual error apparently preceding the semantic one. One example is:

favour → "taste"

which is derived

favour → (flavour) → "taste"

Other examples are:

brought → "buying"
allegory→ "crocodile"

Other characteristics of these patients include a complete inability to read out nonwords.

The usual hypothesis concerning deep dyslexics is that their grapheme–phoneme route is completely disconnected, and the route from the input lexicon to the output lexicon is impaired (note that these patients usually have quite massive left hemisphere **lesions**, so that one would not expect simple deficits). As one would expect, their performance is equally good on irregular as on regular words. We note this in contrast to another type of patient, called **surface dyslexics**, who are characterized as being relatively unimpaired in the grapheme–phoneme route but who have no route directly into semantics. These patients are better at reading regular rather than irregular words. Further, they make just the kind of error on irregular words that one would expect from a grapheme–phoneme system. Examples include:

guest → "just"
shoe → "show"
listen → "liston ... that's the famous boxer"

Homophones may be correctly read, but interpreted incorrectly:

billed → "to build up, buildings"
oar → "that's the ore of metals, the raw materials"

again indicating that the grapheme–phoneme system is independent from the semantics, which, with these patients, can only be accessed via the response buffer, where homonyms are equivalent.

There are a number of other kinds of dyslexia which have been described in terms of models like that in figure 6.7 (see Shallice, 1988). The accounts given of the different kinds of patient have usually been that one or more of the processing elements in the model have been damaged, or that the connections between elements have been cut. Occasionally, other hypotheses have been put forward. For example, Coltheart (1980), has suggested that deep dyslexia reveals right hemisphere functioning. One reason for suggesting this is that many of the deep dyslexics have massive lesions in the left hemisphere, such that the middle one-third of the cortex has been destroyed. Patients with deep dyslexia also usually have other problems with their cognitive functioning. It is common that they are aphasic, for example, (i.e., have problems understanding or producing spoken language). In fact, it is now felt by some to be a little misleading to call the group of patients deep dyslexics rather than saying that the patients have deep dyslexic symptoms (see Morton and Patterson, 1980).

SAQ 6.6 Why might we want to say that learning to read is more difficult than learning to talk?

How do We Learn How to Read?

Learning to read is a fairly complex business. And, compared with learning to talk, it is rather more difficult. Most children do not start to read until they are 5- or 6-years old, and the process of acquisition of an alphabetic script takes some time and passes through a series of overlapping stages which have been described by Uta Frith. The three stages are called **logographic**, alphabetic and, orthographic (Frith, 1985, 1986).

SAQ 6.7 Make a list of the things a child will have to learn in order to become an adult reader. What do you think the order of acquisition of items will be?

The logographic stage

The first words that a child learns to read are more like pictures. They will be the patterns that the child is particularly interested in such as ***MacDonalds***, ***Smarties (or M&Ms)***, the name of the favorite cereal and the child's own name. In our scenario, Zara would recognize "coco-pops" in this way. In the logographic phase, words seem to be recognized independently of each other. Indeed, it seems plausible to think of each word being identified by an idiosyncratic schema. While it is the case that individual letters typically enter into the process of word recognition, it is also the case that not all the letters in a word are crucial and that, in some instances, non-alphabetic information appears to be crucial. Typically, the first letter acts as a salient feature, but irrelevant detail too can be incorporated into the recognition schema. Thus, a child may only be able to read the sign ESSO when the letters are surrounded by the familiar oval. Similarly, a child will respond with "Harrods" when presented with "Hrorasd" or "HaRroDs" (examples taken from Morton, 1988). The extent to which the logographic way of reading is elaborated depends upon the age at which the child is taught to read and by what method. The indifference of the child to quite major changes in the form of the word makes sense of equating words with pictures in the child's mind. Drawings of the same cat can differ greatly from each other as, indeed, the cat itself will undergo major changes in projected form as it moves around. The picture of the cat may show four paws, but there could be only two or three visible. Equally, then, why should it be important that all the letters in a word should be represented all the time? And if a lemon has to be yellow to be a lemon, then why should COCA COLA not have to be red and white to be "coca cola"? The logographic reading stage has not been studied very extensively in normal children, largely because it lasts for such a short time and contains such a small vocabulary. We are currently conducting a study of this phase, and in table 6.1 we present the responses of Thomas, aged 4. It will be seen that his vocabulary is reasonable, though the response of "pull" for ***yellow*** might seem a little extraordinary. However, it is possibly no coincidence that both words have a double ***l*** which may act as the salient feature. In table 6.2 can be found Thomas's responses to the same words misspelled. He almost does as well with these stimuli as with the words, and he can now read yellow. Particularly revealing is his response to *liltle* where the stimulus satisfied the general description of "little" but failed in some particular. Again, presumably not by coincidence for this child, the double letter is salient and the fact that it is now *ltl* rather than *ttl* is just a little disturbing to him. His response is rather like saying that it can't be a cat because it is wagging its tail.

Logographic writing

At some point while the child is building up a vocabulary of logographic words (what many people call a "sight vocabulary"), she begins writing. The child's own name is the most favorite word, and that will be learned by copying from a sample. Later on, children try to write other words, especially ones they have

Table 6.1 The responses of Thomas, aged 4

milk "milk"
child "camel"
house "house"
blue "blue"
grandfather "grandfather"
little "little"
yellow "pull"!

Table 6.2 The responses of Thomas to the same words misspelled

grodftehr "gr ... grandfather"
honse "house"
mlik "milk"
yollwo "yellow"
chld "cat"
bleo "blue"
liltle "little ... no that's not little, not two 't's ... life"!

learned to recognize. Mostly, these words are not perfect copies, and so must be based on representations that the child has derived herself. The form of these representations is likely to be idiosyncratic to a certain extent, but the few children whose early written output has been studied, share the feature that letter order, apart from the initial letter, is variable. One might imagine that the representation consisted of a *collection* of letters (in contrast to a *list* which would be order preserving) which were selected according to some (aesthetic?) criteria.

SAQ 6.8 How would you write computer routines to capture the difference between a collection and a list?

The alphabetic stage

Logographic writing is eventually replaced by alphabetic writing. What is happening during this stage is that the child gains access to her own phonological representations and is able to isolate individual phonemes within them. This is an absolutely essential step because of the way alphabetic writing systems work. Alphabetic scripts represent speech sounds at the level of the phoneme. In non-alphabetic scripts, where the basic elements of the script represent consonant–vowels pairs (as in Hindi), syllables (as in the Kana script of Japanese), or words (as in the Kanji script in Japanese), this stage would be very different.

There are a number of options available to us in accounting for how the child is able to isolate individual phonemes. These options are dependent upon the detail of the model and they will not be discussed further here. All of these options involve processes which follow the output lexicon in speech production. Given that individual phonemes are isolated, they have to be mapped onto written letters. This requires the setting up of phoneme–grapheme rules. Clearly the phoneme–grapheme rules will use the letter knowledge that has been accumulating.

The process of phonemic segmentation is the most important single aspect of learning to read. Rozin and Gleitman (1977) say "The child's insufficient access to the segmented nature of his own or another's speech ... is the major cognitive barrier to initial progress in reading." A large proportion of backward reading children are stuck at the logographic stage, though they can relatively easily be helped over the barrier by a well directed program of remediation (Bryant and Bradley, 1985; Bradley and Bryant, 1985).

The alphabetic reading stage is divided into two sections which are distinguished by the connections that are made with the semantic system. At the beginning of the alphabetic phase the child is beginning to segment words into component letters rather than recognizing them as wholes. Individual letters, then, will consolidate their representations and a set of mapping rules will be set up between them and the system of phonemes. Presumably, the setting up of the **grapheme–phoneme** rules is influenced by the existence of phoneme–grapheme rules. In any case, there is no simple reverse procedure to arrive at the correct rules. Again, however, there does not appear to be information of sufficient detail on the development of individual children to enable such hypotheses to be tested.

SAQ 6.9 Suppose you had a complete list of the phoneme–grapheme correspondences. What kinds of difficulties would there be in deriving the grapheme–phoneme correspondences from them?

Alphabetic writing cannot develop without a minimum of explicit instruction. Similarly, alphabetic reading appears to require some kind of systematic approach. Such instruction would comprise detailing of the rules themselves, such as ***g*** is "guh". Certainly the explicit knowledge that letters map onto "sounds" is a part of any contemporary reading instruction. This would prove difficult for certain theories in recent or current vogue. As we have already seen, analogy theories of adult reading hold that there are no grapheme–phoneme rules at all, merely mappings between input lexical representations and output lexical representations. Apart from problems related to the viability and adequacy of such theories as have been specified (see Patterson and Morton, 1985), it is not clear how these theories would account for children's abilities in the alphabetic reading stage. Most notably, at this stage the only words the child can read correctly are regular words, and nonwords can be read just as well as words. In addition, if the child has learned an irregular word once the alphabetic phase has begun, then she will not attempt to use that knowledge in trying to read a word differing by the initial letter (Marsh et al., 1980).

The problems for current connectionist models are different. McClelland and Seidenburg (1988) do away with lexicons of all kinds, acknowledging only letters, phonemes, and semantic features. Such a model could not cope with the logographic phase at all, and could only begin to operate at the alphabetic phase. However, their model is not currently built to allow direct mapping between elements such as ***d*** and /d/. Rather, the mapping between elements is distributed via a large number of intervening elements ("hidden units") with connection weights. The owner of such a network (i.e., the child in the alphabetic stage) would not be able to offer any insights into its mode of operation in such a way as would serve to help with its establishment. Such observations (as ***b*** means "buh") would be incidental only.

One interesting feature of the child's performance at the beginning of this stage is that they cannot understand the words they read alphabetically. One psychologist reported his son, after having successfully read a list of words, saying "Now you read them so that I know what they mean." This indicates that the child has no functioning feedback from the response buffer to semantics.

The final stages of development are of less interest to those whose primary concern is with dyslexia. By the time a child has mastered most of the alphabetic stages, progress on to the orthographic stage is a simple consequence of the interaction of the activity of reading, linguistic knowledge, and general cognitive-abstractive processes. Very few children fail to make the transition painlessly. What happens is that input representations become established in which letter order is respected and word structure is central.

Phoneme awareness

The tricky stage in this progress towards reading is the establishment of the grapheme–phoneme correspondences. It is at this point that children can get stuck. Clearly, however these correspondences are implemented, some part of the child's cognitive processing has to have access to graphemes and to phonemes. With an alphabetic script the individual letters are clearly separated in print on the page, and there is not much problem in distinguishing letters (once the reversible b/d, p/g have been sorted out). The phonemes are a different problem. While the child will have a full phonological system by this time, the phonological knowledge will be implicit, not explicit, and representations of the individual phonemes will not at first be available to enter into the mapping process.

There is good reason to suppose that a change in the nature of the representation of phonemes is necessary before the alphabetic stage can proceed. There are a number of tasks that signal this "reading readiness." These tasks involve manipulating phonemes. Thus, a child may be asked: "take the first sound away from the word cat, and what do you get?" Tasks of this kind are called **phoneme awareness tasks**, or PA tasks. There are four views on the relation between **phonological awareness** and the acquisition of literacy in the normal child:

1 PA is a consequence of learning to read (Morais, Cary, Alegria, and Bertelson, 1979).
2 PA is a causal factor in learning to read (Bradley and Bryant, 1983).
3 An "interactionalist" view, such that being instructed to read in alphabetic systems develops segmental awareness which, in turn, is crucial for mastering the rules of grapheme-phoneme conversion (Morais, Alegria, and Content, 1987).
4 There is no causal relationship between PA and learning to read, merely an association (Liberman, Shankweiler, Liberman, Fowler, and Fisher, 1977; Cossu, Rossini, and Marshall, 1992).

Cossu et al. proposed settling the issue by establishing that a group of children could acquire reading in the absence of ability to perform explicit segmental analysis tasks. They supposed that this would be prima facie evidence that no necessary causal connection holds between the two skills.

They took a group of Down's syndrome children (mean CA 11.4; mean IQ 44) matched with controls (mean age 7:3) on ability to read aloud regular and irregular words and ability to pronounce of words and nonwords. This step is meant to establish that the two groups are equivalent in relation to both orthographic and alphabetic reading. They then tested the two groups on a variety of **metalinguistic tasks**: phoneme counting, phoneme deletion, oral spelling, and blending of a string of letter sounds. This step is designed to enquire whether the groups are equivalent in the skill of phoneme awareness. On all four tasks the Down's syndrome group performed appallingly. Cossu et al. conclude that: "gross failure on phonological awareness tasks has not prevented . . . (the Down syndrome group) . . . from acquiring reading (as a transcoding skill)," and they conclude "all causal hypotheses relating PA to the acquisition of reading (or vice versa) are false if the connection is taken as a necessary one."

The notion of causal connection between diverse skills is an interesting one. The problem is, in the context of reading acquisition, as elsewhere, that skill is

Box 6.2 Development contingency modeling and causal modeling

These are diagrammatic representations that can be used at any time that you are thinking about development. We normally see that in order to develop a particular skill there are certain prerequisites. In a complementary way, if a certain cognitive ability is missing, then the child will not be able to develop a skill. Thus, adequate hearing is necessary for speech to develop; deafness impairs that process.

We can represent such facts (or theories) diagrammatically. Suppose we say that ability X is required before skill Y can be performed. We would represent this as X—⊂—Y. Equally, the absence of X would mean that Y could not develop (at least through the normal developmental route). We would represent this as ~X→~Y where the ~ (which is called "tilde") means NOT.

We can complicate either model indefinitely and maintain the same logic. Thus, suppose that both W and X are required before Y will develop. Then it follows that if either W or X are missing then Y will be impaired, as shown in figure 6.10.

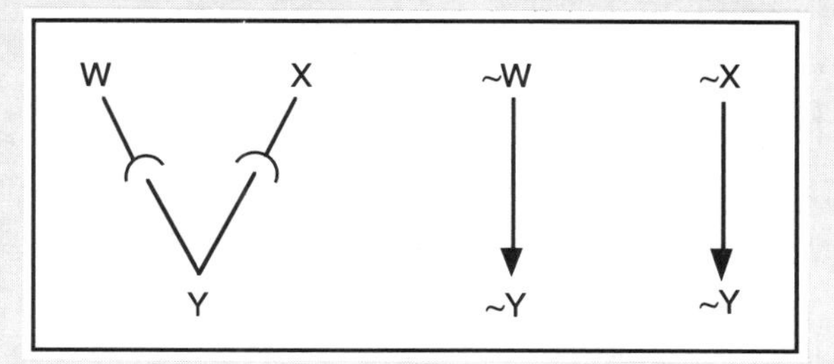

6.10 On the left, a developmental contingency model saying that two abilities, W and X, are required before a third, Y, can develop. On the right are the two corresponding causal models which state that the absence of Y (~Y) can be accounted for developmentally either by an absence of W or by an absence of X.

indexed by performance. In consequence we end up with the traditional conflation of ability and performance that predates cognition. Performance relies on ability, and it is between abilities, defined cognitively, not between skills that cause can be established. Let us represent things diagrammatically, using two kinds of diagram showing **developmental contingency models** and **causal models** (see box 6.2).

Rather than indicating a relationship between PA skills and reading, as in the left-hand part of figure 6.11, we would want to say that reading, as exemplified by grapheme–phoneme rules (GP) requires some phonological representation, P, plus some other abilities, A. The PA skills will also require the phonological representation, P, but will, in addition need some high-level skill, an ability to reflect on language. We call this **meta-linguistic skill**, M. The inability of the Down's syndrome group to perform the PA tasks, can be seen as a consequence of a lack of M rather than a lack of P. The inability of a young normal child, on the other hand, to perform the PA tasks will be attributable to the lack of P. We can now ask questions concerning the development of P and can speculate that certain phoneme tasks could assist in the development of the crucial, P, representations. Indeed, as Maclean, Bryant and Bradley (1987) have shown, knowledge of nursery rhymes at age 3 was specifically

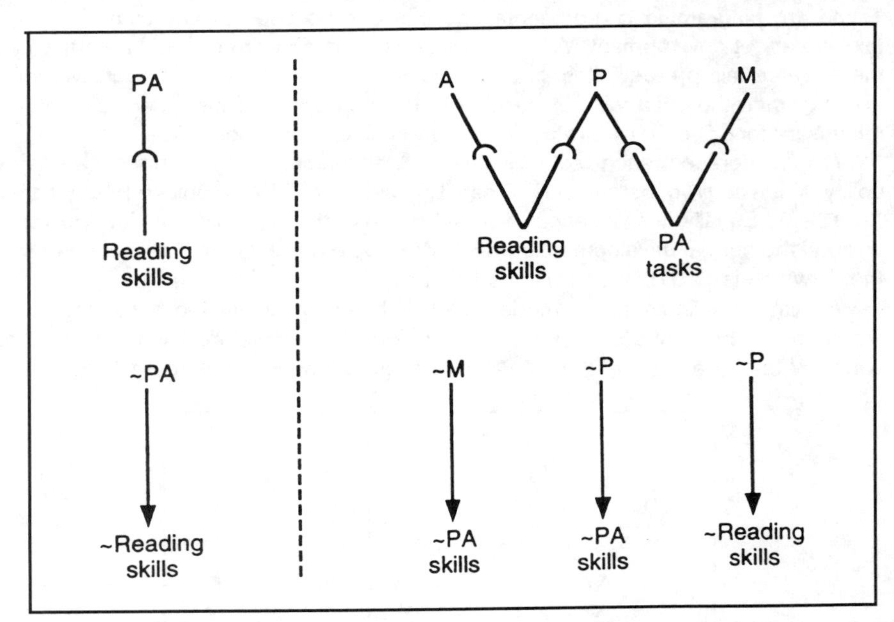

6.11 Developmental Contingency Models (DCMs) of reading acquisition and the corresponding Causal Models (below) of problems in reading. On the left the DCM claim is that learning to read depends on phonological awareness. Correspondingly the causal model says that an absence of ability to perform the phoneme awareness (PA) tasks causes reading failure. On the right the claim is that reading depends on having a phonological skill, P, while PA tasks require both P and a metalinguistic skill, M. In the causal models, Down's syndrome children fail PA tasks because of lack of meta-skills. Normal young children fail PA tasks and reading because of lack of P.

related to reading over a year later. This relationship held after variables such as intelligence and the socio-economic class of the parents had been taken into account. If we assume that Down's syndrome children have an intact P and that the teaching/learning context is satisfactory, then they would develop a normal GP system.

By separating the performance skills from their underlying abilities, we can settle the conflict created by Cossu et al. Further, we see that in thinking about causal relationships we have to take care that, for any skill, we think about all the abilities which are necessary for its development.

Summing-up

We have looked at reading performance by normal subjects, at the breakdown of reading skills as a result of brain damage, and at the process of learning to read. In all three situations, we found that the presence of a lexical basis of converting print into sound as well as a non-lexical basis was either necessary or at least made considerable sense of the phenomena.

Further Reading

For an easy introduction to the topics in this chapter, read *Reading, Writing and Dyslexia* by A. Ellis, (1984), London: Erlbaum. *Reading and the Mental Lexicon*, by M. Taft (1991), London: Erlbaum, gives a more extended account of the field. The best available book on neuropsychology is *From Neuropsychology to Mental Structure*, by T. Shallice (1988), Cambridge: Cambridge University Press. Only a portion of this book directly concerns reading, but his account of methodology in cognitive neuropsychology is superb. For further reading on developmental dyslexia, a good introduction is still *Dyslexia: A Cognitive Developmental Perspective* by M. Snowling (1987), Oxford: Blackwell.

Exercise

1 What other abilities that have not been discussed here would be necessary for the development of the reading of words? If we broaden the nature of reading to include full understanding, what would you add to the list?

CHAPTER

7 The Structure of Sentences

Outline

What does it mean to know a language? How is language acquired? These key questions about human language form the focus of this chapter. We consider the evidence for the view that language is a mentally represented system of rules and that its acquisition and shape is determined to a large extent by a genetically determined component of the mind. We then sketch an approach to the study of language which takes a cognitive view of the language component of the mind and explores certain minimalist assumptions about its design. We conclude with a discussion of the perspective this approach offers on the problem of **language acquisition**.

Learning Objectives

After reading this chapter you should be able to:

- evaluate the evidence for treating language as a rule-based phenomenon
- evaluate the plausibility of the language faculty having two interface levels
- discuss elementary properties of the interfaces of the language faculty
- sketch how the language faculty constructs representations at its interfaces
- describe the role of feature-checking in derivations of the language faculty
- sketch the concepts of representational and derivational economy
- discuss the logical problem of language acquisition and its role in the study of language
- indicate what light is shed on this problem by the minimalist approach to language

Key Terms

- derivation
- economy
- grammar
- language acquisition
- language faculty
- logical form
- phonetic form
- universal grammar

In the scenario that opens this book Mom, Dad, Lucy, and even little Zara all use language as a means of communication. Using language is such an everyday thing to do that it is hard to imagine that it might be based on complex mental processes. Yet, there is no doubt that it is. Each utterance that is spoken and understood has several layers of linguistic information built into it. For instance, in the fragment

Lucy: I'm eating at Jane's tonight, Mom – OK?
Dad: I thought it was your homework evening.
. . .
Mom: She's worked every night, John . . .

Lucy, Dad, and Mom are producing *structurally* correct sentences of English. If Lucy had said *I eating am at Jane's tonight, Mom – OK?*, you would have understood her but you also would have noticed that the structure of her sentence was "wrong." So apparently you can make judgments about which sentences are structurally good or bad. But apart from a structure, sentences have *meanings*. The meaning of a sentence is somehow made up from the meaning of its component parts and the order in which these parts appear. Thus, *I thought it was your homework evening* does not mean the same as *I thought it was your evening homework*, although both sentences contain the same words and are structurally fine. Finally, the fragment also nicely illustrates that the meanings of the component words do not fully determine the *interpretation* of a sentence when it is used in a particular context (in which case we call it an *utterance* rather than a sentence). The obvious interpretation of Dad's response to Lucy's plea to Mom to be allowed to eat at Jane's is not "Dad had a thought, namely that tonight is Lucy's homework evening" but rather something like "Dad thinks that Lucy should stay home tonight to do her homework." But of course that is not at all what Dad *said*. But in the given context, it is the *interpretation* that Dad intended to convey. That this is indeed the interpretation that he has in mind is confirmed by Mom's attempt to support Lucy's request. We are thus led to distinguish between aspects of meaning that seem to be determined by linguistic form (let us refer to this as *linguistic meaning* or *sentence meaning*) and aspects of meaning which depend on the interpretation of an utterance in a particular context (*pragmatic interpretation*).

There is an important difference between the processes of determining the structure and the linguistic meaning of a sentence on the one hand and determining the interpretation of an utterance in some context on the other. Only the latter involves what we might call *knowledge of the world*. For instance, in order to interpret Mom's reply to Dad's sentence in the fragment above, you have to make certain assumptions, such as that Mom believes that Dad is not aware of how diligently Lucy has worked this week. But nothing you know about the world is going to change your judgment that *Dad always Lucy pesters* is a structurally bad sentence. Similarly, you do not need to consider your knowledge of the world to determine that the sentence *women are male* is false. It seems that such judgments depend on linguistic knowledge only.

In this chapter we will be concerned with the nature of linguistic knowledge. We will have nothing to say about how linguistic knowledge is put to use by the human sentence processor in actual sentence comprehension and production (a process often referred as *parsing*: for a brief discussion of computational issues relating to parsing, see box 2.2 in chapter 2). In fact, we will restrict our attention still further to the study of just that part of linguistic knowledge that is concerned with sentence structure (syntax). We will have little to say about the theory of meaning (semantics) and nothing at all about the theory of utterance interpretation (pragmatics). Many of the questions that arise in the study of syntax also arise in the study of sentence meaning. By contrast, utterance interpretation has properties which distinguish it sharply from both syntax and semantics, as you will see in the chapters on meaning and conversation, and on pragmatics and the development of communicative ability (chapters 8 and 9). The discussion of modularity and modularization in chapter 3 deals with the key aspects of this contrast. Chapter 5 discusses a further topic in the study of language, namely speech production and speech perception. Having carved out sentence structure as our area of study, let us now consider some of the basic questions we will have to address.

Every normal human being acquires a natural language and eventually produces and comprehends that language with astonishing ease and speed. This is a remarkable cognitive achievement, which becomes all the more intriguing if you consider in some detail what must minimally be involved in producing that behavior. As we will see, it seems difficult to escape the conclusion that a person who speaks and understands a language has developed a certain system of knowledge, which is somehow represented in the mind. We will call this mentally represented system of rules a **grammar**. The vast majority of present-day research into the nature of this mental object is based on the important work of the American linguist Chomsky (e.g., 1965, 1977, 1981, 1986b, 1993, and many other publications), whose views on language and cognition have had a deep, shaping impact on the field of cognitive science. Other current approaches to the study of the structure of sentences not discussed here include Generalized Phrase Structure Grammar (see Sells, 1985, for an introduction), Head-Driven Phrase Structure Grammar (Pollard and Sag, 1987, 1993), Lexical Functional Grammar (see Sells, 1985, for an introduction), and Categorial Grammar (Steedman, 1993).

Chomsky's linguistics is generally known as **generative grammar**. According to linguists working in this tradition, the following questions are fundamental to the study of language:

1 What kind of knowledge is represented in the mind of a speaker of a language?
2 How did it get there?
3 How is it used in comprehension and production of language?
4 What are the physical mechanisms that serve as the material basis for this system of knowledge and for the use of this knowledge?

The way generative linguists carve up the study of language into different studies at different levels of abstraction is typical of a methodology which is common to both research in the natural sciences and much research into properties of the human mind, and is comparable to the methodological approach advocated by Marr (Marr, 1982), whose views were alluded to in chapter 1. Marr suggested that it is profitable to construct theories of complex information processing tasks (such as vision) at various levels of abstraction, the idea being that we can often specify abstractly what a complex system is doing (its task) by trying to understand what rules govern the mapping from its inputs to its outputs. For instance, in the case of vision we can ask how the visual system manages to map stimuli impinging on the retina to three-dimensional images. We can then propose a solution to this problem (some set of rules, say) which abstracts away from how the mapping is actually computed by the brain. The next step would be to try to determine which reasonably efficient procedure computes the proposed solution and finally to discover how that procedure is realized in the brain.

In much the same way you can think of the language component of the mind of a speaker/hearer as a black box performing a complex information processing task. Let us agree to refer to the language component of the mind as the *language faculty*. The language faculty has two "modes of operation." During production, it maps meanings into speech (or writing or a sequence of signs from a sign language). During comprehension, it maps speech (or written language or sign language) into meanings. In what follows, we will only consider the mapping from sound to meaning and vice versa (chapter 6 discusses some aspects of what is involved in the reading of words). The first of the questions above (what kind of knowledge is represented in the mind of a speaker of a language?) can be answered by proposing a set of rules that specify the mapping from sound to meaning (Marr's task level). To answer question 3 (how is knowledge of language used in comprehension and production of language?) we must propose an effective procedure for using the proposed system of rules (Marr's algorithmic or representational level). To answer question 4 we would have to show the biological mechanisms that serve as the material basis for the system of rules and the procedure that applies to them (Marr's level 3).

SAQ 7.1 Suppose it is true that the speaker/hearer has a language faculty with rules in it. In how many ways could those rules *in principle* have got there?

What about the second question? It is all very well to say that the language faculty contains knowledge of language, but how did that knowledge get there? This question does not fit in Marr's three levels, but it is a problem that is as acute in the study of language as it is in the study of other systems of the mind. Suppose we had found a nice set of rules to represent your knowledge of language but we could not explain how you could ever have acquired some or all of those rules! That wouldn't be any good, would it? An intimately related question concerns the nature of the *initial state* of the language faculty (the state it is in when language acquisition begins). It could be that the initial state imposes little or no constraint on the rule system that is ultimately acquired by the language learner or alternatively that the rule system of mature speakers is predetermined to a large extent.

The requirement that grammars be learnable offers a challenging perspective on the study of language and has had a profound influence on how generative linguists characterize linguistic knowledge. In particular, the conclusion seems inescapable that the initial state of the language faculty imposes quite strict constraints on what constitutes a possible human language.

In this chapter we will be primarily concerned with the first two questions discussed above: what is knowledge of language and how is it acquired? We first sketch a preliminary answer to the first question, where we discuss the arguments for a mentally represented grammar and introduce some basic concepts involved in the study of sentence structure (syntax). We then go into much more detail and consider the minimal properties that must be attributed to the language faculty. We do this by concentrating on a small number of basic properties of natural language sentences which must be represented by the language faculty. We will discover that natural languages have properties which distinguish them sharply from artificial languages, such as programming languages and the like. Finally we reconsider the fundamental question of language acquisition in the light of the characterization of linguistic knowledge that we have previously arrived at.

Language as a Mentally Represented System of Rules

Why Rules?

We can think of the language faculty of the mind as a black box performing a complex information processing task: it maps meanings into speech (or into writing or a sequence of signs in a sign language) and vice versa. The abstract specification of this mapping could take the form of a set of rules.

Let us begin by clarifying the idea that language is mentally represented, that one's linguistic behavior is somehow guided by mentally represented linguistic rules. After all, it would seem that a much simpler account of our language abilities is possible which does not rely on rules at all. Could it not be the case that what you have in your head when you know a language is simply a long list of sentences (with their meanings)?

A particularly down-to-earth problem with this idea is that there is just not enough space in a finite brain to store a list that is anywhere near long enough. Jackendoff (1993) hammers home this point by devising a few simple methods to run up the number of sentences. For instance, there is the rather long list of sentences in (1):

(1) Amy ate two peanuts
Amy ate three peanuts
Amy ate four peanuts
. . .
Amy ate forty-three million, five hundred and nine peanuts
. . .
. . .

There are as many of these sentences as there are nameable integers. The biggest number listed in Jackendoff's *Webster's Collegiate* is a vigintillion (10^{63} in US/French usage; 10^{120} in British/German usage). He concludes that just with this method, it is possible to create more sentences than there are elementary particles in the universe.

SAQ 7.2 See if you can find a few more methods for creating lots of sentences with only a small number of different words. Sample answers at the end of the chapter.

"Well," you might say, "perhaps we can capture all the vigintillion sentences of example (1) with a pattern like *Amy ate X peanuts.*" Quite a clever move! But it is easy to show that you would need an infinite number of such patterns to capture the entire English language. Consider the examples below:

(2) a Bill thinks that Beth is a genius
b Sue suspects that Bill thinks that Beth is a genius
c Charlie said that Sue suspects that Bill thinks that Beth is a genius
d Jean knows that Charlie said that Sue suspects that Bill thinks that Beth is a genius
. . .

Jackendoff points out that you could capture these examples with patterns like those in (3):

(3) a X Verbs that Y is a Z
b W Verbs that X Verbs that Y is a Z
c T Verbs that W Verbs that X Verbs that Y is a Z
. . .

However, there is always a sentence of this type that requires a longer pattern. It follows that just to capture all the sentences of this type you need an infinite number of patterns.

Another piece of evidence for the view that knowledge of language involves rules rather than knowing a long list of sentences comes from the acquisition of language. If knowing a language means having a mentally represented system of rules, then language acquisition must involve the creation of rules. Interestingly, it has been widely observed that during language acquisition children go through phases in which they produce errors as a result of overgeneralizing a rule. For instance, English children go through a phase in which they produce such past tense forms as *comed, buyed, wented* and *bringed.* Such phenomena strongly suggest that language acquisition is a process involving the creation of rules. If language acquisition was simply based on memorizing sentences, then we would not expect language learners to produce forms that are not present in the language to which they are exposed.

SAQ 7.3 Why would the past tense forms just discussed also include forms like *wented*? Our answer at the end of the chapter.

We are in a serious quandary with our language-is-a-long-list hypothesis: we could never store in memory all the sentences of the language, not even with the help of the kinds of patterns we are considering. Moreover, we cannot get away with just storing a finite list of sentences, because then we could not account for the fact that we can understand and produce sentences that we have never heard or produced before. This is often called the creative aspect of language use. Of course it is the case that we use certain stock phrases over and over again (Coffee? – *Yes, please!*), but that should not blind us to the fact that we have the ability to understand and produce novel utterances. In fact, since you started reading this chapter, you have read and understood many sentences you had never seen before.

We can account for our ability to produce and understand novel utterances if we assume that part of our knowledge of language consists of a set of recursive rules. Recursive rules allow an object to occur embedded inside an object of the same type. For example, recursive grammar rules allow a sentence to occur embedded inside another sentence, as in (2). Since this type of embedding can occur an arbitrary number of times, recursive rules are the key to capturing an infinite number of sentences with a finite set of rules.

Hierarchical Structure

Our knowledge of language does not stop with our ability to identify sentences as belonging to our language or not. Our intuitions about sentences appear to be systematically related to how words in the sentence are grouped together. These intuitions can be accounted for if we assume that we mentally represent sentences as having a hierarchical structure. This in turn implies that we have a mentally represented system of rules for building and interpreting such structures and that it is this system of rules which underlies those intuitions.

For instance, speakers of English have no trouble at all replacing the

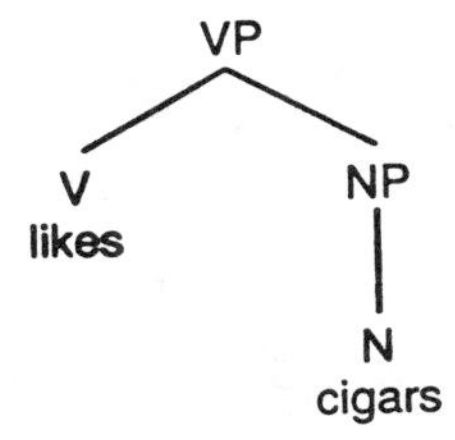

7.1 The tree structure for *likes cigars.*

italicized part in sentence (4a) with its corresponding pro-form (*he*) in sentence (4b). Similarly, speakers of English will uniformly reject the pro-form substitution in (5) (we will mark ill-formed sentences with a star throughout):

(4) a *The tall man from Havana* seems to like cigars
b He seems to like cigars

(5) a The tall *man* from Havana seems to like cigars
b *The tall he from Havana seems to like cigars

We could explain these intuitions as follows. Let us assume that the words in the sentence group together to form larger structural units called *phrases.* For instance, the verb *like* and the noun *cigars* intuitively group together to form the phrase *like cigars.* When two words join together to form a phrase, the grammatical properties of the resulting phrase are always determined by one of the two words. In this case, the resulting phrase has verb-like properties, as it can follow the infinitive marker *to* just like a verb can (*he seems to sleep*) but a noun cannot (**he seems to cigars*). We therefore call *like cigars* a verb phrase (or VP). We will call the VP a **projection** of the verb and the verb the **head** of the VP. A head is a node which is itself not projected from anything. Crucially, then, we distinguish between a head and its projection. What about a word that does not appear to project at all? For instance, given what we have said so far, *cigars* in the phrase *likes cigars* does not project at all, because when we combine *likes* and *cigars* it is the verb that projects to form a VP. Since *cigars* is itself not a projection of anything, it is certainly a head. Let us assume here and in what follows that in fact every head projects a phrase, so that *cigars* is the head of a noun phrase (NP) and the structure of likes *cigars* is as in figure 7.1. The group of words *the tall man from Havana* in (4) also forms a phrase. First the preposition *from* and the noun phrase *Havana* form a prepositional phrase (PP) *from Havana.* The noun *man* and the PP *from Havana* form a noun phrase (NP) *man from Havana.* This phrase groups together with the adjective phrase (AP) *tall* to form another noun phrase *tall man from Havana.* As you see, we assume that a phrase can project still further to form a larger phrase. The head of the NP *tall man from Havana* is still the noun *man.* Finally, this NP groups together with the determiner *the* to form the determiner

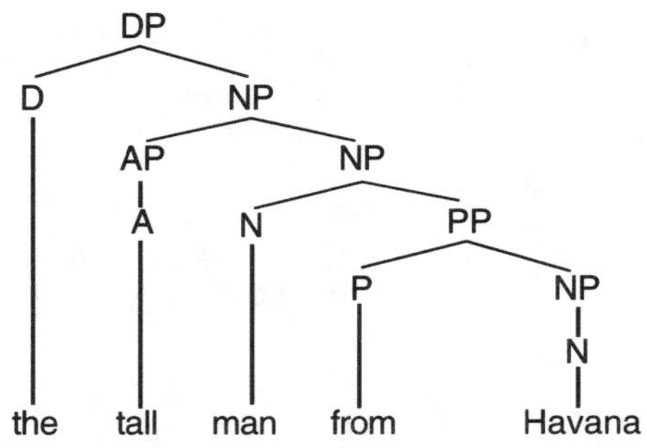

```
                DP
        D              NP
               AP            NP
               A        N          PP
                               P         NP
                                         N
       the    tall     man    from     Havana
```

7.2 The tree structure for *the tall man from Havana.*

phrase (DP) *the tall man from Havana.* The full structure is given in figure 7.2. We will call such structures *tree structures.*

We can now formulate the following rule of pro-form substitution:

(6) *Pro-form substitution*
A pro-form can be substituted for a phrase of the same category-type.

Thus, if we assume that the pronoun *he* is a DP, then it follows that it can substitute for the DP *the tall man from Havana* but not for the noun *man* or the NP *tall man from Havana.* Let us consider another example. The pro-form *do so* substitutes for a VP. This explains why (7b) is good, because *do so* substitutes for the VP *likes books* whereas (8b) is bad, because *do so* substitutes for the verb *likes* (and although the verb is the head of a VP, it is not a VP itself).

(7) a John likes books and Mary *likes books* too
b John likes books and Mary does so too

(8) a John likes books and Mary *likes* books too
b *John likes books and Mary does so books too

It should be clear that if we assume that hierarchical structures of this kind are mentally represented in the language faculty of speaker/hearers, then we can account for these systematic judgments. By contrast, if we do not make this assumption, then these facts about language users remain a mystery.

Let us look at one more piece of linguistic knowledge that depends on sentences having a hierarchical structure. Every speaker of English intuitively knows that words like *himself* are dependent on another part of the sentence (and speakers of other languages have the same intuitions about similar words in their language). Let us call such a dependent word an *anaphor* and the word or phrase upon which it depends its *antecedent.* Thus, (9a) is fine because the anaphor *himself* has *John* as its antecedent, but (9b) is not well-formed because there is no potential antecedent for *himself* at all:

(9) a John likes himself

b *Himself sleeps

As the examples in (10) show, the antecedent on which *himself* depends cannot just occur anywhere:

(10) a *Himself likes John
b *John's mother likes himself

You might conclude from the contrast between (9a) and (10a) that what is at stake here is just linear order, in other words that the antecedent of *himself* must precede it. But example (10b) suggests that something rather more complex is going on. Careful study of these and other examples in English and a large number of other languages have led linguists to the conclusion that the antecedent of an anaphor must always enter into a particular structural configuration with it. This relation, which is called *c-command* (constituent command) is defined in (11):

(11) *C-command*
Node A c-commands node B in a tree structure if and only if (i) A does not dominate B and B does not dominate A; and (ii) the first branching node dominating A also dominates B.

(12) *Dominance*
Node A dominates node B in a tree structure if and only if A is higher up in the tree than B and you can trace a line from A to B going only downwards.

As an example consider the relation between the D(eterminer) *the* and the N(oun) *man* in figure 7.2. D does not dominate N and N does not dominate D. Also, the first branching node above D (DP) also dominates N. Therefore, the D c-commands N.

SAQ 7.4 In figure 7.2:
Does the P *from* c-command the N *Havana*?
Does the N *man* c-command the D *the*?
Does the DP-node c-command the N *Havana*?
Answers at end of chapter.

SAQ 7.5 Draw tree structures for examples (9a) and (10b). Does *John* c-command *himself* in (9a)? And in (10b)? Answers at the end of chapter.

We could therefore formulate the following (incomplete) rule for the interpretation of anaphors:

(13) *Anaphor Interpretation*
An anaphor depends on a c-commanding antecedent

We can of course only define something like the rule of anaphor interpretation in (13) if we assume that sentences are associated with a hierarchical tree structure.

To conclude, we have found several arguments against the language-is-a-long-list hypothesis and in favour of the view that knowledge of languages is rule based. We also found strong evidence that these rules range over tree structures formed from the words in a sentence and therefore provide evidence for the view that sentences are mentally represented as having a hierarchical structure.

Grammars

The set of rules that specify a speaker's knowledge of language is usually called a *grammar.* We use the terms *grammar* and *grammar rule* in this sense throughout this chapter. There are other uses of the word *grammar* that you may be more familiar with which are irrelevant to the present discussion. For instance, most of us have been taught rules of "correct grammar" in school. A moment's reflection should be enough to convince you that these "rules of correct grammar" and the mentally represented rules for the language you speak cannot be the same. They cannot be simply because you could already speak your mother tongue before anyone had ever taught you a "rule of correct grammar."

Grammar can be divided into a set of rules about the structural well-formedness of words (morphology) and a set of rules about the structural well-formedness of sentences (syntax). Morphological rules determine that you can use a noun like *sleep* and an affix like *-less* to make the adjective *sleepless.* Similarly, morphological rules determine that the plural form of *house* is *houses* and not **hice.* Syntactic rules determine that the sentence *Who did Mary see* is well-formed, whereas the sentence **Who did Mary see John* is not. When we say that speakers know a language, we are saying that they have the command of a system of rules (a grammar) which allows them to make judgments about the well-formedness of sentences. This is an odd use of the word *know,* because we have no introspective awareness of grammar rules at all. Our knowledge of language is *tacit* knowledge: we have it and we use it but we cannot become conscious of it. Some people find the concept of tacit knowledge difficult to grasp or consider it a downright con, but there is nothing terribly special about it. After all there are lots of other things going on in your brain that you cannot become conscious of either. Jackendoff (1993) draws the following analogy:

> Think about getting from an intention such as "I think I'll wiggle my fingers now" into commands to be sent to the muscles, so that our fingers wiggle. Just

> how do we do it? From the point of view of introspection, the experience is immediate: we decide to wiggle the finger, and the finger wiggles, unless there is some obstruction or paralysis. How the mind accomplishes this is entirely opaque to awareness. In fact, without studying anatomy, we can't even tell which muscles we've activated. So it is, I want to suggest, with the use of mental grammar. (p. 19)

There is an important sense in which knowing grammar rules and using that knowledge are quite distinct. For instance, you might well know that the word for "small rechargeable battery-operated vacuum cleaner" is *dust-buster* but if you're tired or after you've had a few cans of beer you might well pronounce it as *bust-duster.* Normal speech is full of these and other errors, such as hesitations, half sentences, and what have you. We should therefore distinguish a speaker's tacit knowledge (his *competence*) from the way that knowledge is put to use (his *performance*). The latter may be influenced by all kinds of external factors, such as tiredness, drunkenness, drugs and distractions, whereas we take the underlying system of knowledge to be more or less constant. Grammatical theory is concerned with discovering the nature of a speaker's competence and abstracts away from performance factors. This emphasis does not reflect a value judgment, rather it seems difficult to study performance without first studying competence. Basically you first have to know what speakers know about their language before you can study the effects of external factors on the use of this knowledge.

The first task for a linguist, then, is to construct a rule system (called a grammar) that captures the input/output characteristics of a speaker's language faculty. If this is successful, then the grammar will specify exactly the sentences that the speaker deems to be part of his language. But this is a very limited goal, because a grammar which captures the input/output characteristics of a speaker's language faculty might do so in a way which sheds no light whatsoever on the linguistic knowledge of that speaker. We would also want the proposed system of rules to capture important generalizations of the language which form the basis for the intuitions of native speakers about the well-formedness of sentences of the language. We call a grammar *descriptively adequate,* if it meets this further requirement.

The Theory of Universal Grammar

But given our overall project, namely to gain an understanding of the human language faculty, descriptive adequacy is still quite a moderate goal. What we really want is a theory about what constitutes a *possible* grammar of a human language. It would tell us which properties of natural language grammars are non-accidental, in the sense that they are determined by the initial state of the language faculty itself. Such a theory is generally referred to as the *theory of* ***Universal Grammar*** (or UG, for short). UG is the theory of the initial state of the language faculty. Formulating a theory of UG is a daunting task, not least because it has to satisfy conflicting criteria. On the one hand it must be *universal*: that is, since we are taking the initial state of the language faculty to be a genetically determined property of the species, it must allow us to specify the grammar of any human language. And intuitively there are considerable differences among languages. But on the other hand, UG must be sufficiently *restrictive*: that is, it must be constrained enough to specify *only* the grammars of natural languages (as opposed to artificial languages or animal communication systems). Apart from the conditions of universality and restrictiveness, UG must also satisfy the requirements *of explanatory adequacy* and *learnability*.

UG is explanatorily adequate if it can explain why natural language grammars have the properties that they do. For instance, it might well be the case that certain properties of natural language grammars can be naturally explained as a consequence of the fact that the language faculty must generate "sound representations," that is representations which can be interpreted by the articulatory–perceptual mechanisms of the mind. To put it differently, it is not inconceivable that the articulatory–perceptual mechanisms impose certain "design" constraints on the language faculty (cf. Armstrong et al., 1995). Note that "perceptual" here refers to the very earliest (sensory) stages of language processing, as distinct from the way in which Fodor uses the term when he talks of the language input system (see chapter 3).

UG satisfies the learnability requirement if for any natural language it makes available a grammar which can be learned. You might think that this is a pretty trivial requirement. After all, children just mimic the language around them and that's how they pick up the rules. Or do they? More careful consideration shows that this simple view of language acquisition is misplaced. Recall for instance that language learners produce forms that are not present in the language to which they are exposed (cf. *comed* etc. above). When we say that a speaker knows a language, what we mean is that he or she knows which sentences are part of the language and which sentences are not. The evidence on which the process of language acquisition is based, therefore, might seem to fall into two categories. On the one hand, the language learner must have access to *positive evidence*, showing which sentences are possible in the language, and on the other, access to *negative evidence*, showing which sentences are impossible. The trouble is that the language input to the language learner only shows which sentences are possible but not which ones are impossible. This would perhaps not constitute a problem if children's language mistakes were regularly corrected and if the language learning child could in general put the correction to some use, but extensive research has shown that neither condition is met. Parents typically correct their child's language not if the sentence produced is ungrammatical but if what the child says is not true. Furthermore, if the correction concerns a child's grammar, then parents' good advice is typically ignored, as illustrated in this well-known example:

child: Want other one spoon, Daddy.

father: You mean, you want the other spoon.
child: Yes, I want other one spoon, please, Daddy
father: Can you say "the other spoon"?
child: Other . . . one . . . spoon.
father: Say "other."
child: Other.
father: "Spoon."
child: Spoon.
father: "Other spoon."
child: Other . . . spoon. Now give me other one spoon?

The conclusion seems to be that the process of language acquisition is based on positive evidence only. This creates what is known as the *logical problem of language acquisition*: how does the mature speaker end up knowing which sentences are ungrammatical in the absence of relevant evidence? It appears that we are forced to the conclusion that the missing evidence must be attributed to the initial state of the language faculty (UG). Linguists working in the Chomskyan framework consider UG to be crucially involved in the process of language acquisition. Indeed it is claimed that language acquisition should be impossible for any organism that is not equipped with UG. It seems that humans are unique in having a Universal Grammar, since no other animal is capable of acquiring anything even remotely resembling a natural language. In the past years there have been sustained efforts to teach various kinds of apes some of the rudiments of natural language. As Chomsky (1988, p. 38) points out, not without humour:

> . . . it is now widely recognized that these efforts have failed, a fact that will hardly surprise anyone who gives some thought to the matter. The language faculty confers enormous advantages on a species that possesses it. It is hardly likely that some species has this capacity but has never thought to use it until instructed by humans. That is about as likely as the discovery that on some remote island there is a species of bird that is perfectly capable of flight but has never thought to fly until instructed by humans in this skill.

In what follows we first take a closer look at the language faculty and try to uncover some of its basic properties. The discussion will have two broad aims: to focus on some distinctive properties of natural languages as opposed to artificial languages and to see to what extent UG may be said to explain some of these. Finally, we return to the question of how UG is involved in language acquisition.

Properties of the Language Faculty

Interfaces

We said earlier that the language faculty specifies a mapping from meanings to sounds (and vice versa). Let us now take a cognitive science perspective and consider this mapping from the point of view of the overall architecture of the mind. Suppose we think of the language faculty as specifying representations at two interfaces to other systems of the mind. In other words, we assume that for each sentence, it determines a pair {meaning-representation, sound-representation}.

At one interface, the language faculty specifies a representation of *linguistically determined* aspects of the meaning of a sentence. Following widespread usage in generative grammar, we will refer to this interface of the language faculty as **Logical Form** (LF) and to the representation specified at this interface as the LF-representation. You will recall from the introduction that we distinguish between the linguistic meaning of a sentence, which is independent of context and knowledge of the world, and the interpretation of a sentence, which is the result of extra-linguistic interpretive processes. In what follows, we will take the meaning-representation produced by the language faculty to encode linguistic meaning only. The LF-interface delivers these representations to interpretive processes which make use of extra-linguistic information (knowledge of the world) to arrive at the intended representation of sentences. There exists substantial evidence that the construction of LF-representations does not involve access to extra-linguistic knowledge. In fact, the language faculty as a whole appears to be informationally encapsulated in the sense of Fodor (1983). Some evidence for this view was discussed in chapter 3 (but cf. Marslen-Wilson and Tyler, 1987, for a contrary opinion).

The language faculty must also construct a representation at the interface to the articulatory–perceptual systems. We will refer to this interface as **Phonetic Form** (PF) and to the representations specified at this interface as PF-representations. Put simply, in production the PF-interface specifies instructions to the motor system to produce speech (or signs), whereas in comprehension based on auditory input (or visual input, as in the case of signs), it specifies whatever we take to be the output from the perceptual system. Recall that we take the articulatory–perceptual interface to be neutral between speech and signing, even though we restrict attention here to the former. In what follows we will have relatively little to say about the PF-interface of the grammar. The LF-interface has been widely investigated and much is known about its properties. Let us look at two small examples of interpretive principles of the LF-interface.

The first of these we have already briefly discussed: the rule of anaphor interpretation, which we gave as (13), repeated here for convenience:

(13) *Anaphor Interpretation*
An anaphor depends on a c-commanding antecedent.

Although this principle as it stands needs quite a bit of refinement, it captures a basic property which must certainly be captured at the LF-interface of the language faculty. An anaphor has as an inherent property that it cannot be "understood" in the absence of a c-commanding antecedent.

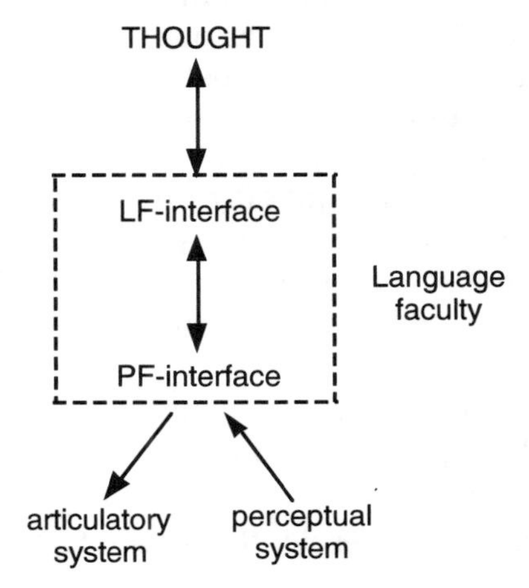

7.3 Information flow in comprehension and production.

SAQ 7.6 What further refinement to the principle of Anaphor Interpretation given in (13) is suggested by the following examples?

John likes himself
*John said that Mary likes himself

Answer at end of chapter.

The following examples illustrate another fundamental property of the LF-interface, often referred to as the θ-Criterion (pronounced *theta criterion*):

(14)	a	Dad sleeps
	b	*Dad sleeps the newspaper
(15)	a	Zara likes coco pops
	b	*Zara likes

While we have no problems interpreting the examples in (14a) and (15a), the examples in (14b) and (15b) are uninterpretable. Why should this be the case? There isn't much to say, but what there is to say is absolutely vital. It is part of the meaning of *sleep* that it specifies a "scenario" which involves just one "role," so to speak: the role of the "sleeper." The meaning of *like*, by contrast, specifies a scenario which involves two roles: the "liker" (the one/ones doing the liking) and the "likee" (the object/objects being liked). The example in (14b) is not interpretable because it contains two phrases wanting a role (*dad* and *the newspaper*), whereas *sleep* has only one role to give away. In (15b) exactly the opposite is the case: *likes* must give away two roles but there is only one player on the scene (*Zara*). In generative grammar the roles specified by a verb are called *θ-roles* and the phrases acting as players are called ***arguments***. It is a basic interpretive property of sentences that every θ-role must be given to an argument and that every argument must be given a θ-role. More precisely, there must be a one–one mapping between θ-roles and arguments. It is exactly this interpretive property that is captured by the θ-criterion.

There are also phrases which do not act as players and therefore do not take a θ-role. These are called *adjuncts* (in traditional grammar often referred to as *modifiers*). Adjuncts indicate time, place or manner:

(16) a Dad sleeps under the newspaper
b Dad sleeps every evening

Because adjuncts do not need a θ-role, they can be "stacked", as in (17).

(17) Dad sleeps under the newspaper every evening

Building Interface-Representations: Computations of the Language Faculty

Now that we have some idea of the output of the language faculty, namely PF–LF pairs, we can start thinking about the minimal components and mechanisms that it must contain which allow it to construct these paired representations. It stands to reason that the two representations must be intimately linked: they are, as it were, the meaning and the sound of the same expression. We therefore expect the language faculty to construct *one* object, which can serve as the basis for both the LF- and the PF-representation.

It is intuitively entirely obvious how this could be done. We may assume that the language faculty has access to a store of words which have attached to them a representation of their meaning (a semantic specification) and of their sound (a phonological specification). We will refer to this abstract store of words as the lexicon (see chapter 6 for a psychologically informed discussion of the mental lexicon and chapter 8 for a discussion of word meaning). Suppose further that the language faculty contains an operation which combines a selection of words into a tree structure along the lines discussed earlier. Then the object so formed can be mapped to LF by extracting from it those parts relevant to the LF-interface and it can be mapped to PF by extracting from it the parts relevant to the PF-interface. Let us call this mapping operation Spell-Out, to capture the intuition that it is the first step in the process of producing the PF-representation. It is only a first step, since we must assume that a number of non-syntactic operations are carried out between Spell-Out and the PF-interface. For instance, there will be phonological rules responsible for producing an intonation contour and for phonological processes between words, such as *wanna*-contraction, which produces (18b) from (18a):

(18) a I want to eat coco-pops
b I *wanna* eat coco-pops

It would be more accurate, then, to say that Spell-Out maps an object to the *PF-component* of the language faculty. This component of the language faculty eventually produces a PF-representation. We will return later to the question whether any operations take place between Spell-Out and the LF-interface.

Although this is a simplified picture, it shows us a number of things. First, the grammar must contain a computational procedure for combining words into larger units. We already came across this procedure, which we will call **Merge**, in the discussion of pro-form substitution. Merge takes two syntactic objects (either words or phrases) and combines them by *projecting* one of the two. Let us consider a simple example, say the construction of a structure for the string *Zara likes coco-pops.* Merge first combines *likes* and *coco-pops*, projecting *likes* to form a VP. As before we assume that every head projects a phrase, so that the head *coco-pops* has projected to NP (figure 7.4). Merge then combines *Zara* and the structure already formed, projecting the verb further to create a still larger VP (figure 7.5). We will refer to a head and the nodes projected from it as the projection of that head. Thus, the V-node and the two VP nodes which project from it together form the projection of V.

Since *likes* has to assign two θ-roles, it requires the presence of two arguments. This is a lexical property of *likes*, not shared for instance by the verb *sleeps.* It is therefore natural to assume that the arguments of a verb are associated with their θ-roles within its projection. Strengthening this assump-

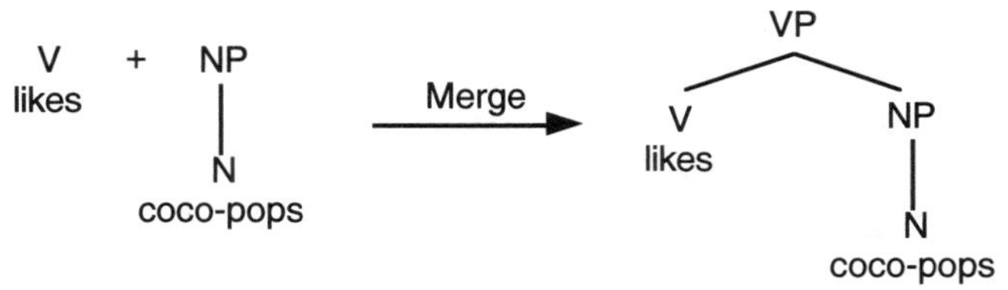

7.4 Merge combines *likes* and *coco-pops*, projecting *likes* to form a VP.

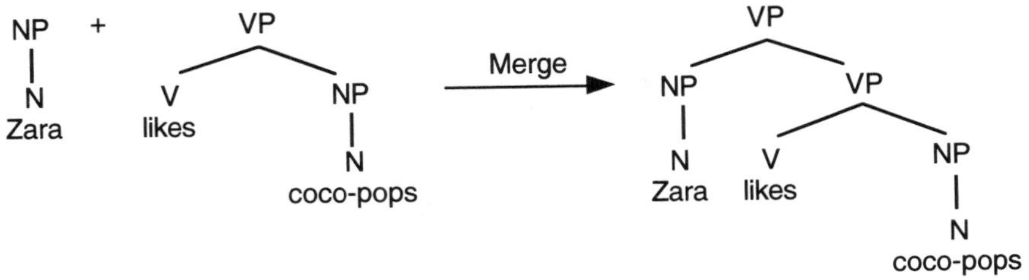

7.5 Merge combines NP *Zara* and VP *likes coco-pops*, projecting V.

tion still further, we could say that the verb *assigns* a θ-role to the argument with which it Merges.

We will refer to the sequence of steps involved in constructing a tree structure as a *derivation.* We will refer to a phrase which is a sister of a head as a complement. For instance, the NP *coco-pops,* in the tree structure formed by the two applications of Merge in the example above, is a sister of the head *likes* and therefore its complement. We will refer to the phrase which is attached to the projection immediately after the **complement** as the **specifier**. For instance, in the structure above, the NP *Zara* is a specifier of *likes.* We see that θ-role assignment is a strictly local process in that a head can only assign a θ-role to the complement and the specifier within its own projection.

Second, it can be seen that words do not just contain a semantic (meaning-related) and a phonological (sound-related) specification. For instance, *coco-pops* is also specified to be of category N and *likes* is specified to be of category V. The category specification is neither semantic nor phonological but determines *syntactic distribution.* For instance, it allows us to express the generalization that the category determiner can take a noun phrase as a complement (i.e., as its sister) but not a verb phrase. Or it allows us to state the fact that the complements of the English verb *put* must be phrases of the category D or N and P respectively. No other categories and no other orders are allowed. Some examples illustrating this are given below, using labeled bracketing to indicate a string of words forming the constituent given as the label:

(19) a Lucy put [$_{DP}$ the bread] [$_{PP}$ in the frying-pan]
b *Lucy put [$_{PP}$ in the frying-pan] [$_{DP}$ the bread]
c *Lucy put [$_{DP}$ the bread] [$_{DP}$ the frying-pan]

We can think of words, then, as containing three kinds of specifications: semantic, phonological and syntactic. Each of these specifications can be captured in terms of *features.* A feature has a name and a value. For instance, the lexical entry for the word *coco-pops* contains a feature with the name *category* and the value *N.* We will write this as [cat = N]. The word *coco-pops* is therefore

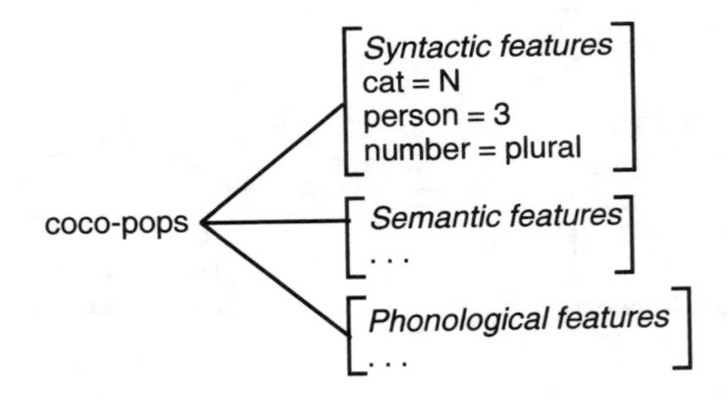

7.6 The entry in the lexicon for *coco-pops.*

associated with three groups of features, of which in the schematic representation in figure 7.6 only the syntactic features are specified.

Staying still with the example *Zara likes coco-pops,* we see that *Zara* and *likes* share certain features. In particular, both are specified as [person = 3] and [number = singular]. These features show up on the verb in the form of the ending *-s.* We say that *Zara* and *likes* agree on these features. Although languages may differ strongly in this respect, it is very common for many kinds of syntactic features to be phonologically realized in some way.

SAQ 7.7 Find two other examples of syntactic features which are phonologically realized in English. Sample answers at the end of the chapter.

In our example it is not the case that *Zara* and *likes* agree in person and number features accidentally; it appears that the subject and the verb must agree in these features or the sentence is ungrammatical:

(20)	a	Zara likes coco-pops (**agreement** in person and number)
	b	*Zara like coco-pops (agreement in person only)
	c	*You likes coco-pops (no person or number agreement)

Two questions arise at this point:

I Does the verb always agree with the subject?
II Second, in those cases in which the verb must agree with the subject, how does the language faculty *enforce* agreement?

We will deal with the second question in the following section. As for the first, notice that the subject only agrees with the verb in a *tensed* sentence (i.e., one in which the verb is in the present or in the past) and that no such requirement holds if the verb is in its infinitival form (preceded by the infinitive marker *to*):

(21)	a	Dad believes [Zara to like coco-pops]
	b	Dad tried [to like coco-pops]

In (21a) the verb *believe* takes a sentence as a complement of which *Zara* is the subject. Clearly, *Zara* does not agree with the verb *like* and the sentence is grammatical. In (21b) the verb *try* takes a sentence as a complement which appears not to have a subject at all.

SAQ 7.8 Given the θ-criterion, which we discussed earlier, is it really possible for *like* in example (21b) not to have a subject at all?

But given our earlier discussion about the interpretive requirement captured by the θ-criterion, it seems that there should be two players available to take the two roles of *like*. It is clear that *coco-pops* plays the role of the *likee*. Now notice that it cannot be *Dad* who takes the role of the *liker* because *Dad* already has the role of *tryer*. And since we must have a one–one mapping of roles onto players, *Dad* is not a candidate for the role of *liker*. Following this reasoning to its logical extreme, we are led to the conclusion that there must be another, invisible (or silent) argument in example (21b). In our terms, an "invisible" constituent is simply a word or a phrase which does not have any phonological features. As a result it is not pronounced at the PF-interface: it is silent. Thus, the structure of (21b) is really something like (22), where *ec* indicates the "empty category":

(22) Dad_1 tried [ec_1 to like coco-pops]

Although *ec* does not have phonological features, it does have semantic and syntactic features. In particular, it must carry the features [person = 3] and [number = singular], since it is interpreted as referring to the same person as *Dad* (indicated here by co-subscripting *Dad* and the *ec*). In terms of agreement the example is therefore completely parallel to (21a): *ec* does not agree with the verb.

SAQ 7.9 How does the following example lend support to our conclusion that infinitival clauses may have silent subjects?

To cut oneself is painful

Answer at end of chapter.

Summarizing the discussion so far, we have sketched a minimal specification of the language faculty, consisting of a lexicon of words specified for semantic, phonological, and syntactic features and the operation Merge, which, given a selection of words from the lexicon, constructs a syntactic object which forms the basis for both the LF- and the PF- representation of a sentence. We call this process a *derivation*. We also discovered the existence of agreement relations within the tree structure constructed by the derivation. We observed that a tensed verb must agree with its subject, but we did not resolve the question how the language faculty manages to enforce this agreement relation. We return to this problem after the following two sections, in which we discover and discuss a second fundamental property of the language faculty, which will turn out to interact with agreement in important ways.

Displacement Phenomena: the Operation Move

One question concerning agreement that was left unresolved in the previous section is how the language faculty manages to enforce agreement between a

Devil's Advocate Box 7.1 Avoiding empty categories

In (22) we saw that an empty category was required to fill the subject of *like.* There is substantial dispute in the literature, however, about the psychological reality of such empty catergories, and several linguistic formalisms (e.g. Generalized Phrase Structure Grammar (GPSG: Gazdar et al. 1985), Head-Driven Phrase Structure Grammar (HPSG: Pollard and Sag 1993) and Categorial Grammar (CG: Steedman 1993)) avoid the use of them completely. In these theories, the complement of *tried* in (22) is taken to be exactly what it appears to be – a verb phrase in its infinitival form.

The trick is "binding" the subject of *tried* to the subject of the embedded verb (in this case *like*) via a method called *unification.* In a formalism like HPSG, for example, the lexical entries for *Dad* and *coco-pops* are represented as sets of features:

PHONOLOGY	<dad>
CATEGORY	*noun phrase*
SEMANTICS	*dad*

PHONOLOGY	<coco-pops>
CATEGORY	*noun phrase*
SEMANTICS	*cocopops*

The entry for *like* is more complex:

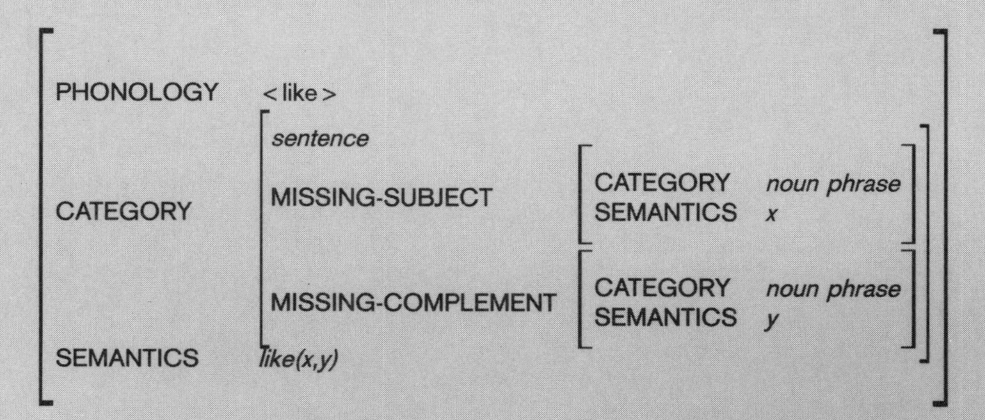

This is taken to mean that *like* requires a subject noun phrase with semantics *x* and a complement noun phrase with semantics *y*, and when it gets those noun phrases it will become a sentence whose meaning can be represented by *likes (x,y), x* and *y* are variables, which are filled when the verb combines with its missing subject or missing complement.

Devil's Advocate Box 7.1 continued

We can combine *like* with *coco-pops* to get:

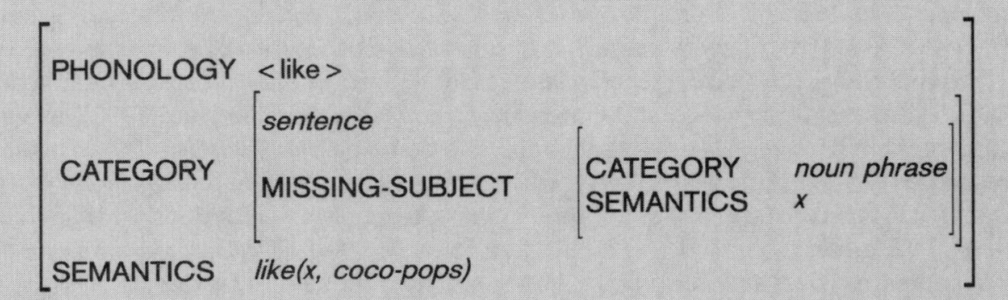

The entry for *tried* is quite complex:

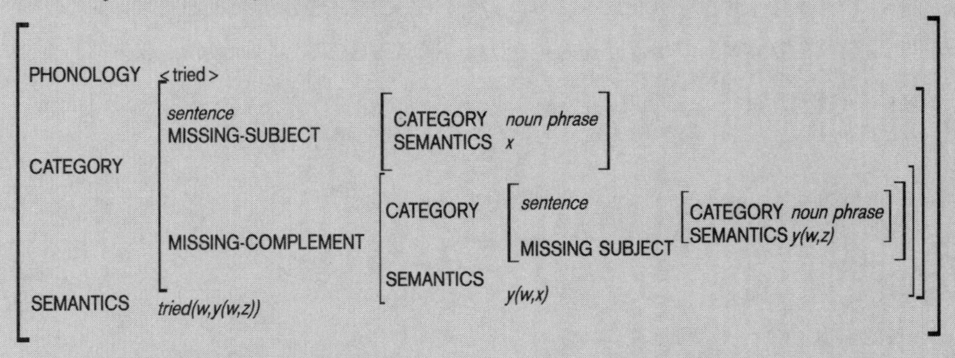

This says that *tried* would be a sentence with meaning *tried(w,y(w,z))* if it has a subject noun phrase with meaning *w* and a complement that was itself a sentence missing a subject noun phrase with meaning *w*. Ignoring the infinitival marker *to*, we can discharge one of these requirements by combining *tried* with *to like coco-pops*, giving:

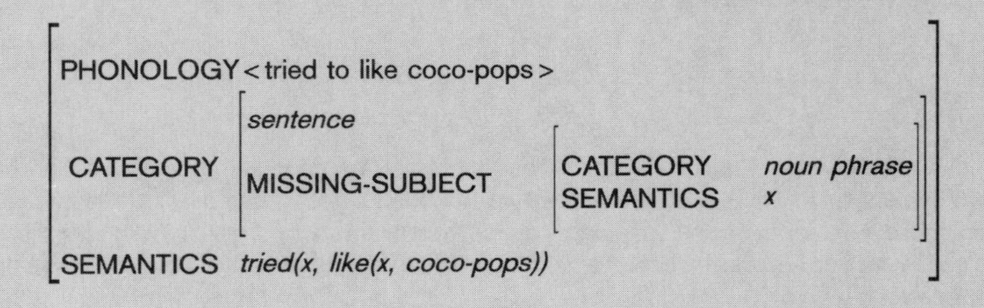

In filling in the complement, we have automatically bound the variables in matching positions, so *y* has become equal to *like*, *z* has become equal to *coco-pops*, and *w* has become equal to *x*. Now when we combine this sentence fragment with the subject noun phrase *Dad*, we bind *x* to *Dad* and end up with:

PHONOLOGY	<dad tried to like coco-pops>
CATEGORY	*sentence*
SEMANTICS	*tried(dad,like(dad,coco-pops))*

tensed verb and its subject. Given the very simple example that we were looking at (*Zara likes coco-pops*), you might hypothesize that a subject simply agrees with the verb of which it is an argument, but the following example, when properly analyzed, shows that that cannot be correct:

(23) Zara seems to like coco-pops

At first sight you might think that (23) is just like (22), but they are in fact quite different.

We begin by making the observation that arguments have the property that they refer to someone or something in the world. Thus, the argument *Zara* picks out an individual in the world named *Zara*. Now notice that *seem* can take a subject which does not refer in this way:

(24) It seems that Zara likes coco-pops

(24) is a paraphrase of (23). In other words, these two sentences express the same meaning. The pronoun *it* in (24) does not appear to be an argument: it does not behave in the same way as the pronoun *it* in *It likes milk*, where we can interpret *it* as referring to, say, the cat. If *it* in (24) were an argument, then the fact that (23) and (24) are paraphrases would be a great surprise. The *it* in (24) is known as an *expletive*.

We now make the next step: if *it* is not an argument, then it also does not have a θ-role. This follows directly from the well-established interpretive requirement that there is a one–one mapping between arguments and θ-roles. But although *seem* does not give a θ-role to its subject, *try* does. For this reason the non-argument *it* cannot appear as its subject:

(25) *It tried Dad to like coco-pops

The ungrammaticality of this example (on the intended reading, with *it* a non-argument) follows immediately if we assume that a non-argument with a θ-role is uninterpretable at the LF-interface.

We have evidence that *Zara* in (23) does not appear in the position in which it receives its θ-role. As example (24) suggests, *Zara* is an argument of *like* and receives its θ-role in the V-projection of which *likes* is the head. But *Zara* surfaces as the subject of *seems* and agrees with it, even though it is not its argument. It therefore appears that *Zara* in (23) has been displaced from its original position. With this discovery we have hit upon what is undoubtedly the most striking property of natural languages, namely that the surface word order of sentences does not always reflect the underlying argument structure: arguments may show up in a position in which they are not interpreted. Perhaps the most obvious example of displacement occurs in direct questions, such as (26) below:

(26) What does Zara like?

Here *what* is interpreted as the object of *like* (it corresponds to the *likee*) but it surfaces in sentence-initial position.

We are led to the conclusion that we must enrich the computational procedure of the language faculty, since it must include a mechanism that is responsible for displacing constituents. Suppose, therefore, that in addition to Merge there is an operation called **Move**. What does Move consist of? Recall that Merge combines two constituents by projecting one of the two. Move could be very similar to Merge, except that one of the two constituents it Merges is taken from within the constituent that projects. Thus, given a constituent A, Move takes B, a subconstituent of A, and Merges B with A, projecting A:

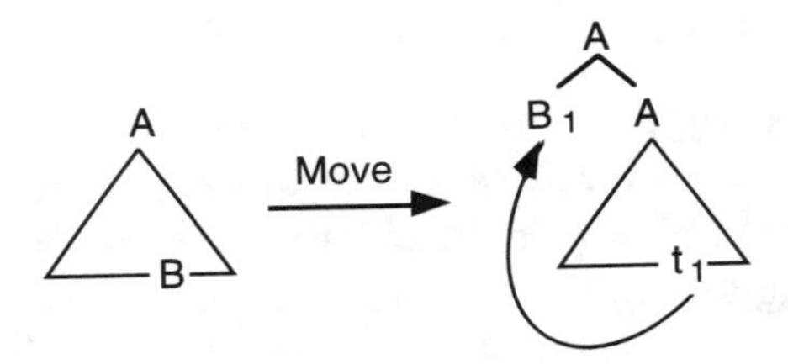

7.7 The operation Move.

Suppose we assume that Move leaves behind a copy of the constituent it displaces, but the copy lacks phonological features and is therefore silent. We will refer to the silent copy left behind by Move as a *trace* (indicated as a subscripted *t* in figure 7.7 and in the examples below). The moved constituent and the trace together form a *chain*. At the LF-interface a chain is the abstract representation of an argument. Therefore, the moved element at the head of the chain will be interpretable if its chain contains exactly one position which receives a θ-role. As examples, consider the schematic LF-representations of (23) and (26), given in (27) below:

(27) a Zara$_1$ seems [to t_1 like coco-pops]
b What$_1$ does Zara like t_1

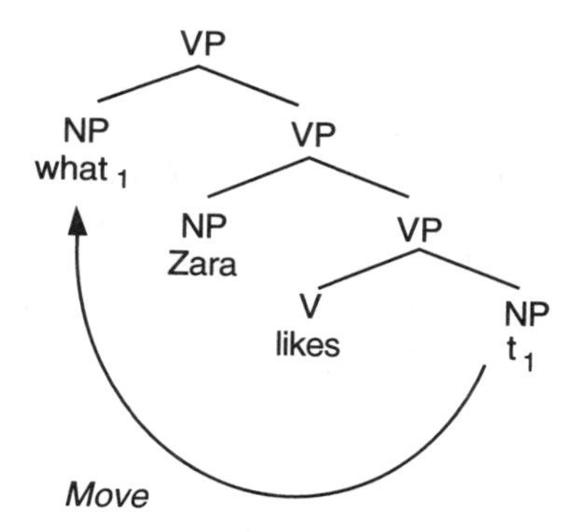

7.8 The structure of (27b) (abstracting away from the auxiliary *do*).

The trace of *Zara*$_1$ in (27a) is indicated as t_1. The identical subscripts indicate that *Zara* and its trace form the chain <Zara,t>. In this chain, the element *t* is in a position which receives a θ-role. *Zara* is not. Therefore, the chain <Zara,t>, which represents the argument *Zara* has exactly one θ-role, as required. Similarly, the chain (what$_1$,t$_1$) of (27b) is the chain representing *what* and carries exactly one θ-role, namely the one assigned by *like* in the position of t_1.

Notice that if Move moves an argument to a position where it receives a second θ-role, the resulting structure will be uninterpretable at the LF-interface. Thus we cannot have the sentence *John hit* with the structure

(28) John$_1$ hit t$_1$

and the interpretation *John hit himself*, because the argument-chain <John,t> contains two θ-roles, thereby violating the interpretive requirement that an argument must have one and only one θ-role.

SAQ 7.10 It seems reasonable to suppose that the structure of a passive sentence like *John was hit* is something like:

John$_1$ was hit t$_1$

After all, *John* corresponds to the person who gets hit. What does this indicate about the θ-role assigning properties of passive *hit*? Answer at end of chapter.

In summary, we have found that the surface word order of sentences does not always reflect the underlying argument structure. We postulated the existence of an additional computational operation of the language faculty, Move. As a result of the application of Move in the course of a derivation, arguments may show up in a position other than the one in which they are interpreted.

Why does Move Exist?

The existence of "displaced" constituents shows up a fundamental difference between natural languages and what are often called artificial languages, such as programming languages. Every theory of the syntax of natural languages must somehow get to grips with this phenomenon. As it turns out, this poses a severe challenge, which is only partially met by existing theories. The basic problem is twofold. First of all, displacement only occurs in certain environments. For instance, Move does not move a question word like *what* across another question word:

(29) *What$_1$ do you wonder who saw t$_1$

We therefore face the task of specifying these environments, preferably in a way that will give us some insight into the workings of the language faculty. Second, languages differ considerably in the extent to which displacement occurs. For instance, Move moves a question phrase to the beginning of the sentence in English but not in Chinese:

(30) a What$_1$ do you remember that John bought t$_1$
b Ni jide [Zhangsan mai-le shenme]
You remember Zhangsan buy-ASP what

Our theory about displacement must also be able to capture these cross-linguistic differences.

Having attributed the displacement property of natural languages to the existence of the computational operation Move, we may begin by asking why Move should exist in the first place. Notice that this question does not arise with respect to Merge: without it the language faculty would not be able to construct interpretable objects at all. A string of words like *Zara likes coco-pops* would not have the interpretation that it has at the LF-interface if Merge did not construct a structured object in which the θ-roles of *likes* could be associated with the arguments *Zara* and *coco-pops*. There is nothing more to say. But the operation Move moves constituents *away* from the position in which they are interpreted. What purpose could that possibly serve?

Recent work in generative grammar carried out in the so-called **Minimalist Program** attempts to understand the application of Move in terms of a number of "minimalist" assumptions about the language faculty, the leading idea being that both its representations and its computations are geared towards *economy*. What is meant by economy?

Let us start with representations. Recall that, for each sentence, the language faculty specifies two interface representations: an LF- and a PF-representation. Representational economy requires that each of these representations only contain elements that are *interpretable* at the relevant interface. For instance, an LF-representation must not contain any phonological features, because these only have an interpretation at the PF-interface. Similarly, a PF-representation must not contain any semantic features, because these only have an interpretation at the LF-interface. Thus, the requirement of representational economy is that interface levels be *fully interpretable*:

(31) *Full Interpretation*
Interface representations must contain interpretable elements only.

If a derivation of the computational procedure of the language faculty produces two interface representations that are fully interpretable, we say that the derivation *converges*, otherwise it *crashes*. We can also speak of a derivation converging/crashing at LF or at PF.

SAQ 7.11 Given Full Interpretation, is it necessary to have the θ-Criterion as a separate principle holding of LF-representations? Answer at end of chapter.

What consequences does Full Interpretation have, in concrete terms, for the computational procedure of the language faculty? Recall that we are assuming that Merge combines items which are specified for semantic, phonological and, syntactic features. Let us refer to the set of words which form the tree structure with the symbol *N*. Once the derivation reaches the point where all the words in *N* have been combined into one tree structure Σ, it must be the case that some operation extracts from this object all and only the parts relevant to the PF-interface (forming the structure Σ_{PF}), leaving behind an object consisting of features only relevant to the LF-interface (Σ_{LF}). This is the operation Spell-Out, which we have already informally referred to earlier (see figure 7.9). If Spell-Out does not apply, the derivation cannot converge, because it fails to produce representations that can satisfy Full Interpretation at the LF- and PF-interface.

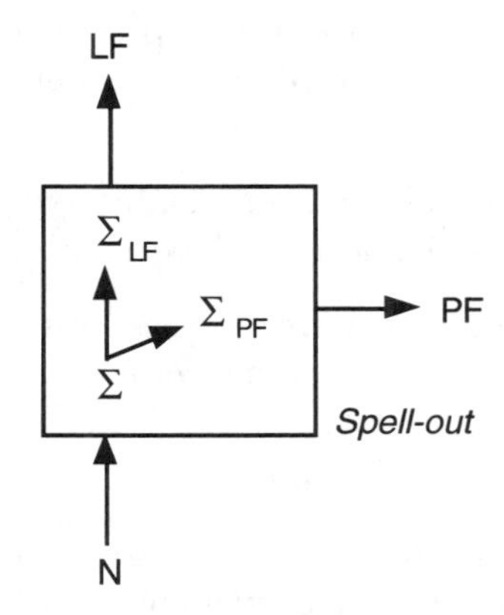

7.9 Spell-Out produces Σ_{LF} and Σ_{PF} from Σ.

For concreteness, let us assume that Spell-Out creates two syntactic objects: one which does not contain any phonological features (to be mapped to the LF-interface) and one which does not contain any semantic features (to be mapped to the PF-interface by the PF-component). We must certainly assume that Spell-Out leaves the syntactic features of the object mapped to LF intact, because many of these have an interpretation at the LF-interface. For instance, the syntactic feature *number* affects the interpretation of a noun which carries it (for instance, *boy* and *boys* do not mean the same thing). Similarly, the LF-interface must distinguish between syntactic categories, such as nouns and verbs, because these have radically different interpretive properties. We may also assume that syntactic features play a role in the PF-component of the grammar. For instance, the phonological rules determining intonation contour are sensitive to the presence of major constituent breaks. It follows that there must also be rules of the PF-component which remove syntactic features: since these features have no interpretation at the PF-interface (i.e., they do not correspond to articulatory instructions), their presence at the interface would lead to a violation of Full Interpretation.

What does economy mean with respect to computations? Economy considerations do not apply to Merge, basically for the reasons already given: Merge is the minimal procedure that given a set of words constructs a single syntactic object. It could not be more economical than it is. With respect to Move, economy requires that Move apply *as little as possible.* What this means in concrete terms is that in the course of a derivation, Move may only apply if failure to do so would cause the derivation to crash.

Given Full Interpretation, it is possible to formulate a coherent story about why Move should exist. Suppose that the operation Merge regularly introduces syntactic features into the emerging tree-structure that have no interpretation at the LF-interface. Suppose furthermore that an application of Move can somehow eliminate such syntactic features. If an application of Move can eliminate an uninterpretable feature and thereby save the derivation from crashing, then it must apply. If an application of Move cannot eliminate an uninterpretable feature then it may not apply, since, by the economy requirement on derivations, Move must apply as little as possible. We must now answer the following two questions:

I What constitutes an uninterpretable syntactic feature?
II How can an application of Move eliminate such a feature?

The following section addresses these questions by exploring the intricate relationship between movement and agreement.

Agreement, Move, and Feature-checking

We now return to the question how the language faculty enforces agreement between a subject and a verb. In particular, we will investigate the idea that movement and agreement phenomena are inextricably linked together.

From the discussion of the example *Zara seems to like coco-pops* we have reason to believe that the agreement relation between a subject and a verb is not established inside the V-projection, where the θ-role is assigned. Recall that the surface position of *Zara* in this sentence (in which it shows *agreement*) does not correspond to the position in which it receives a θ-role (i.e., the position in which it is *interpreted*). Recent work in generative grammar, carried out in the Minimalist Program, suggests that both the subject and the object of a verb enter into a number of different agreement-like relationships and that all these relationships are established *outside the domain in which* θ-role assignment takes place (the V-projection). In other words, the domain for θ-role assignment and the domain for agreement relations are disjoint. It seems not unreasonable to assume that this is a fundamental property of the language faculty. It has therefore been proposed that the V-projection is contained in a number of agreement-related projections. These projections are referred to as the *checking domain* of V, because each of them is involved in checking an agreement relationship involving the verb. The arguments of the verb and the verb itself move to positions in the checking domain in order to enter into agreement relationships. Here we consider just one of the projections in the checking domain of V, namely the one involved in checking subject–verb agreement, but the general mechanism for checking features which we are about to discuss holds quite generally across all agreement relations.

We begin by making the assumption that the agreement relation between the subject and the verb is represented outside the V-projection. In particular, we assume that the agreement features of the subject and the verb are not just present on the subject and the verb themselves but also as two sets of agreement features in a projection outside the V-projection called I (short for Inflection) of which the V-projection is a sister, as indicated in figure 7.10.

We now think of the agreement relation between the subject and the verb as being mediated by the two matching feature sets in I. On this view, subject–

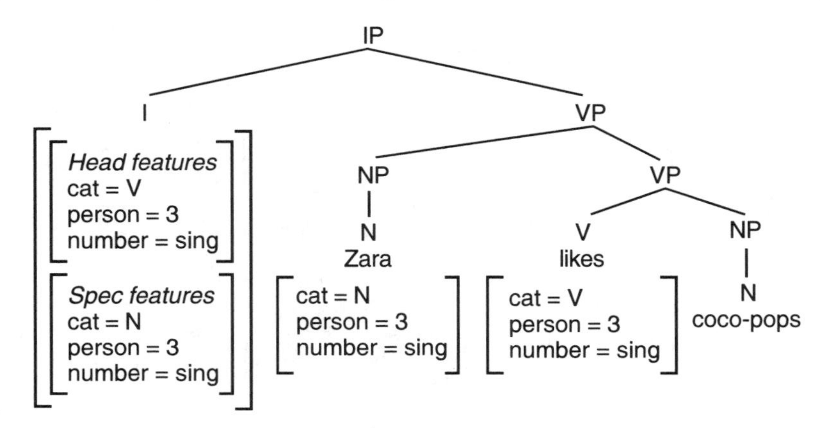

7.10 The agreement features of the subject and the verb are represented by two sets of features in the head I: the head-features of I must match those of the verb and the spec-features must match those of the subject.

verb agreement consists of *checking* that the specifier-features of I match the agreement features on the subject and that the head-features of I match the agreement features on the verb.

The next step is to specify how feature-checking takes place. The minimal assumption is that it is a strictly local process, just like the assignment of θ-roles, and that it therefore has to take place within the I-projection. Suppose then that Move applies to bring both the subject and the verb inside the I-projection, so that feature-checking can take place locally. The specifier-features of I will be checked by the subject when Move makes the subject the specifier of I, and the head-features of I will be checked by the verb when Move attaches V to I in the manner indicated below (we say that V is *adjoined* to I). In summary, for a sentence like *Zara likes coco-pops* all the features of I can be checked in the configuration in figure 7.11.

We assume that for some reason these checking relations must be established or we cannot explain how the grammar enforces agreement between the subject and the verb. Now recall that by economy of derivation Move can only apply if failure to do so would cause the derivation to crash. In the example at hand, Move applies to create a configuration for feature-checking. But why should failure to check the features in I cause the derivation to crash? Suppose we make the following two assumptions: (i) the agreement features in I are uninterpretable at the LF-interface and (ii) when a feature is checked it deletes. It follows that failure to check the features in I will cause the derivation to crash at the LF-interface.

Let us now briefly review some of the examples discussed earlier, beginning

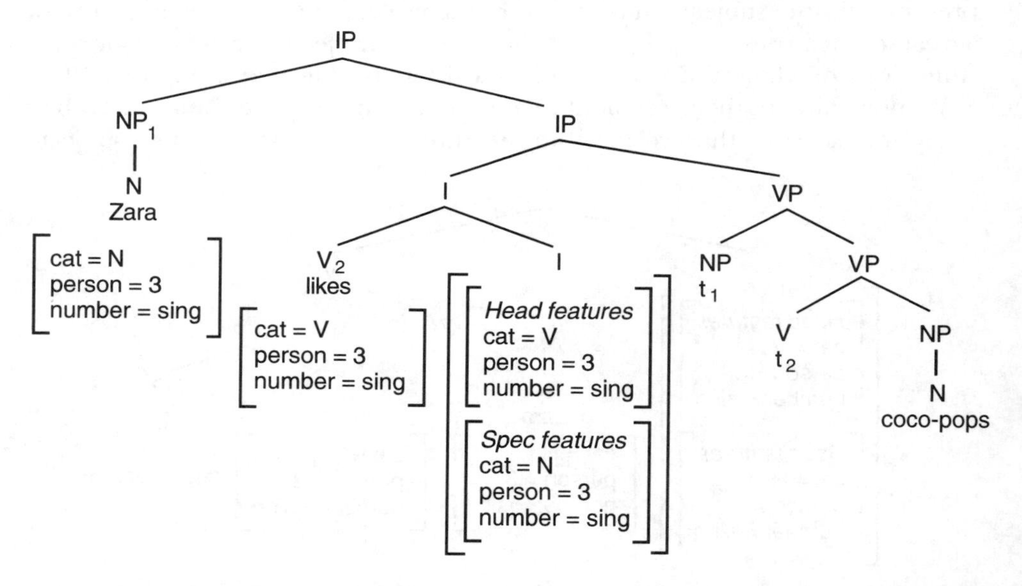

7.11 Both V in its position adjoined to I and the NP in the specifier of I are in a strictly local relation with I itself, thus allowing V to check its head-features and the NP *Zara* to check its specifier-features.

with (23), to which we attributed the schematic LF-representation (27a), both repeated here for convenience:

(23) Zara seems to like coco-pops
(27a) $Zara_1$ seems [to t_1 like coco-pops]

Leaving aside the question whether *Zara* checks any features in the embedded clause in this sentence, Move must raise it to the specifier of the I-projection in the main clause. If Move does not apply, the uninterpretable specifier-features of matrix I (i.e., the I of the main clause) will not delete and the derivation will crash at the LF-interface. Similarly, *seems* must raise to I to check the head-features of I. What happens in the example (24), repeated below:

(24) It seems that Zara likes coco-pops

Since both the verb in the embedded and in the main clause have agreement features, both clauses have their own I-projection. *Zara* checks the specifier-features of the embedded I, while *likes* checks its head-features. In the main clause, *seems* checks the head-features of I. Since *seems* does not assign a θ-role to a phrase in its specifier, there is no subject in the specifier of the main VP which Move can raise to check the specifier-features of I. For reasons which go beyond the present discussion, Move cannot raise the embedded subject *Zara* to check the specifier-features of matrix I. Therefore, if nothing else happens, the derivation will crash. There is one possibility left, however. Merge may apply and merge non-argument *it* with matrix I, projecting I. Thus, *it* ends up as the specifier of I and checks its specifier-features. We now see that the occurrence of non-argument *it* in this example is not an accident: if Merge failed to introduce it, the derivation would not converge (see figure 7.12).

When we started our discussion of Move, it was pointed out that in accounting for displacement phenomena, we face two tasks. The first of these was to specify the environments in which displacement occurs, preferably in a way which sheds light on the working of the language faculty. Our account of agreement phenomena is a first step towards achieving this goal. We have characterized the environment in which Move applies as involving feature-checking. The phenomenon of displacement is fully determined by the interaction of the presence of uninterpretable features and economy conditions on representation and derivation. In the following section we turn to the second task, which was to account for the fact that languages differ considerably in the extent to which they exhibit the displacement phenomenon.

Covert Syntax

We have so far tacitly held to the assumption that all applications of Merge and Move happen prior to Spell-Out. But there are good reasons to assume that this assumption is false. What is more, if we assume that Move can apply both

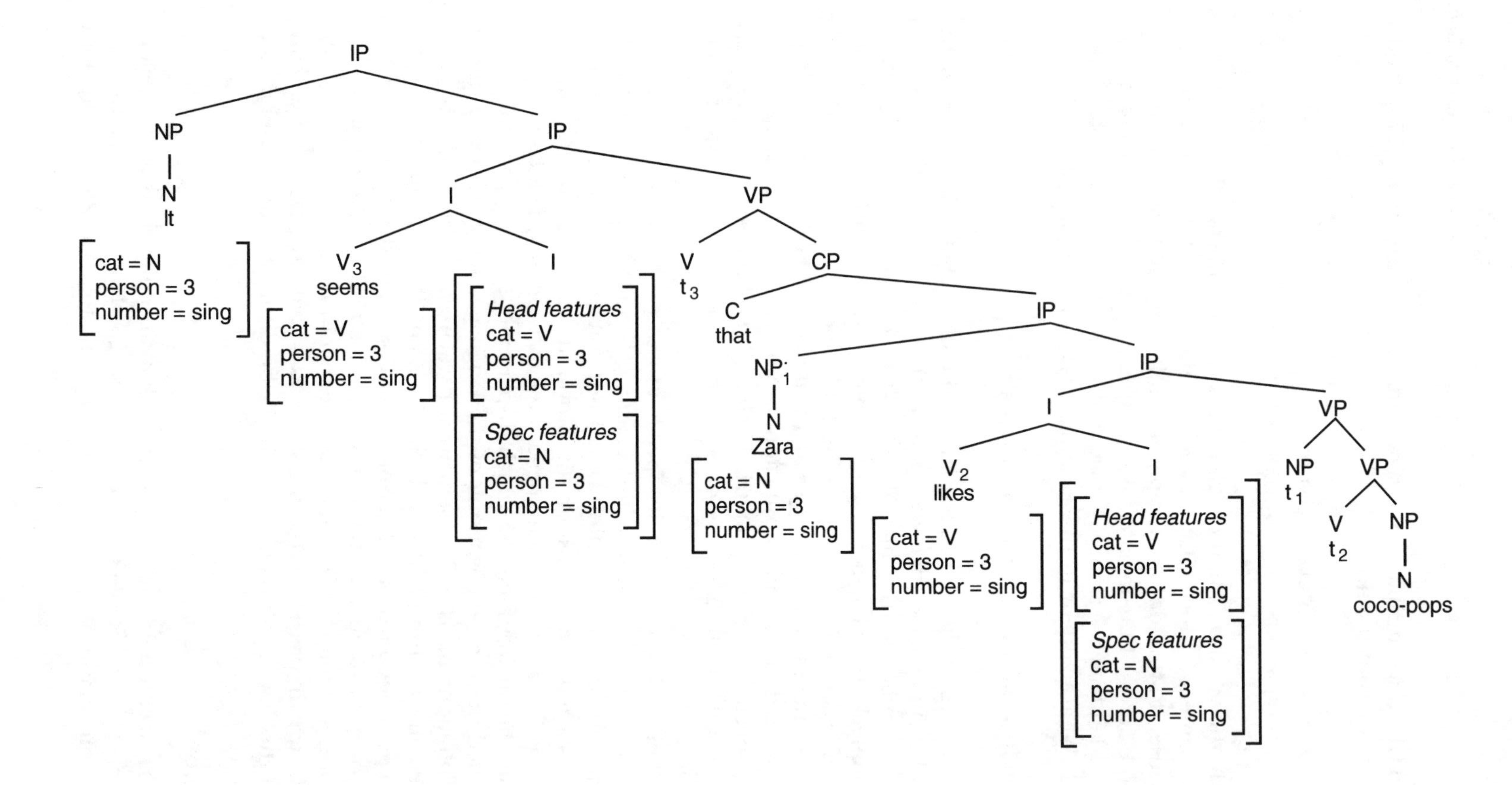

7.12 The tree structure for example (24). Move has placed NP *Zara* in the specifier of the embedded I-projection, where it checks the specifier-features of embedded I. In the matrix clause it is Merge which has placed expletive *it* in the specifier position of the I-projection, where it checks the specifier-features of matrix I.

before and after Spell-Out, then we may be able to account for the fact that languages appear to differ in the extent to which they show displacement. Suppose an instance of Move applies *after* Spell-Out. We cannot "hear" such a movement, because only computations which apply prior to Spell-Out can have any effect on the order of elements in the PF-component of the grammar. This opens up the possibility that if a language does not appear to show a certain type of feature-checking, then this is just because Move creates the configuration for feature-checking after Spell-Out. In other words, different languages may diverge as regards the way they divide up a derivation between overt and covert applications of Move. We will refer to the part of the derivation which happens prior to Spell-Out as the *overt syntax*, and refer to the part of the derivation after Spell-Out as the *covert syntax*.

SAQ 7.12 Would it be possible for Merge to apply after Spell-Out as well?

Returning to our earlier examples, we can in fact find some evidence that in English Move does not move V to I overtly. We might expect, for instance, that if *Zara* and *likes* were in the strictly local checking relation of specifier–head agreement in (15a) (as in figure 7.11), repeated here as (32a) for convenience, then nothing would be able to intervene between these two words. But (32b) shows that this is not true:

(32) a Zara likes coco-pops
b Zara always likes coco-pops

Our suspicion is deepened by the fact that in Dutch the subject and the verb cannot be separated in this way, suggesting that in Dutch the verb does move in the overt syntax:

(33) a Zara eet coco-pops
Zara eats coco-pops
b *Zara altijd eet coco-pops
Zara always eats coco-pops
c Zara eet altijd coco-pops
Zara eats always coco-pops

We might also expect to find languages in which the verb moves in the overt syntax, but the subject is only moved in the covert syntax. In such a language the verb would precede the subject in a normal declarative sentence like *Zara eats coco-pops*. And indeed this is what we find in Celtic languages like Welsh (examples from Awbery, 1976):

(34) a Gwelodd y dyn y ci
Saw the man the dog

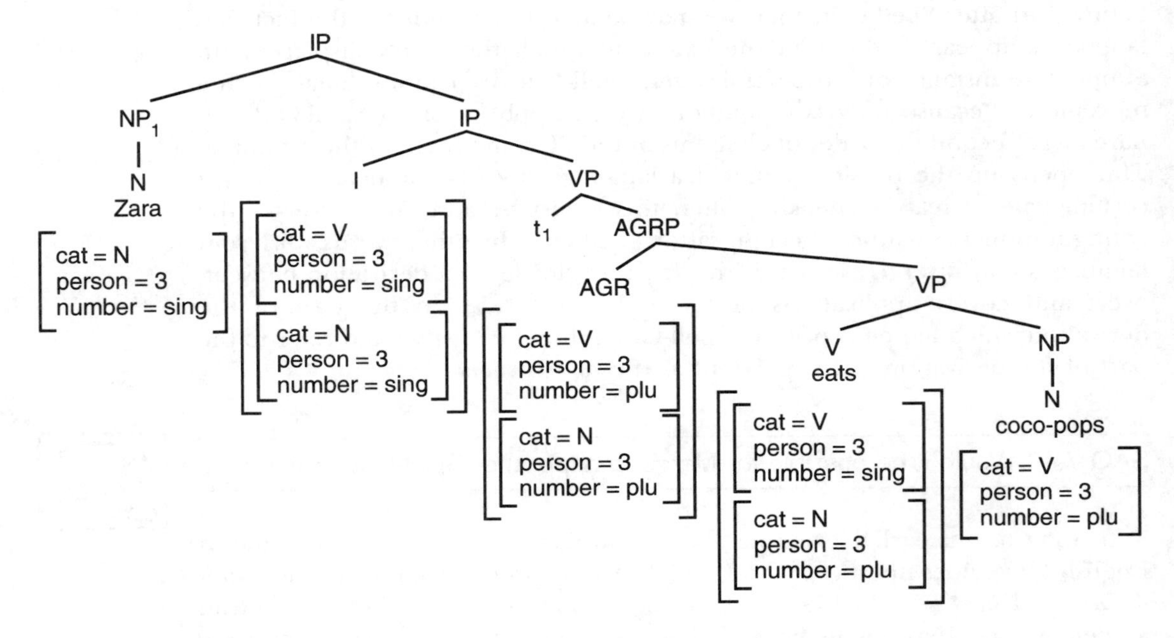

7.13 The structure of *Zara eats coco-pops* at the point in the derivation at which Spell-Out applies. The agreement features of the object are represented by two sets of features in the head AGR: the head-features of AGR must match those of the verb and the spec-features must match those of the object. The verb is still inside the VP.

b Rhoddodd y dyn y ffon i'r ci
Gave the man the stick to the dog

SAQ 7.13 In many languages the object precedes the verb, whereas in others, like English, the object follows the verb. How would you propose to deal with this word-order difference in the theory outlined here?

We could develop a similar account for the relative order of the object and the verb. In many languages the object and the verb show agreement just like the subject and the verb do in English. Suppose that even in languages in which object–verb agreement is not phonologically realized, there is object–verb agreement in the syntax. More precisely, suppose we assume that a simple sentence like *Zara eats coco-pops* has the structure in figure 7.13 at the point in the derivation at which Spell-Out applies. When Spell-Out takes place the verb and the object are still inside the VP, which results in Verb–Object word order at PF. After Spell-Out, the object NP *coco-pops* attaches to AGRP to check the Specifier-features of AGR. The verb attaches to AGR to check its head-features and the whole AGR–V complex then moves to I, which results in a configuration in which the verb can check the head-features of I as well.

If this story is on the right track, then we expect to find languages in which the object attaches to AGRP overtly, while V moves to I covertly, resulting in Object–Verb word order at PF. This is exactly what appears to be the case in Dutch embedded clauses:

(35) . . . dat Zara coco-pops eet
. . . that Zara coco-pops eats

We will not explore the possibilities afforded by the overt–covert distinction any further here, but will focus on two key questions which are raised by this approach to word-order variation:

I If Move can apply covertly, can Merge apply covertly as well?
II What will determine whether an operation of the language faculty applies overtly or covertly in a given language?

Let us begin with the first of these. If Merge could apply covertly, then we might end up with an LF-representation which is not "correctly" related to the PF-representation: it might contain many more words than appear at PF. This is clearly not what we want. We must therefore assume that Merge cannot apply covertly. Fortunately, there is no need to stipulate this. Recall that Spell-Out strips away phonological features because these are uninterpretable at the LF-interface. But if Merge gets a word out of the lexicon *after* Spell-Out, then this word will contain phonological features and therefore cause the derivation to crash at LF. It follows that Merge cannot apply covertly without causing the derivation to crash.

The second question asks how we can arrange for some movement to happen overtly in one language and covertly in another. One way in which we might approach this question is to attribute overt movement to some property of the PF-interface. An answer along these lines could take the following form. To begin with we modify our economy condition on Move, which said *Move as little as possible*, to *Move as late as possible*. Let us call this condition *Procrastinate*:

(36) *Procrastinate*
Move as late as possible.

The intended interpretation of Procrastinate is: put off moving for as long as you can and if possible do not move at all. Changing the economy requirement in this way has the effect that covert movement is preferred above overt movement. It follows that if overt movement is not necessary for convergence, then it does not happen. Our next step is to assume that an uninterpretable feature set may be either *strong* or *weak*. A strong feature set has the property that it causes the derivation to crash at PF. It follows that a strong feature must be eliminated before Spell-Out. In other words, a strong feature set triggers movement in the overt syntax. A weak feature set, by contrast, does not cause the derivation to crash at PF. It follows that such a feature set can be eliminated after Spell-Out, in the covert syntax. Since this is possible, it is preferred, by Procrastinate. For instance, in English the specifier-features of I are strong and therefore trigger overt movement, but the head-features of I are weak and, therefore, V moves to I covertly. In Welsh the head-features of I are strong and the specifier-features of I are weak, so that only the verb moves overtly.

SAQ 7.14 Which feature sets are strong and which are weak in a Dutch embedded clause? Answer at end of chapter.

In summary, we have seen that word-order differences between languages may be captured in terms of the contrast between overt and covert feature-checking. We put forward the idea that features in the checking domain may be either strong or weak and that strong features must be eliminated by overt movement. All features in the checking domain must have been eliminated when the derivation reaches LF. The economy condition Procrastinate determines that overt movement is allowed if and only if it is necessary for convergence.

Language Acquisition in the Minimalist Program

We began this chapter with a number of questions that we took to be fundamental to the study of language. The first of these was "what kind of knowledge is represented in the mind of a speaker of the language?" The Minimalist Program answers this question by saying that knowledge of language consists of the combination of a lexicon and a computational procedure. The computational procedure has the properties that we have outlined. In particular, it consists of two operations, Merge and Move, which build two interface representation, LF and PF. Both the computation itself and the interface representations satisfy certain economy principles in an "optimal" way.

We must now address the second question that we posed, namely, how is this knowledge acquired by a native speaker? The Minimalist Program gives a radical answer to this question. The computational procedure and the conditions that apply to it are part of the initial state of the language faculty (i.e., innately specified) and determine to a large extent what constitutes a possible human language. The only source of variation among languages is to be found in the lexicon. The association of a particular word-meaning with a particular word-form is arbitrary and therefore unpredictable, but other kinds of variation in the lexicon are likely to be quite restricted. We do not, for instance, expect to find languages in which nouns can be inflected for tense. Similarly, the kinds of features that can be carried by nouns and verbs vary within narrow bounds. It is therefore to be expected that substantial parts of the mental lexicon are innately specified as well. Putting aside arbitrary lexical variation, the Minimalist Program restricts the possibilities for language variation to simple choices, called *parameters*, which are associated with items in the lexicon. An example of such a parameter is provided by the strength of features. We may think of the strength property as a binary valued parameter associated with features: a feature or features set is either strong or weak. Thus, the child acquiring English must find out that the specifier-features of I are strong rather than weak, as they are in Welsh. The child can set the strength parameter on the basis of positive evidence, on the assumption that the word order in the sentences that the child hears suffice to determine which phrases have moved where (see the section *The Theory of Universal Grammar* for discussion of the notions of positive and negative evidence).

Summing-up

We have sketched an approach to the study of language which postulates the existence of a richly structured component of the mind dedicated to language, called the language faculty, which consists of a lexicon and a computational procedure. The computational procedure selects words from the lexicon and constructs two interface representations: a "meaning"-representation (LF) and a "sound"-representation (PF). We discovered that natural languages have the curious property that phrases do not always show up in the position in which they are interpreted. This "displacement" of phrases was attributed to the existence of an operation called Move, whose application is driven by the requirement to eliminate uninterpretable features before the derivation reaches the interface levels. Both the interface representations and the computational procedure satisfy certain economy principles in an optimal way. According to this characterization of human language, knowledge of the structure of sentences is almost entirely innate with language variation restricted to parameters associated with items in the lexicon.

Further Reading

The Language Instinct by S. Pinker (1994), London: Allen Lane/The Penguin Press, is a very well-written, comprehensive and informative introduction to the subject of linguistics. *The Twitter Machine* by N.V. Smith (1989), Oxford: Blackwell, and *The Chomsky Update* by R. Salkie

(1990), London: Unwin Hyman, are enjoyable introductory texts providing insight into the nature of the scientific study of language. If you would like to find out more about the Minimalist Program and the framework it developed from, then a good place to start is *Government and Binding Theory and the Minimalist Program* by G. Webelhuth (ed.) (1995), Oxford: Blackwell. Those interested in phonology will find a very basic introduction in *Phonology: A Cognitive View* by J. Kaye (1989), Hillsdale, NJ: Lawrence Erlbaum and an accessible but more comprehensive introduction in *English Sound Structure* by J. Harris (1994), Oxford: Blackwell. A very accessible, non-technical introduction to semantics can be found in *Realms of Meaning* by Th.R. Hofmann (1993), London: Longman. For those who are looking for a somewhat more demanding read, there is *Formal Semantics* by R. Cann (1993), Cambridge: Cambridge University Press, which is an excellent introduction to mainstream formal semantics. *Understanding Utterances* by D. Blakemore (1992), Oxford: Blackwell, is a good introduction to the study of pragmatics.

Answers to SAQs

SAQ 7.2

There are many possible answers to this question. Here are two examples:

By forming conjunctions :

Bill left
Bill and Harry left
Bill, Harry, and Susan left
Bill, Harry, Susan, and Jack left
Etc.

By embedding one relative clause inside another :

This is the linguist who knows Bill
This is the linguist who met a man who knows Bill
This is the linguist who met a man who met a man who knows Bill
This is the linguist who met a man who met a man who met a man who knows Bill
Etc.

SAQ 7.3

Because the rule for forming past tense verbs attaches the ending *-ed* indiscriminately to any lexical item of the category V, including *wented* .

SAQ 7.4

(i) The P *from* c-commands the N *Havana* , because P does not dominate N, N does not dominate P and the first branching node dominating P (PP) also dominates N.
(ii) The N *man* does not c-command the D *the* , because the first branching node dominating N (NP) does not dominate D.
(iii) The DP-node does not c-command the N *Havana* , because it dominates it.

SAQ 7.5

Draw tree structures for examples (9a) and (10b). Does John c-command himself in (9a)? And in (10b)?

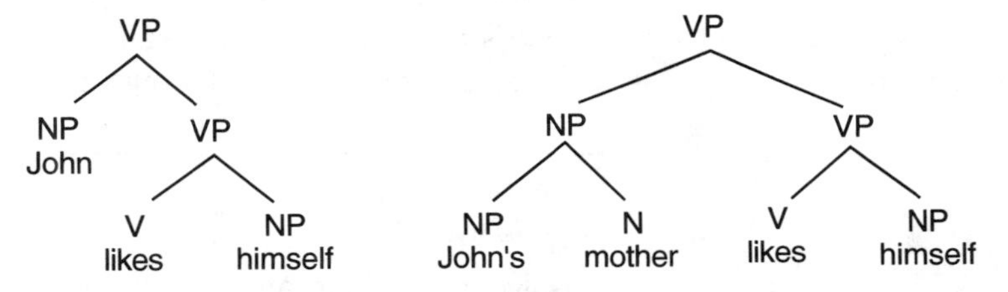

7.14 The tree-structures for *John likes himself* and *John's mother likes himself.*

Later on in the chapter these somewhat simplified tree structures will be replaced by more complex ones, but the c-command relations between *John* and *himself* are unaffected by this. *John* c-commands *himself* in the tree structure for (9a) on the left but not in the tree structure for (10b) on the right.

SAQ 7.6

What further refinement to the principle of Anaphor Interpretation given in (13) is suggested by the following examples?

> *John likes himself*
> **John said that Mary likes himself*

It appears that the antecedent of an anaphor must be local in some sense. From the examples given here one might draw the preliminary conclusion that "local" means "in the same clause."

SAQ 7.7
Past tense is realized on verbs by the ending *-ed* .
Progressive aspect is realized on verbs by the ending *-ing* .

SAQ 7.9
How does the following example lend support to our conclusion that infinitival clauses may have silent subjects?

To cut oneself is painful

Recall that an anaphor such as *oneself* needs a local c-commanding antecedent. If the infinitival clause has an empty subject, so that its structure is:

ec to cut oneself is painful

then the *ec* can serve as the anaphor's antecedent and the principle of anaphor interpretation is satisfied. Without the *ec*, the sentence violates this principle and would be incorrectly ruled out.

SAQ 7.10
Passive *hit* assigns only one θ-role, while active *hit* assigns two.

SAQ 7.11
Given Full Interpretation, is it necessary to have the θ-Criterion as a separate principle holding of LF- representations?
If we assume that an argument lacking a unique θ-role (i.e., lacking one altogether or having more than one) cannot be interpreted at LF, then it follows that such an argument violates Full Interpretation. Similarly, if we assume that a θ-role which has not been assigned to an argument cannot be interpreted at LF, then an unassigned θ-role violates Full Interpretation as well. Therefore, on these assumptions, the θ-Criterion can be dispensed with.

SAQ 7.14
Which feature sets are strong and which are weak in a Dutch embedded clause?
Example (35) suggests that both the subject and the object are

outside the VP. The object precedes the verb, so must have moved out of the VP. The subject precedes the object, so it must have moved as well. It therefore must be the case that the specifier-features of I and AGR are strong. It is not clear from this example whether the verb has raised to AGR or not. But it certainly has not raised to I or it would precede the object. Therefore, the head-features of I must be weak.

Exercises

Exercise 1

It is often claimed that chimpanzees are capable of learning human language. Read up on this subject and evaluate the evidence for this claim. The following references should help you get going.

Gardner, R.A. and B.T. Gardner (1969). Teaching Sign Language to a Chimpanzee. *Science* 165: 664–672.

Patterson, F.G. (1978). The Gestures of a Gorilla: Language Acquisition in another Pongid. *Brain and Language* 5: 56–71.

Wallman, J. (1992). *Aping Language*. New York: Cambridge University Press.

Exercise 2

Draw a tree diagram for the sentence *John is intelligent*. In many languages John and intelligent would show agreement in a sentence like this. Consider the following examples from French:

(i) Jean est intelligent
(ii) Marie est intelligente

How would you revise your structure for *John is intelligent* in view of these facts? What movements must be assumed to take place?

Exercise 3

Consider the following examples:

(i) John_1 likes him_2
(ii) * John_1 likes him_1
(iii) John_1's mother likes him_1
(iv) John_1 said that Peter likes him_1

Propose a Principle of Pronoun Interpretation, which will account for these facts. What conclusions can be drawn about the distribution of anaphors and pronouns?

Exercise 4

We have seen that the operation Move is subject to the economy condition *Procrastinate*, which says that categories should move as late as possible. Now consider the following examples:

(i) * John seems that it is certain to win
(ii) * What do you wonder to whom John gave?

Propose a condition on Move which will ban the derivation of examples like these.

CHAPTER

8 Meaning and Conversation

Outline

In this chapter we first consider the nature of the meanings of words. Our approach is guided by the need to ensure that language, mind, and the world connect up. We adopt the view that language provides a means to build mental models based on a set of innate and basic elements. The language individuals use should allow ready contact with their models of the situation. We consider how individuals develop a shared language to allow effective reference by exploring two experimental tasks. The intended meaning of an utterance frequently diverges from what is literally said. This poses a theoretical challenge and we discuss the nature of the interpretation processes involved in understanding metaphor and irony.

Learning Objectives

After reading this chapter you should be able to:

- appreciate the contrast between semantics and pragmatics
- outline a theory of word meaning
- understand the concept of grounding in conversations
- appreciate the role of a mental model in the understanding and production of language
- consider how individuals coordinate their use of terms
- appreciate the processes involved in understanding metaphor and irony

Key Terms

- contribution model
- coordination
- grounding
- irony
- metaphor
- relevance theory
- subconcepts

Introduction

> A common language connects the members of a community into an information-sharing network with formidable collective powers.
>
> (S. Pinker *The Language Instinct* (1994, p. 16))

Conversation is of fundamental importance: as social agents, we learn from others and coordinate with them to achieve certain tasks. In the scenario, for instance, the family acts together to have breakfast. In doing so, the participants take turns in speaking. Some utterances are based around the task directly: "Oh, can you get her some coco-pops – I'm seeing to the toast." Others request information: "Where did she go last year? I've completely forgotten." The reply serves to construct a shared memory. Other talk aims to achieve effects indirectly. Dad says "Toast! Lucy! It's Etna in here!" Clearly he does not mean that Mount Etna is in the kitchen. Other talk involves establishing the legitimacy of certain actions (see Potter, Edwards, and Wetherwell, 1993, for research on talk as a form of social action). Lucy wants to eat at Jane's place. Dad is seemingly not persuaded by Lucy's attempt to influence him. In this chapter we address the following questions: what do we mean by meaning? How do individuals achieve their goals through conversation? Why do they use the language they do?

The Conversational Problem: Code and Inference

What problem or problems do conversationalists face? In order to communicate linguistically individuals need to share a common code – they have to know the meanings or senses of words and their sounds. A linguistic code pairs senses with sounds. It is at least arguable that the system which achieves this mapping is a module in Fodor's terms (see chapter 3). However, in order to understand an utterance, its **context** also has to be considered. The same words can convey many different thoughts – which one was intended?

SAQ 8.1 Consider the utterance "I'll go" from the scenario. Imagine it uttered in three different contexts. What do you take it to mean in each case?

Understanding the nature of the processes that fill out the linguistic meaning of a sentence is the domain of pragmatics and such processes cannot be modular in Fodor's terms because they draw on general knowledge (chapter 3). Let us consider another example from the scenario so the contrast between the meaning of words and the **intended meaning** or significance of an utterance is clear. Mom says: "She's worked every night, John." In context, it is clear Dad will take Mom to have communicated that Lucy has done schoolwork every night of that week and should therefore be allowed a night off to see her friend Jane. However, the linguistic string "She's worked every night" most certainly does not encode all that! In another context, it might be taken to refer to another female person who has worked every night in some other way and imply that this person is a workaholic. The decoded content of Mom's utterance, that is the "meaning" or semantics of the sentence she used, is a product of the meaning of the individual words and the syntactic structure in which they are combined: this meaning is invariant across different contexts of use. The meaning actually conveyed is, of course, highly dependent on context, including previous utterances in the conversation and particular items of knowledge available to participants at the time. In order to understand how conversations work, we first have to consider the meanings of words. We can then go on to consider some aspect of the pragmatic processes. Further background and discussion of the semantic/pragmatic distinction can be found in chapter 9. What do we mean by the meaning of words? In order to think about the meaning of words, we have to address a deep question: how do language, the mind, and the world connect up?

The Meaning of Words

There are two partial answers which we record only to dismiss. First, it might be felt that the meaning of a word is what it refers to in the world. But this cannot be right. The meanings of "the young child in the scenario" and "Lucy's sister" are not the same though they both refer to Zara. Nonetheless, what an expression specifically refers to is very important as we shall see. Second, we might think that the meaning of a word is equivalent to the mental image that it evokes. One problem with this theory is that an image is particular. On reading about a "cat" one person might imagine a sleepy Siamese and another a little tabby. But does the word "cat" really mean something different for the two individuals? We need to distinguish the sense of a word from its reference and from individuals' associations. The sense of a word is the concept associated with the word. The reference of a word is a set of things in the world to which it can be applied. The concepts of a "young child" and "sister" are different though both can refer to the same individual.

In order to develop an appropriate view of word meaning we have to consider what individuals are trying to do by uttering such expressions. Individuals are trying to get their **addressees** to pick out some entity. Once you have read the scenario, then the expression "The young child" should allow you to pick out Zara uniquely because the referent falls under the concepts of "young" and "child" and is the only member of that class in the context as required by "the." As an addressee you can then search your mental representation for this unique referent. The next question is this: what is the nature of concepts such as "child" and "sister"?

There is a long tradition according to which lexical items can be defined in terms of certain basic concepts or semantic components which fully define them. For example, certain nouns like "mother" do seem to be definable in terms of a set of more basic components. We might define mother as adult, female, parent. Likewise, we could define bachelor as unmarried, adult, male. Certain verbs also seem to be readily decomposable. For example, the verb "kill" could perhaps be defined as "cause to become not alive."

SAQ 8.2 Try to formulate definitions of "triangle" and "sell." The concepts you use should each be necessary properties of anything to which the lexical items can apply and they should be jointly sufficient in that they should exclude anything to which the words cannot be correctly applied.

However, this sort of enterprise has met with great problems (see Fodor, Garrett, Walker, and Parkes, 1980), not least of which is that the vast bulk of words in the language do not seem to be definable in terms of necessary and sufficient components. Consider, for example, the meanings of words used in the scenario such as "coffee," "toast," and "milk." Nouns with more abstract meanings seem even less obviously definable in such a fashion (consider "thought" and "surprise"). Verbs generally fare no better – it is difficult to extract a complete set of components.

There are a number of possible responses to such an outcome. One possibility is to suppose that there are no basic concepts at all. Words stand in a one-to-one relation with concepts (Fodor et al., 1980). On this account, understanding the meaning of an utterance involves the translation of sentence content into some mental language with tokens for each concept. Consider Lucy's question: "Has anyone seen the tray?" It might be translated into a representation of the basic idea (a proposition, see chapter 2: box 2.1) consisting of tokens of concepts SEE (ANYONE, TRAY). But this translation can only be half the story because it provides no contact with the world: it does not identify the specific referents in the world. In understanding the question, other members of the family recognize a request to search their memory or the immediate environment for an object with certain properties.

How then shall we proceed? One step we can make is to acknowledge that there are different kinds of concepts (Johnson-Laird, 1993). Some concepts are analytic and so can indeed be defined in terms of necessary and sufficient components of some kind (e.g., triangle). However, other kinds of concepts (e.g., animals and other natural kinds) are not like this. In such cases, individuals may know how to recognize these instances and their knowledge of these concepts can be of three sorts: they may know which properties are necessary (if any); they may know which properties normally occur (default properties); and they may know those properties that vary quite freely. A prototypical cat has all the necessary and default values of "cat." Another class of concepts is exemplified by "games" or "tables," man-made objects or inventions (constructive concepts). What counts as an example of a particular natural kind or constructive concept depends not only on the word's meaning but also on how the concept contrasts with other related concepts (Miller and Johnson-Laird, 1976). Whether a person calls something a "cat" depends on its similarity to a typical cat as well as its relationship to other related concepts such as a typical dog or lion. So the upshot of this is that there can be vagueness in specification. Expertise can be invoked where necessary. In consequence, the failure to find necessary and sufficient components for all concepts should not be taken to mean that the search for fundamental conceptual components is vacuous. On the contrary, it makes both evolutionary and computational sense to suppose that our thoughts are assembled as a set of basic components (see also chapter 10, "How are Concepts Represented in Memory"). It is vital though that these components or **subconcepts** (Johnson-Laird, 1993) relate to the world in some way.

Subconcepts are the means by which people think, that is construct propositions about the world. Such propositions allow them to create mental models that are specific and explicit instances of their thoughts. A mental model, as described in chapter 1, corresponds to the structure of the situation described (the entities, their properties, and relations) rather than to the linguistic structure of the sentence. Language, mind, and the world interrelate then because language allows the construction of a mental model. This model can contact the models created through perception. We view the process of understanding a sentence as a process by which individuals come to know the situation truthfully described by it.

Let us consider a concrete example. To understand an utterance such as "The circle is on the right of the star" a proposition is assembled that can be used to build a model of the situation. This model can be used to contact a perceptual model and hence to determine the truth or otherwise of the assertion. But what is the meaning of the relation "on the right of." This is based on a set of subconcepts that locate an object in space. Given three dimensions: left–right; front–back; and up–down, then, 1 0 0 indicates that the object should be located by incrementing its value by one on the left–right dimension, but holding its value constant on the other two dimensions. "On the right of" is part of a system of spatial relations that includes "in front of" and "above" based on the same set of subconcepts (Johnson-Laird, 1993). Table 8.1 displays this system of relations.

We assume a language processor uses a set of subconcepts to assemble the meaning of the sentence. On hearing "The circle," the subconcept underlying circle is used to build a model:

○

Given the sentence, "The circle is on the right of the star," subconcepts are combined to yield the meaning of the sentence represented by the proposition

Table 8.1 System of spatial relations based on subconcepts

	Dimension		
	Left–right	Front–back	Up–down
on the right of	1	0	0
on the left of	–1	0	0
in front of	0	1	0
behind	0	–1	0
above	0	0	1
below	0	0	–1

Adapted from Johnson-Laird, (1993).

in which the relation "on the right of" is captured by the subconcept (100) and the subconcepts for circle and star :

((100) (○) (*))

This representation can then be used to construct a mental model of the state of affairs:

* ○

On this view, everyday concepts are akin to instructions in a high-level programming language: understanding and thinking requires compiling them into subconcepts. These subconcepts are used in procedures for constructing and manipulating models and so link up directly with the primitive mental operations used in perceiving, attending, and intending (see Clark and Clark, 1977, p. 442). The task of determining the set of subconcepts is immense but it can be guided by the need to ensure that subconcepts are useful in constructing mental models. The truth of the utterance "The circle is to the right of the star" or rather the proposition expressed by it, can be determined by embedding the model based on the utterance in the perceptual model of the situation. The objects and their relations can then be compared.

Just one complication we have glossed over: an assertion such as "A circle is to the right of a star" is true of an infinite number of different situations (how far is the circle to the right of the star? What is the size of the circle and star?). Of course, an infinite number of models cannot be constructed, nor could they be maintained in a finite memory. An effective solution to this problem (Johnson-Laird, 1983) is to suppose that individuals construct just one model which serves as a representative example of the set of models and, indeed, for the model the speaker had in mind. If the model is incorrect it can nonetheless be revised. So, for example, an initial model might contain default values for an entity which might turn out to be incorrect.

What is the evidence that understanding involves both linguistic/propositional representations and model representations? There is considerable experimental support for the position (see, e.g., Stevenson, 1993). We illustrate, here, with some linguistic examples. Listeners must retain some information about the form of an utterance if they are to interpret certain kinds of utterance correctly. Consider, for example:

The boys were chasing the girls
The dogs were too

The interpretation of the elliptical utterance "The dogs were too" depends on the form of the first utterance. If it had been "The girls were being chased by the boys," then the boys would have been chasing the dogs as well. In the case of Mom's utterance: "She's worked every night" the pronoun "she" refers not to a form of words but to a representation in a mental model. Each new utterance is interpreted with respect to that model and amends the state of that model. Since it is a model of the world it contains representations of people, things, and context. A reference in an utterance or text using a definite pronoun ("she"/"he") refers to a token or element that has been or is to be treated as being in the model constructed to date. In this case, "she" refers to a mental element or token standing for Lucy in the model. The following example (Garnham and Oakhill, 1992) may help to make the point clearer:

Last night we went to hear a new jazz band.
They played for nearly five hours.

In this example, there is an indefinite noun phrase (a new jazz band) which introduces a single referent (the band). However, the next utterance, refers to it using the pronoun "they." This pronoun does not refer to the band as such, but to the related concept of members of the band. Experimental research indicates that individuals establish a token for band from the start indicating that it comprises more than one individual.

The kind of account we have outlined here provides a sketch of the way in which language, mind, and the world connect up. The explicit content of an utterance is usually just a point of departure to capture a state of affairs. The addressee has to fill out the details. The appropriate domain of such an account is not natural language sentences, but propositional representations which permit the construction of models. These models can then make direct contact with models of states of affairs in the world.

Informative and Communicative Intentions

We have said a little about how we view the meanings of words. What broad factors need to be considered if we are to understand how individuals determine the intended meaning of an utterance? Suppose Lucy wanted to communicate to her friend Jane that she had a sore throat. If she had one, she could simply speak and Jane could realize it. She would do so, most probably, even if Lucy did not intend her to realize this fact. Suppose that Lucy wanted to inform Jane that she had a sore throat yesterday. She cannot give her direct evidence since she no longer has one. Instead she could say: "I had a sore throat yesterday." Her utterance is directly caused not by yesterday's sore throat, but by her intention to inform Jane of this fact. So her utterance is direct evidence of her informative intention. This utterance, an example of ostensive-inferential communication (Sperber and Wilson, 1986, p. 50–64), consists of two intentions: an informative intention and a communicative intention. Information is pointed out by saying something (this is the informative intention) and this intention is intended to be recognized (this is the communicative intention). Conversations are prototypical instances of ostensive-inferential communication.

According to **relevance theory** (Sperber and Wilson, 1986), individuals pay attention to the most relevant phenomena, that is, phenomena which are likely to have **cognitive/contextual effects** without costing too much **processing effort**. By issuing an utterance speakers claim the listener's attention and thereby suggest that the information is relevant enough to be worth attending to. Ostensively communicated information, in this view, comes with a guarantee of relevance and this is a crucial factor in determining intended meaning. We shall presuppose that utterances are treated as relevant and will consider some implications of this account in a later section on figurative language. First though, we try to capture the nature of the conversational process.

Conversation

What is needed for successful conversation? One possibility, the autonomous view (Schober and Clark, 1989), is that each participant in a conversation should produce the right utterance at the right time. Sperber and Wilson (1986, p. 43) envisage that the responsibility is indeed asymmetric. They use a ballroom analogy (one partner takes responsibility for leading). It is the speaker's responsibility to assess the contextual information and to choose an appropriate utterance to elicit the desired effect. Such a view can be contrasted with one where the participants work together to establish the mutual belief that the addressee has achieved satisfactory understanding of the speaker's meaning. This process of collaboration is termed **grounding** (Clark and Schaefer, 1987). Consider the following excerpt from Svartvik and Quirk (1980; cited in Schober and Clark):

A: well wo uh what shall we do about this boy then –
B: Duveen?
A: m
B: well I propose to write uh saying. I'm very sorry I cannot – uh teach at the institute

Here B checks the intended referent of A's question which A then acknowledges. Only then does B continue. Now although individuals may try to design their utterances so that they are mutually understood, they may need to perform additional checks to ensure that what has been said has been understood. This is an essential part of the grounding process and what makes conversations acts of collaboration. Consider B's utterance, "Duveen?", in the earlier sequence: B did not continue until he/she had checked the intended referent and received confirmation. Participants do not seek perfect understanding, rather they seek the mutual belief that the partners have understood what the contributor meant to a criterion sufficient for current purposes (Clark and Wilkes-Gibbs, 1986). So in the context of a task where there is considerable danger if an error is made, checking will be more rigorous than in a situation where precision is not important.

How might this process work? Clark and colleagues (e.g., Clark and Schaefer, 1989) proposed a **contribution model** based on work on turn taking and repairs in conversations (e.g., Sacks, Schegloff, and Jefferson, 1974). In this model, contributions divide into two phases: a presentation phase and an acceptance phase.

Presentation phase
A: well wo uh what shall we do about this boy then –
Acceptance phase
B: Duveen?
A: m
B: well I propose to write uh saying. I'm very sorry I cannot – uh teach at the institute

This excerpt, establishes a difference between merely uttering some words and achieving a communicative effect, such as asking a question about a specific person. Notice that the acceptance involves a further sequence "Duveen?", which itself is a presentation followed by an acceptance "m." This sequence is embedded within the acceptance phase of the main contribution. You might assume then that typically, individuals only need to be on the look out for evidence that they have been misheard or misunderstood. In the absence of such feedback they simply assume by default that they have been understood.

According to the contribution model, however, individuals often search for positive evidence. For example, speakers might seek so-called back-channel responses ("yeah," "uh-uh," "m"). A second form of positive evidence is the initiation of a relevant next turn such as in the answer to a previously posed question. In the scenario, John says: "Florida" in response to a question about where Granny went for her holiday last year. If the answer is relevant to the question, then it is also evidence that the question has been understood. Indeed this might typically be the way of signaling acceptance. The third form of positive evidence is continued attention, but this could also deliver ambiguous information: individuals might be hoping that the utterance will make sense to them, eventually, as they listen further.

SAQ 8.3 If conversations are collaborative, what other kinds of evidence of the attempt to coordinate might you expect to find?

Are Conversations Collaborative?

If conversations are really acts of collaboration, then we should see evidence of it in the way that speakers formulate their utterances. Speakers in real conversations may have limited time for planning. Accordingly, they may start an utterance then repair it; they may hesitate so as to allow their addressee to complete the utterance. According to Clark and Wilkes-Gibbs (1986), therefore, the appropriate way to capture the workings of the process of presentation and acceptance is the principle of least collaborative effort: participants try to minimize the cognitive work both do from initiating to accepting an utterance.

Such a view contrasts with one where it is the speaker's responsibility to create utterances which addressees will understand without difficulty (cf. Sperber and Wilson, 1986). According to this model, an instance of the autonomous view mentioned above, reference is expressed via three types of noun phrases: a proper noun (Ronald Reagan, King Kong); a definite description (that book, next week); or a pronoun (we, they). The noun phrase is used to allow the addressee to identify the referent uniquely against the common ground (i.e., the mental models of the participants), and merely issuing the noun phrase completes the intention. The speaker alone according to this model controls the conversational process. In fact, speakers do issue improper utterances (Clark and Brennan, 1991) as will become clear from work on **referential communication**. But this is not to deny that there can be occasions where individuals choose their words most carefully.

Referential Communication

Experimental research provides an important way to create circumstances in which to observe collaborative processes in conversation. In one procedure, developed from earlier work by Krauss and Weinheimer (1964), Clark and colleagues required pairs of subjects to communicate about the order of a set of abstract figures (e.g., Clark and Wilkes-Gibbs, 1986). Some examples of these Tangram figures are shown in figure 8.1. One subject (the director) has a set of 12 such figures arranged in two rows of six. Separated by an opaque screen, the other subject (the matcher) has the same set of figures randomly allocated to the two rows. The director describes each figure so that the matcher can order his or her Tangram figures in the same way. Having completed the task once, the materials are randomized over the two rows and the procedure repeated.

What expectations does the collaborative view suggest? It is plausible to expect that more words will be required to agree a figure on the first trial

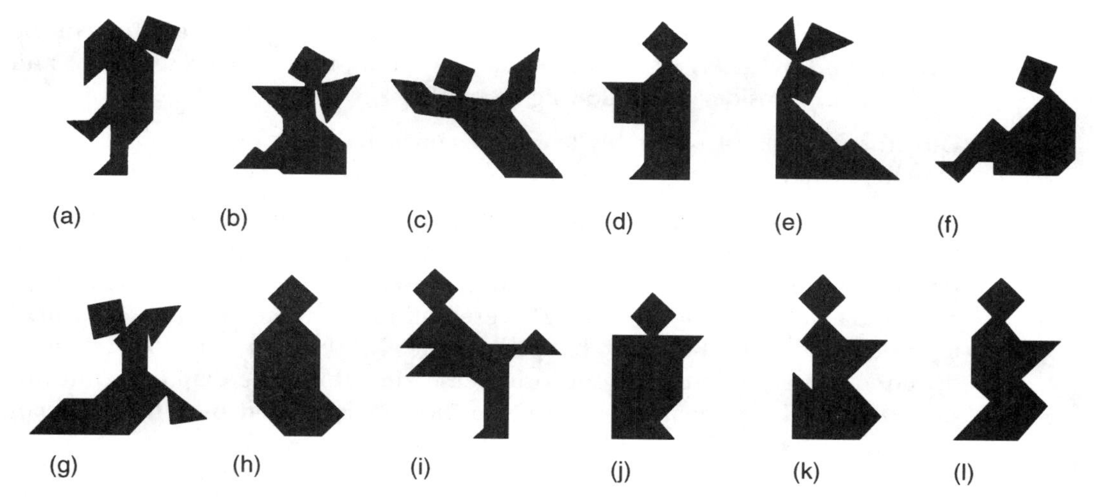

8.1 Tangram figures used in a referential communication task by Clark & Wilkes-Gibbs (1986). Reproduced with permission of the author and publisher.

because the director may have to use non-standard phrases. Subsequently the director can assume that the matcher can recall a previous description. This can then be referenced in a more standard fashion. In addition, later trials in the experiment might be expected to yield fewer turns from acceptance to presentation.

Both of these expectations were borne out by the data. An example of a simple sequence of changes over the six trials is shown below:

All right, the next one looks like a person who's ice skating,
except they're sticking two arms out in front.
Um, the next one's the person ice skating that has two arms?
The fourth one is the person ice skating, with two arms.
The next one's the ice skater.
The fourth one's the ice skater.
The ice skater

This sequence illustrates that directors initially describe the figure with an indefinite reference ("a person who ... "). In subsequent trials they refer to the figure with a definite description and move from non-standard noun phrases (e.g., trial 2) to standard noun phrases (e.g., "the ice skater"). In fact, an average of 41 words per figure was required to ensure acceptance on the first trial compared to an average of only eight words in the last trial. The number of turns to secure acceptance of each figure also declined from just below four turns for the first trial to one turn by the last trial. The participants come to rely on descriptions mutually accepted on previous trials.

Let us look at the process in more detail by considering the utterances of the director and matcher: we should expect to see a process of presentation and acceptance. Consider the following example:

Director: Number 4 is the guy leaning against the tree
Matcher: Okay

In this exchange the matcher uses "Okay" to assert that she believes that she has identified the correct figure and placed it correctly, and in doing so she presupposes acceptance of the director's presentation. There is no need here for the parties to rework the initial presentation. Such an exchange is termed a basic exchange. Basic exchanges are likely, therefore, to be relatively infrequent on the first trial, but to increase after that. The data bore out this expectation; the percentage of basic exchanges increased over the six trials (18, 55, 75, 80, 88, 84).

The experiment also indicates more about the process of mutually accepting a reference consistent with a collaborative view. Three processes can be discerned: initiating a reference; refashioning a reference; and evaluating the presentation. We will illustrate just one of these: refashioning.

Refashioning

Self-repair, expansion, and replacement are different ways to refashion. In the communication task, self-repairs (such as, "Um, next one is the guy, the person with his head to the right but his legs are, his one leg is kicked up to the left") declined over trials. There were also instances of covert repairs: "Okay, number, uh, 4 is the, is the kind of fat one with legs to the left – er, I mean, to the right." In repeating "the" the director may have been repairing something they were about to say. An uttered phrase may also be expanded. Such expansion might be initiated by the director as in:

Director: Okay, number 1 is just the kind of block-like figure
with the jagged right-hand side. The left side looks like a square.

or they might be triggered by the matcher signaling the need for expansion as in:

Director: Okay, the next one is the rabbit.
Matcher: Uh –
Director: That's asleep, you know, it looks like it's got ears and a head pointing down?
Matcher: Okay.

The third form of refashioning involves the matcher presenting a noun phrase of her own, as in the following example:

Director: Okay, and the next one is the person that looks like they're carrying something, and it's sticking out to the left. It looks like a hat that's upside down.
Matcher: the guy that's pointing to the left again?

Director: Yeah, pointing to the left, that's it! (laughs)
Matcher: Okay.

In this case, the two parties accepted an alternative description of the figure presented by the matcher.

Referential Descriptions

The communication task provides a rich source of data on another front. What kinds of content do people use to secure reference? The initial description provided a perspective on the figure which could then be mutually accepted. This choice, Clark and Wilkes-Gibbs supposed, serves to minimize collaborative effort because unlike referring to a common object, such as a chair or shoe, no common perspective could be assumed to be part of the common ground of the participants.

A particular figure can be described by identifying the linkage between its component parts (a segmental perspective: "Um, it's a, oh, hexagonal shape, and then on the bottom right side it has this diamond"), or the object can be conceived of as a whole (an holistic perspective). For example, individuals may use an analogy to natural objects (" . . . looks like a girl dancing sort of"). Which kind of perspective should be preferred if participants seek to minimize collaborative effort? Holistic perspectives are likely to be shorter. In fact, on the first trial directors chose an analogy for a figure (to reflect the way they saw it) but also described the component parts. By the sixth trial, however, most figures were described from an analogical perspective alone.

Logically, a definite description can be made in one of two ways: individuals could refer to previously established perspectives ("the rabbit"; "the person with his arms up") or they could refer to procedures associated with the previous reference (e.g., "the one we got confused on last time" or "the first one from last time"). Which kinds of descriptions should be preferred? From the point of view of minimizing collaborative effort, participants should prefer referring to the permanent properties of objects rather than to their temporary properties (such as their position in the sequence), which change over trials and so may become confusable. Overall, 90 percent of references in the Clark and Wilkes-Gibbs study were based on permanent properties such as shape and appearance. This allows the tokens in the mental models to have these properties too.

In a nice extension to the Tangram study, Schober and Clark (1989) examined the performance of matchers compared to that of overhearers of the same conversation. According to the collaborative view, individuals in a conversation actively collaborate to ground their utterance. On this view, individuals overhearing a conversation should be at a disadvantage because they have not been able to participate in this process. Whereas the matcher in the communication task can reach understanding through collaboration, the overhearer can only do it by conjecture. The overhearer cannot signal further refashioning nor can the overhearer contribute to the process of framing the perspectives. If it is not a cogent perspective for her – too bad. Over trials such difficulties may accumulate. Schober and Clark confirmed this prediction.

This experimental task provides a rich picture both of the process of referential communication and of the content of the definite descriptions. Of course, the task itself imposes some constraints and may not be suited to studying the development of referential skills (see chapter 9). Certain cues such as eye gaze and facial expression cannot be used to convey agreement or disagreement. But what it demonstrates unequivocally is that pairs work together to secure a definite reference. They collaborate to ensure the mutual belief that the matcher has understood what the director meant so as to allow the correct figure to be identified. The processes involved in grounding are central not only in these experimental situations but also in other task-oriented dialogues such as telephone calls to directory enquiries (Clark and Schaefer, 1987) and in ordinary conversation (Clark and Schaefer, 1989). Collaboration establishes a suitable referential language unique to the conversation. Such collaborative effort may reduce the need to draw on large amounts of general knowledge since the language used, more or less guarantees correct choice of referent within the mental model.

Coordination and Mental Models

Individuals work to refine their language, then, to suit their communication goals and change the nature of the descriptive language used over the course of the task in order to access their models of the situation as effectively as possible. This process has been studied in detail in another task by Garrod and colleagues (Garrod and Anderson, 1987; Garrod and Doherty, 1994), and this provides further insight into the nature of coordination. In their task individuals had to cooperate in order to find their way through a computer-controlled maze which consisted of box-like structures connected by paths, along which players could move their position markers. Players had to move these markers to the goal position. In a particular game both players had the same type of maze but each player could only see her own start position, goal, and move marker.

The cooperative nature of the game arose from the fact that some of the paths contained gates which blocked movement and some of the box-like structures contained switches. Gates and switches were distributed differently for the two players. In order to overcome obstacles, verbal cooperation was required since the principle of the game was that if a player (A) entered a box with a switch marked on B's screen, then the entire configuration of B's gates would change: all paths previously open would be gated or paths previously gated would be open. In order to change a gate, therefore, B has to guide A to a switch box, which is only marked on B's screen. The two participants have to verbally cooperate in order for both of them to get to their own goal positions. A game consists of the players attempting to reach their respective goals with

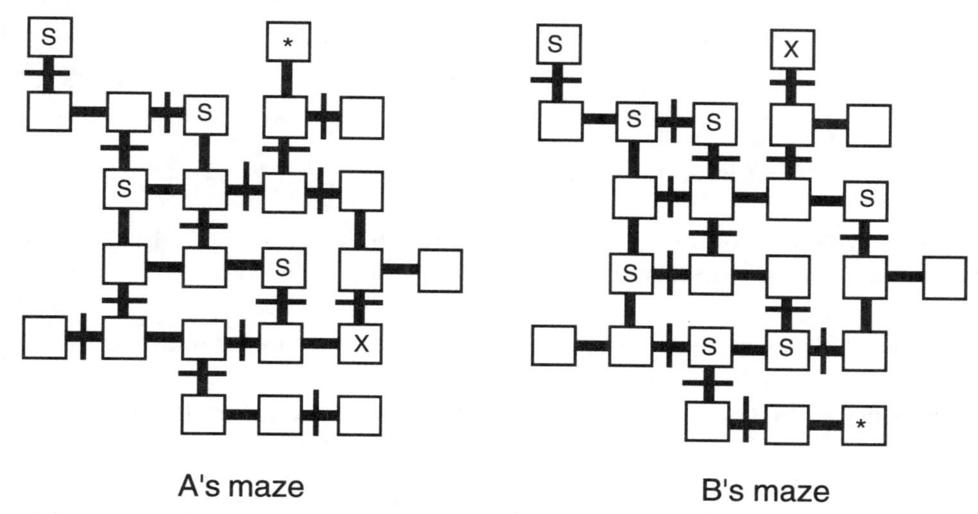

8.2 Schematic diagram of the mazes used by Garrod & Anderson (1987). Boxes containing Ss indicate switches, whereas those containing an X indicate the players' current positions and asterisk the players' goal positions. Reproduced with permission of the author and publisher from S. Garrod and D. Doherty (1994). Conversation, co-ordination and convention: an empirical investigation into how groups establish linguistic conventions, *Cognition*, 53, 181–215.

dialogue intervening between moves. This dialogue contains descriptions of the players' current positions, switch locations, and goal positions.

In order to coordinate, participants in the maze game have to reach a way of describing the mazes – a descriptive scheme. Unlike the Tangram task, an overall scheme is needed and this must reflect an appropriate model of the maze.

SAQ 8.4 Consider a number of ways in which a location in one of the mazes could be described.

In a path description, the speaker invites the listener on a small tour until they reach the target point, as in the example below:

> See the bottom right, go two along and two up. That's where I am.

The next most common type established an abstract frame of reference. The maze was viewed as a kind of grid as in the next two examples:

> I'm on the third row and fourth column
> I'm at C4

In the latter case, letters were used to designate vertical lines, and numbers used to designate horizontal lines. A related type designated a position by referring to a line, and then described the position relative to that line as in:

> Third bottom line, third box from the right.

A final type also occurred. Individuals identified a figure in the box-like structures and then indicated their position with respect to it:

> See the rectangle at the bottom right, I'm at the top left-hand corner.

The data revealed a number of patterns. First, speakers within the same game rapidly adopted the same scheme and this process increased as the dialogue progressed. Second, there was a shift in the descriptive scheme used. Path and figural descriptions were common in the first game whereas by the second game, line and grid schemes predominated. Individuals moved from a perceptually based scheme to a more abstract scheme which provided greater flexibility.

Mental Models and Descriptive Schemes

What underlies the different descriptive schemes? The different schemes reflect different underlying mental models of the maze. Each model captures different aspects of its spatial organization and functional properties. In order to produce and to understand descriptions individuals have to establish a restricted set of terms which may be mapped onto their mental model of the maze. Presentation and acceptance sequences are also evident in these materials.

Consider a dialogue in which participants refer to the maze as comprising a set of figures:

A: You know the extreme right, there's one box
B: Yeah, right, the extreme right it's sticking out like a sore thumb
A: That's where I am
B: It's like a right indicator?
A: Yes, and where are you?

The participants subsequently described positions with respect to the various "right indicators" that they perceived in the figure. They developed a special language suited to their initially adopted model. In the next sequence B is trying to describe his position to A based on this model.

B: I'm er, well you know how, well it's difficult ehm, you know
where the middle right indicator is?

A: Yes
B: Well count that middle right indicator as a box
A: Mm
B: Then move to the left, that's one box, two boxes, three boxes,
four boxes, five boxes, right?
A: Yes
B: I'm in the fourth box
A: I'm lost, I'm lost, I'm lost

In short, the adoption of an initial model leads to specific types of description which can have consequences for the ease of conveying information. In a path-type description, the underlying model is one in which positions are related in a network corresponding to the actual paths in the maze. This means that certain points which are physically close may be difficult to describe because they are remote from one another in terms of traceable paths. Consider, for instance, a dialogue in which the path description scheme is adopted. The description starts by identifying a salient point and traces a path to the desired location.

A: Right, see the bottom left-hand corner
B: The bottom left
A: There's a box and then there's a gap
B: Uh-huh
A: And there there's a box and then there's another box
B: Uh-huh
A: I'm right there

In reporting their position, which is just adjacent to the other one, the description still follows the paths rather than ignoring them:

A: I'm one to the right, then one up, then there's a gap, right?
B: Uh-huh
A: I'm just in the box above the gap

In another description scheme (line description) speakers specify a position by referring to the horizontal or vertical line of points on which it is located. Here the maze is interpreted with respect to a more abstract model, which sees it as consisting of a set of horizontal rows and vertical columns related in various ways. This model gives rise to a restricted set of language terms. As the researchers noted, a horizontal line could be referred to exclusively as a "row," "line," "level," or "floor" depending upon the dialogue pair.

Most economical was a description based on the coordinate model comprising row elements and column elements. In this case only relations between columns and rows need to be represented as in a matrix. One pair develop an economical description by referring to the columns as A, B, C, D, E, F and the rows as 1, 2, 3, 4, 5, 6. Having established the model and the description language, they could then share positional information.

A: . . . I'm presently at C5 OK

B: E1

This model and language scheme yields economical solutions to the problem of coordinating action to achieve goals.

How in such dialogues can participants minimize collaborative effort? Minimizing collaborative effort means that participants should formulate their utterances so that they avoid unnecessary time and effort. One way to achieve this is to "formulate your utterance according to the same principles of interpretation as those needed to understand the last utterance from the other participant." Garrod and Anderson term this principle output/input coordination. Such a principle implies that both participants will be locally consistent. Note that in dialogue each participant takes the roles of both speaker and listener. If individuals interpret the input against some model of the situation then their production may be naturally guided by the same model-description scheme. However, if turn-taking involves being locally consistent then no modification to the language scheme can be introduced. But this means that participants cannot adapt their practices in the face of difficulties. What possibilities exist? One participant can act as the conversational leader and the other can adapt to whatever schemes are introduced. This suggests at least one circumstance where individuals might deliberately introduce asymmetry into the communication task and is consistent with the earlier suggestion of Sperber and Wilson (1986). Alternatively, participants may actively negotiate a model and descriptive scheme.

Literal and Non-Literal Uses of Language

We have seen that in using language in referential communication tasks, individuals base their descriptions on an underlying model or perspective. Individuals, for instance, used analogies ("looks like an ice skater") to capture the perceptual properties of a figure or developed a language suited to their underlying model of a maze. We noted at the outset of this chapter that the

Devil's Advocate Box 8.1

You might argue that the experimental situation of the maze game is unrepresentative of normal language situations. Hence, the data carry no implications for normal language use. Are there any language situations which correspond to this? People give directions to strangers. In this case they have to describe landmarks and provide route descriptions. This is a normal use of language. The experiments would seem to model best situations of language change: change occurs in the context of stress between existing language and current task. In this sense identifying positions on a maze stretches resources. As Glucksberg and Danks observed (1975) language flexibility leads to the development of specific vocabularies within social groups: "photographers speak of hypo; psychologists of shaping; and skiers of powder."

same words can convey different thoughts. The **literal meaning** of an utterance can differ from its intended meaning. In this section we examine aspects of the interpretation of figurative language and return to notions proposed in relevance theory.

"Please shut the door" and "It's cold in here" mean different things. Yet, the latter can be uttered as a request for the door to be closed. According to Grice (1975; see chapter 9 for further details) intended meaning, or what is implicated, is largely determined by maxims of conversation. Conversation works according to Grice because two or more people mutually agree to cooperate in accordance with certain maxims, namely: the maxim of quantity (say as much as and no more than is needed); quality (only say what you believe to be true); relation (be relevant); and manner (be clear).

According to Grice, an inference or implicature is invited when one maxim is deliberately flouted. In such circumstances, the hearer assumes the overarching cooperative principle is still being adhered to and so tries to make sense of the violation by assuming that the intended sense is different from the actual sense. The utterance: "It is cold in here," in a context where it is clear that everyone is shivering, is either a statement of the self-evident or can be construed as a request for someone to close the door.

This position treats literal meaning as the priority in processing. Listeners derive this meaning, test its fit with context and if necessary devise some informative non-literal meaning (Clark and Lucy, 1975). However, research has established that in understanding idioms (e.g., "throw in the sponge") or indirect requests, individuals may not derive any literal meaning (Gibbs, 1984). As Récanati (1995) points out, a figurative interpretation of an utterance does not mean that its literal interpretation has been computed. If a non-literal interpretation of a constituent fits the context the interpretation may be retained and its literal interpretation suppressed. He considers the utterance: "The ham sandwich left without paying," said by a waiter of a customer who had ordered and eaten a ham sandwich. Although the description, "ham sandwich" may receive a literal interpretation, it may invoke a representation of the customer, that is, a reference to a token of a customer in a mental model, and the customer is clearly a better subject for the argument "has left without paying." As a result, the literal reference might be suppressed and no complete literal interpretation of the utterance computed.

The "literal precedes figurative" view also has problems explaining why it is that subjects seem unable to ignore the figurative meaning of an utterance. Thus, in deciding whether a given statement is literally true or false, subjects found it much more difficult to decide that "My job is a jail" is literally false compared to deciding that "My job is a snake" is literally false (Glucksberg, Gildea and Bookin, 1982). The former is rated a better **metaphor** than the latter. Yet this difference should carry no implications for processing in terms of the serial model.

How are metaphors such as: "It's Etna in here" or "Cigarettes are time bombs" understood? One traditional theory supposes that individuals treat metaphors as if they were implicit comparison statements. Understanding a metaphor involves the same process as understanding statements such as "Copper is like tin." How might this process work? Individuals could retrieve information about both metals and compare their properties or features. But such an account cannot be general because (a) we can use metaphors to attribute properties to a substance, person, or animal that are new. To hear that "Jack is a tiger in union meetings" is to learn something new about Jack. We may not have known before that he could be verbally fierce. A second problem with the comparison view is that it predicts that similes (explicit comparison statements of the form A is like B) could be expressed as nominative metaphors (statements of the form A is B), but this is incorrect. The statement "copper is tin" is simply false.

An alternative view is that metaphors are what they seem to be: class-inclusion statements rather than disguised similes (Glucksberg and Keysar, 1990; see also Rubenstein, 1972). Glucksberg and Keysar point out that literal class-inclusion statements often have figurative meanings too. Consider the statement: "Dogs are animals". Suppose you hear this in the course of a conversation. It's likely that the speaker knows that you know this fact. Hence, you might reasonably suppose that the speaker means that dogs lack human qualities, they cannot be treated as if they would behave like a human being. Pragmatic information has to be used to understand literally true statements as well as figurative statements.

Understanding a metaphor requires consideration of the context. Is a literal or non-literal category statement intended? In the case of the statement "My job is a jail" we understand it to mean that the person feels constrained and lacking in autonomy. The category "JAIL" refers to a type of object that constrains one against one's will. A jail was chosen presumably because it was prototypical or an exemplary instance of such a category. This category contains instances, which include jails and my job. This proposal seems to extend nicely to cases such as "She's no angel" or "He isn't exactly a dynamo." In these cases, the statement is literally true, but like "Dogs are animals" there is a figurative meaning.

SAQ 8.5 Outline an analysis of the expression "Boys will be boys" according to the class-inclusion account of metaphors.

Consider next, the metaphor: "Cigarettes are time bombs." According to a class-inclusion analysis, the predicate "time bombs" refers to objects designed to explode and cause harm. It also refers to a prototypical type of object that can have lethal and unpredictable effects. The word "time bomb" is used to name this latter category. Such an account also explains truisms such as "Boys will be boys." The predicate "boys" functions in two ways: it refers to instances of young human males and it refers to a category of person who act in boyish ways. In the scenario, Dad says: "It's Etna in there." The referent of "Etna" is both the volcano (i.e., a specific instance) and a type of object which yields smoke and hazard. The actual utterance serves a number of functions: it makes a figurative assertion and thereby signals a concern, and it makes an indirect request (for Lucy to open the window or switch off the toaster). Why was it chosen? Perhaps the family spent a holiday in Sicily last year and so the reference appealed to their common ground.

SAQ 8.6 To what extent can class-inclusion be a general account of the processing and representation of metaphors? Does it apply to all metaphors?

Understanding any utterance, literal or non-literal, requires understanding the communicative intentions of the speaker. **Ironic** utterances present a more complex example of this requirement. Consider an utterance such as "Another wonderful day" said while looking out at torrential rain when the forecast had promised sun. The statement is counter to fact and according to the Gricean cooperative principle, it should lead the listener to infer that the intended meaning was opposite to that expressed. The listener infers that the speaker has deliberately flouted the maxim of truthfulness. But such an account leaves unexplained why a person should choose to say the opposite of what they intended. What is the relevance of it? According to relevance theory (Sperber and Wilson, 1986; Wilson and Sperber, 1992), it is designed to communicate the speaker's attitude towards a thought, view or, utterance, which is attributed to someone other than the speaker. The attitude in this case being one of derision or mockery. Kreuz and Glucksberg (1989) suggest that it serves to remind the listener of what was expected (see Clark and Gerrig, 1984; also Sperber and Wilson, 1986, for different views). Not all ironic utterances refer to specific views and thoughts, and express an attitude towards them. In some cases they reference a generalized expectation within the community and comment on that (see Sperber and Wilson, 1981; Jorgensen, Miller, and Sperber, 1984; Gibbs, 1986). Such a view explains the asymmetry between positive and negative ironic statements. Thus, a statement such as "You're a fine friend!" uttered in the context of a friend failing to help, requires no prior explicit remark to invoke a view. It is a cultural norm to help a friend in need. In contrast, the utterance: "You're a terrible coward!" cannot be used ironically when someone has been courageous unless the person had previously said that they were cowardly. Kreuz and Glucksberg confirmed that negative ironic statements ("You're a terrible coward!") were rated more sensible in the context of an explicit remark. In contrast, positive ironic statements ("You're a fine friend!") were rated equally sensible in the absence of an explicit remark. Now it may be that the notion of reminding requires further development because it seems odd to speak of reminding in the context of an ironic comment on an utterance just made. Perhaps the ironic remark makes the prior thought the focus of a listener's attention.

Of more direct interest is the relationship between metaphor and **irony**. Grice (1975) makes no distinction between the difficulty of understanding metaphor and irony. They both pose the same problem: to infer a speaker's meaning that fits the context when the speaker has blatantly said something false (i.e., flouted the maxim of truthfulness). However, the discussion above suggests a crucial difference between the two figures of speech as far as relevance theory is concerned. Understanding an ironical utterance is a more complex process because the addressee has to infer that the speaker is attributing a thought to someone other than herself and is expressing her own non-supportive attitude towards it. That is to say, whilst understanding any utterance involves attributing an intention to the speaker, understanding an ironical utterance compared to a metaphorical utterance requires an extra level of representation.

It is natural to expect, then, that certain groups of individuals who fail to develop a normal understanding of mental states will have problems understanding certain kinds of figurative expressions. Autistic children display certain specific deficits in this area. In the false belief task devised by Wimmer and Perner (1983) children watch a puppet play, during which a marble is moved from a basket to a box, while a character is absent. Baron-Cohen, Leslie and Frith (1985, 1986) found that normal children around 3-and-a-half years old and Down's syndrome children of below average intelligence could attribute the false belief to the character (that the marble was in the basket), and could therefore predict where the character would look for the marble on her return, namely in the basket. In contrast, 80 percent of the autistic children of normal intelligence failed the test and supposed that the character would look in the box. They were unable to understand another character's false belief. Leslie (1987) suggested that representing such a mental state (the child's belief about another belief) requires meta-representation. Normal children around 6–7-years old are also able to understand nested beliefs. They can make attributions about what one person thinks another character thinks ("Mary thinks that John thinks the ice-cream van is in the park"). That is, they can think recursively (Miller, Kessel, and Flavell, 1970). In a test of such second-order belief attribution, Baron-Cohen (1989; but see also 1993) found that none of the 15-year-old autistic children who passed the first-order test passed the second-order test, whereas the majority of non-autistic 7-year-old children did.

SAQ 8.7 What predictions from relevance theory can be made about the performance of difference groups of autistic children in understanding literal, metaphorical, and ironical utterances?

In an effort to assess this issue, Happé, (1993) examined the performance of autistic individuals on their understanding of similes ("The dog was so wet it was like a walking puddle"); metaphor ("The dancer was so graceful. She really was a swan"); and irony ("What a clever boy you are David!") given a story where the character has done something stupid. The supposition was that all groups should be able to understand similes because understanding them requires determining in what respect two entities are alike. In contrast, understanding metaphor requires the ability to distinguish between what is literally the case and what is figuratively the case. In fact, according to the view outlined above, it requires an individual to recognize that the name of an object or entity ("swan") can be used as the name for a category of things ("graceful actions") of which the object is prototypical. Happé, drew an analogy with the first-order false belief test where the person has to distinguish between the reality of the situation and the character's belief about that reality. Individuals who fail the first-order belief test should fail to understand metaphors. Irony, should be even more difficult to comprehend according to relevance theory (but not according to Grice) because it requires the ability to recognize a second-order belief. An attitude of derision or mockery towards an attributed thought, for example, in "What a clever boy you are David!", the addressee is being reminded of a possible thought (some action is clever), which is being mocked. Individuals who fail second-order belief tests should also fail to understand ironic statements.

The results supported relevance theory. Individuals who failed the first-order false belief test, failed to understand metaphors but showed a much better understanding of similes. Only individuals who passed the second-order test appreciated ironic utterances. This difference also occurs for normally developing children: only children who passed the second-order test showed an understanding of irony. Hence, these results argue for a strong relationship between the degree of ability to construct higher-order mental representations (meta-representational ability) and the degree of communicative competence. There are further interesting examples (e.g., ostensible invitations; see Isaacs and Clark, 1990), which we have not considered, and complexities of theory (see Sperber, 1994b), which we have not addressed. We do need though to explore the possible means by which listeners derive the intended meaning. How do people home in on the right kind of inference to make? We have seen how individuals work to establish the mutual belief that the intended meaning or significance of an utterance is understood. But the relevance of information may also be a factor that constrains inference and so leads to the determination of intended meaning.

Relevance: Cognitive/Contextual Effects and Cognitive Effort

As noted at the beginning of this chapter, an important notion here is the idea that individuals seek to maximize the relevance of all information, including, of course, communicated information (Sperber and Wilson, 1986; forthcoming). Relevance theory also recognizes that cognitive/contextual effects are achieved at a processing cost. Information is relevant in a given context and to the person processing it in this context to the extent that it yields effects that could not have been derived by considering either the context or the new information alone. An example of an effect might be a new belief implied by this information and context. The greater the cognitive effects the greater the relevance of the information. Deriving cognitive effects is not without cost or effort and, all else being equal, a given cognitive effect achieved at greater processing cost will be less relevant.

Let us consider an example (Sperber, Cara, and Girotto, 1995). Suppose you wish to fly to New York as soon as possible but do not know the flight times. Information that the next plane is at 5 p.m. is relevant to your concerns. You can infer that this is the earliest time of departure and this could lead you to inferences regarding your schedule. Informed that the first plane is after 4 p.m. is also relevant, but not to the same extent: knowing that the plane leaves at 5 p.m. implies it leaves after 4 p.m. Knowing that it leaves after 4 p.m. does not imply that it leaves at 5 p.m. Its cognitive effects are fewer. Consider next, a remark that varies the cognitive effort involved. Suppose you were informed that the first plane would leave 7200 seconds before 7 p.m. Given you could calculate the number of hours from the number of seconds, then you could work out that the next plane departs at 5 p.m. and so derive the same range of cognitive effects. However, given the processing effort involved, this remark is less relevant. We can imagine that a number of factors affect the amount of processing effort: these will include the nature of the words used; the syntactic complexity of the utterance; the ease with which we can access a suitable context.

Assessments of relevance, therefore, depend upon two factors (1) cognitive effects and (2) the processing effort required to derive these effects. So this theory makes a claim about the nature and functioning of the cognitive system that subserves language understanding. It is designed to attend to the most relevant information and to process it as relevantly as possible. Sperber and Wilson summarize this in their "First (or Cognitive) Principle of Relevance": Humans automatically pay attention to information that seems relevant to them – and the more relevant the better (Sperber and Wilson, *Relevance and Meaning,* forthcoming). Such a principle applies to our processing of information in conversations as well as information from the environment. It defines part of the task of the cognitive system.

Communicated information differs from other environmental information in that it raises a definite expectation of relevance. Sperber and Wilson define a notion of **optimal relevance** which is meant to specify what a listener is looking for in terms of cognitive effort and cognitive effect. An utterance on a given interpretation is optimally relevant if and only if it achieves enough effects to be worth the addressee's attention and puts the addressee to no unjustified effort in achieving these effects. The "Second (or Communicative) Principle of Relevance" (Sperber and Wilson, forthcoming) spells out the assumption: every utterance creates a presumption or expectation of its own optimal relevance.

Consider, by way of example, the following exchange (adapted from Sperber, Cara and Girotto, 1995):

Dad: Do you want to go to the party at the Smiths?
Mom: They came to our party.

By asking the question, Dad signals what would be relevant to him. Mom does not reply "Yes." Her reply is indirect but it carries a presumption of optimum relevance. Dad has to look for ways in which the information given explicitly might answer his question as to whether or not Mom wants to go to the party. According to this relevance theoretic account, Dad searches for a contextual assumption which together with the stated information allows him to make an appropriate inference (see Carston, 1988; Stevenson, 1993). An accessible assumption in this case might involve the social norm of reciprocity. If Dad is reminded of this rule by Mom's utterance, he will presume its relevance. He can then infer that she wants to go to the party but not necessarily because she expects to enjoy it but because she considers that it is important to observe social niceties. Such cognitive effects could not have been achieved by her answering the question with a simple "Yes." Her answer allowed Dad to infer her reason for thinking they should go to the Smith's party.

SAQ 8.8 Consider the following exchange:

Dad: Do you want coffee?
Mom: It will keep me awake.

What conclusion did Mom want Dad to reach? What contextual assumption must he derive to reach that conclusion? If she just wanted him to know that she did not want a coffee would her reply have been consistent with the communicative principle of relevance?

So for Sperber and Wilson a rational comprehension strategy is one in which possible interpretations are examined in order of their accessibility. Such interpretations include intended contextual assumptions, propositions expressed, and intended **contextual effects**. The process stops when the expected level of relevance is achieved. Only one interpretation at most will meet the criterion of consistency with the expectation of optimal relevance. One implication of this, illustrated in the example above and in SAQ 8.8, is that more effort implies extra effect. Any element of indirectness creates additional processing effort and via the communicative principle of optimal relevance encourages the hearer to search for additional effects. Such a relationship may also underlie people's use of figurative language. These figures are more effective than their literal paraphrases because they achieve more cognitive effects to offset the greater effort they require. Consider metaphor again. Someone says: "John is a lion." Given cultural stereotypes of lions, individuals might derive the contextual implication that John is brave. But if that is all the speaker meant to communicate, why not say so? Why put the addressee to the additional effort of accessing the contextual assumption that lions are brave and deriving John's bravery as an implication? The speaker must have aimed for additional effects according to relevance theory. Perhaps she intended to communicate that John is brave in the way that lions are brave: he is physically courageous. It is also plausible that metaphors evoke certain affective and imaginal consequences. Recent work using brain imaging (Bottini et al., 1994) indicates that a number of sites in the right hemisphere are additionally activated when individuals process metaphors compared to literal statements.

Sperber and Wilson suppose that the critical process of determining a contextual effect is deductive and is based on the application of logical rules. However, current evidence (see chapter 11, "Mental Models") favours the manipulation of mental models as a way to fill out the propositions expressed by an utterance. The idea that interpretation is guided by seeking an interpretation consistent with the principle of optimal relevance is likely to prove fruitful regardless of the procedures invoked to derive intended meaning.

Summing-up

Understanding an utterance involves assembling a proposition by combining subconcepts according to the structure of the sentence. On the basis of this proposition, individuals construct a mental model of the state of affairs described. Conversations work because the partners in the conversation collaborate to achieve certain outcomes, including establishing the use of terms to refer to tokens in their mental models. It was suggested that the principle of minimizing collaborative effort might emerge out of the operation of more basic processes to do with mapping language on to a mental model. Fundamental to the success of communication lies the ability to recognize the speaker's communicative intention and to infer their informative intention. We considered the processes involved in understanding metaphor and irony and concluded that these two figures differ in psychological complexity in a way predicted by relevance theory. This theory also provides a way to explore why individuals use figurative language. It suggests some critical dimensions on which to consider the process of understanding in the context of conversations and may, therefore, offer a route to building a computational account of conversation.

Further Reading

Chapter 1 in *Pragmatics*, Cambridge,: Cambridge University Press by S. Levinson (1983) discusses different concepts of semantics and pragmatics. *Semantics and Cognition*, Cambridge, Mass.: MIT Press (1983) and *Semantic Structures*, Cambridge, Mass.: MIT Press (1991), both by

R. Jackendoff, provide studies of conceptual structure and how it is encoded into natural language expressions. Chapter 2 in *Human and Machine Thinking*, Hillsdale, N.J.: Lawrence Erlbaum Associates by P.N. Johnson-Laird (1993) provides a lucid account of the theory of subconcepts. *Relevance: Communication and Cognition*, Oxford: Blackwell by D. Sperber and D. Wilson (1986) provides a comprehensive statement of the principles of relevance theory and *Understanding Utterances*, Oxford: Blackwell, by D. Blakemore (1992), provides an excellent introduction to this theory.

Exercises

1 Next time you are in a social gathering make notes on the different ways in which individuals ground their utterances. What factors affect the explicitness with which they do so?

2 Garrod and Doherty (1994) compared the performance on the maze game of individuals who played a number of different games with the same partner with that of individuals who played the same number of games with a different partner each time, thereby creating a virtual community. What expectations do you have about the eventual degree of coordination of these two different groups, that is the extent to which they converge on an efficient descriptive scheme?

3 Why might individuals issue invitations that they expect the addressee to decline? For some interesting material see Isaacs and Clark (1990).

CHAPTER

9 Pragmatics and the Development of Communicative Ability

Outline

The study of pragmatics is the study of language usage. In this chapter we will look at the development of pragmatic competence in young children. We will first look at various definitions of pragmatics. As you will see, researchers have struggled with how to take the context and meaning of utterances into account in such definitions and provide a clear demarcation between what is considered the area of pragmatics and what is considered the area of semantics (the study of meaning). Children's developing pragmatic competence will be explored with respect to two fundamental types of inferences that arise when we converse with others: inferences about communicative intent and inferences about shared background knowledge. In each case, we will also briefly examine how the knowledge on which these inferences are based may be represented mentally. Finally, we will examine what happens when people have difficulty with the pragmatic aspects of language and look at some possible explanations for pragmatic disability.

Learning Objectives

After reading this chapter you should be able to:

- highlight some of the difficulties related to defining pragmatics
- discuss the role of intention in a definition of communication
- identify some of the developments in the ability of children to use their language to achieve an intended effect and to perceive the intentions of another speaker
- describe how our knowledge of how to make inferences about a speaker's intention may be represented
- identify some of the pragmatic skills involved in making inferences about shared knowledge
- describe what knowledge structures may underlie our ability to make inferences about shared knowledge

- describe what happens when people have difficulty with the pragmatic aspects of language
- discuss some possible explanations for pragmatic disability

Key Terms

- conversational implicature
- functional approach
- Grice's maxims
- mutual knowledge
- pragmatics
- speech acts
- theory of mind

Introduction

Imagine for a moment that you have decided to learn a second language. What are the things you would need to learn to be able to speak this language? Well, first, you would probably start by learning some of the basic vocabulary (the lexicon) that you would need in everyday situations. Once you had begun to master some of the vocabulary, you would move on and learn some key phrases, expressions, idioms, and jargon that would make you sound more like a native speaker. At this point, in order to begin to put together sentences on your own, you might need to learn something about the grammar (the syntax) of this second language as it might be quite different from your first language. For example, the language you have chosen to learn may form its sentences with a different word order than your first language does.

But, as an adult, there are many things about *using* a language to communicate with other people that you would *not* have to learn about when learning a second (or any other) language. For example, it's probably a good bet that, anywhere you travel in the world, talking about the weather is a good, friendly way to start a conversation with a stranger. How do you know this? And what else do you know about communicating with other people that will be similar no matter what language the community speaks? The following list highlights only a few things you know:

- if you are giving information to someone, you should provide them with enough information and not too much
- when telling someone a story you should take into account what may be familiar or unfamiliar to your listener
- if someone asks you a question, you should provide an answer relevant to the topic at hand
- if you are the speaker you should refer to yourself using the word for "I" and not the word for "you"
- a useful, friendly topic for casual chitchat is the weather

As noted in the previous chapter, the understanding of *how to use* language (and, of course, any other means of communication) in interactions with other people is referred to as understanding the **pragmatics** of language. Pragmatic knowledge is viewed as separate from knowing the syntax (grammar) or lexicon (words) of a language.

Now, let us consider a child such as Zara in the scenario, who is learning her first language. In addition to having to learn the syntax and lexicon of English, Zara *will* have to learn how to use her language in interaction with others. Pragmatic factors may seem second nature to adults, but they are not to children. Zara will have to learn how to make her responses relevant to questions that are posed to her, how to be as informative as another person needs her to be, and how to introduce a new topic of conversation without confusing a communicative partner. In this chapter, we will be taking a look at how some of these pragmatic skills develop in children.

Why Study the Pragmatic Skills of Children?

There are good reasons to study the development of pragmatic skills in children even if your main interest is in the adult forms of this knowledge. Take, for example, the ability to tailor a conversation to the knowledge state of a communicative partner. Adults are constantly doing this in conversation with others, but how? For instance, when Zara's mother says to her husband "Oh can you get her some – I'm seeing to the toast," what has made her think that he will know what she means by "some"? Clark and Marshall (1981) have suggested that communicative interactants take the following four kinds of **mutual knowledge** into account:

1 *Community membership*: things that everyone in a community knows and assumes that everyone else in the community knows (e.g., in a community of Canadians you could assume that everyone knows Toronto lies on the East coast; in a community of native Londoners you could assume people know what the sign for the underground looks like).
2 *Physical copresence*: things that we assume another person knows by virtue of the fact that that person was physically copresent in a certain situation (e.g., Zara's mother assumed that her husband had just heard Zara ask for "Coco Pops").
3 *Linguistic copresence*: things that people assume others know as a result of what was previously communicated to them (e.g., upon saying *I met my friend Ann yesterday*, you will assume that your listener now knows that Ann is being talked about and will not be confused if you refer to her in your next utterance using the pronoun *she*).
4 *Indirect copresence*: things that people assume a communicative partner knows indirectly as a result of being physically or linguistically copresent. (e.g., it may be universally known among your circle of friends that Ann is very short and so you could assume that in mentioning Ann you are also securing the (indirect) mutual knowledge of her height and so will not be misunderstood if you say *Did you hear about Ann's blind date last night – 6ft 5!*).

Although these four types of knowledge may seem quite basic to you on first reading, they actually rely on having a sophisticated understanding of what other people can be expected to have remembered or forgotten from the past, to have noticed along with you, to have inferred from previous conversations and so on. In fact, adults' ability to communicate with others rests on assumptions about other people that are much more basic than those put forth by Clark and Marshall. Child cognitive and language development researchers have recently shown that some very basic assumptions that we make about people are not acquired by children until they are about 3 or 4 years of age. For example, research into *children's theory of mind* (e.g., Astington, Harris, and Olson, 1988) has examined children's understanding of the existence of mental states such as knowledge, desires, and beliefs and their ability to attribute these states to other people. This research has highlighted that toddlers and preschool children only slowly develop such understandings about people as the following:

- people can remember things over a fairly long period of time
- people's attention can wander
- people acquire knowledge through sensory experiences such as seeing, hearing, or feeling something
- people can acquire different types of knowledge from different sensory experiences (e.g., you can learn the color of something by seeing it but not by feeling it)
- people conceptualize and construct their world in roughly similar ways
- people have mental states

Child developmental research has been instrumental, therefore, in highlighting some of the very fundamental assumptions we make about other people that are crucial to the ability to communicate. As adults we take this kind of knowledge about other people for granted and do not realize that there may have been a time in our lives when we did not know these things and needed to acquire this knowledge.

SAQ 9.1 Imagine for a moment that you are a child who does not yet understand all the things about other people in the list above. How might this affect your ability to communicate with others? (We will come back to this question later in the chapter).

Coming to understand the above "facts" about people constitutes part of the early foundation upon which sophisticated, adult, pragmatic abilities rest. So, for example, let's say we wanted to explain how the mother in the scenario knows when she needs to use a proper name (e.g., Granny) and when she can use a pronoun instead (e.g., she). Not only would we want to postulate that the kinds of knowledge suggested by Clark and Marshall (1981) are involved, but we would also want to add to this another, more precursory, level of under-

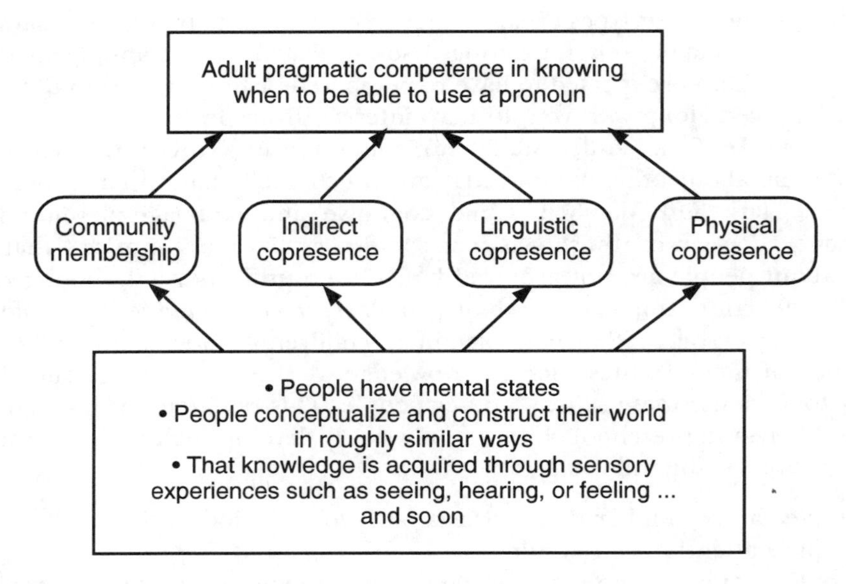

9.1 Factors underlying adult pragmatic competence.

standing, that contributes to the mother's pragmatic competence, as shown above:

Definition of Pragmatics

Successful communication is not only the result of knowing what the words of a language mean (semantics) and how to put the words together in grammatically appropriate ways (syntax), but also of knowing how to use the language in discourse with others. For example, your ability to communicate messages that are informative, relevant, polite, rude, or sarcastic relies on more than your simply knowing how to put words together into a sentence.

A single definition that encompasses the scope of what the field of pragmatics is all about does not exist. One reason for this is the fact that pragmatics is studied not only by linguists, but by anthropologists, philosophers, psychologists, sociolinguists, psycholinguists, and computer scientists. Each of these groups have different ideas about what to include and not include within the boundary of pragmatics, and of what aspects of pragmatics are to be the focus of study. So, for instance, although discussions of pragmatics among philosophers may often focus on the problem of how people are able to make sure that their conversational partners know what they are referring to (i.e., how people establish *mutual reference*), sociolinguists are usually more concerned with pragmatic issues having to do with how our communicative interactions may be affected by the social class, race, and gender of the participants.

As we have seen, a very general approach defines pragmatics as the *use* of language, and thus separates it from the areas of syntax (grammar) and semantics (meaning). If one tries to define pragmatics any further, the next step is usually to say that it has to do with our understanding of how to take the *context* of an utterance into account. But it is at this point that it becomes very difficult to pin down a precise definition that captures what the field of pragmatics is all about. Two issues, in particular, have been problematic with respect to defining pragmatics, namely:

(1) how to include *context* in this definition; and
(2) how to separate the domain of *semantics* from the domain of pragmatics.

Attempts to solve these two problems have led to numerous definitions of pragmatics, a few of which will now be discussed as we take a closer look at the nature of these two problems (for a detailed review of the field and the different definitions of pragmatics see Levinson, 1983).

Context

Let us consider first the relation of context to the study of pragmatics. One problematic issue for pragmaticians has been determining the degree to which aspects of context are or are not encoded within the very structure of the language (i.e., the words themselves) and how necessary it is to go beyond the language structure to determine the meaning of a sentence or utterance. As implied in the previous chapter, the prevailing view today is that language usage and understanding is influenced by contextual factors that have little or nothing to do with actual words used in a sentence or utterance. This was illustrated clearly by Grice (1975) who demonstrated that in conversation with others, we readily and ubiquitously make inferences about what a speaker meant – **conversational implicatures**, as Grice called them – that are based on information that is not encoded in the speaker's utterance. Consider this famous mini-dialogue from Grice:

Speaker A (to a passer-by): I've just run out of petrol.
Speaker B: Oh; there's a garage just around the corner.

In this dialogue, Speaker B does not explicitly utter the information "You can get *gas* at the garage just around the corner." But this information can (and one could say should) be inferred from Speaker B's utterance by Speaker A. A definition of pragmatics must take into account these types of conversational implicatures, by which the meaning of a speaker's utterance relies on aspects of context outside the structure of a language (e.g., information accessed via visual perception, memory).

Meaning

Meaning is generally viewed as the domain of semantics. However, aspects of meaning are also involved in pragmatics. The challenge for pragmaticians, therefore, has been to define what aspects of meaning are going to be covered by semantics and what aspects are going to be covered by pragmatics.

Grice (1968, 1975) observed that in many cases, as with the mini-dialogue above, a distinction can be made between the meaning of the sentence uttered itself and the meaning the speaker intended to communicate using the utterance. So, for example, someone might say *Is there any coffee left?* (sentence meaning = *Is there any coffee left?*) but really mean for the addressee to interpret this as a request for more coffee (utterance meaning = *Can I have some more coffee?*). Grice equated semantics with *sentence meaning* and pragmatics with *utterance meaning*, which he viewed as involving the sentence plus its context. Another way to conceptualize this distinction is to view semantics as concerned with those elements of meaning that can be directly decoded from the words of the sentence itself, and pragmatics as concerned with those elements of meaning that depend on contextual information beyond the words of the sentence itself and on the interpreter's inferential abilities (Carston, 1993). Both of these ways of defining pragmatic meaning bring us back, however, to problems relating to context. It remains to be explained precisely how communicators take the context into account to determine the meaning the speaker intended. This is the topic of the next section.

Approaches to Incorporate Context in a Definition of Pragmatics

What features of the context are important when determining what a speaker meant to say? Grice (1975) proposed one feature of the context that may play an important role, namely, a set of overarching assumptions that participants in conversations are aware of and that guide the conduct of participants in conversation. He identified four conventions (or **maxims** as he called them) that together represent a general **cooperative principle** that participants in a conversation try to observe:

1 *Maxim of Quantity*: Make your contribution as informative as is required, but not more informative than is required.
2 *Maxim of Quality*: Try to make your contribution one that is true. That is, do not say anything you believe to be false or lack adequate evidence for.
3 *Maxim of Relation*: Make your contribution relevant to the aims of the ongoing conversation.
4 *Maxim of Manner*: Be clear. Try to avoid obscurity, ambiguity, wordiness, and disorderliness in your use of language.

SAQ 9.2 How would you account for the talk of a con artist? Or more to the point, how do these maxims help you to explain the success of con artists?

The important thing to note about these four conversational conventions is not that speakers follow them to the letter in conversing with others but rather that, whenever possible, addressees will interpret what speakers say as *conforming* to these maxims on at least some level. So, the existence of these maxims will cause addressees to generate inferences beyond the semantic content of the sentences uttered (i.e., to generate conversational implicatures). Moreover, conversational implicatures will not only be the result of a speaker adhering to the maxims, but will also come about as a result of the speaker flouting these maxims at the level of what is said (i.e., literal meaning), as in the case, for example, of making a sarcastic comment by flouting the maxim of quality (e.g., *Wasn't that a nice dinner*).

Grice's maxims represent only a first step, however, towards explaining how aspects of the context are taken into account when we are determining the meaning of a speaker's utterance. Consider the maxim *be informative.* It is one thing to say that speakers should strive for a certain level of informativeness, but it is altogether a different question to ask *how* speakers are able to determine the level of informativeness at which to pitch a contribution to a conversation. In conversation, listeners and speakers are clearly using aspects of the context to make inferences about the knowledge shared by themselves and the other participants, but how are they doing so?

This view of pragmatics – one that is particularly concerned with the making of inferences that connect what is said to what is mutually assumed or what has been said before – has been the focus of much research by psycholinguists, cognitive psychologists, and researchers in artificial intelligence. This work is often guided by a definition of pragmatics that focuses study on the relations between language and context that are basic to an account of language understanding (e.g., Charniak, 1972). The biggest problem facing this approach has been how to limit the context producing the inferences to anything smaller than *all* the world knowledge possessed by speakers and hearers (note that this is an instance of the "frame problem," outlined in chapter 3). We have already seen one example of this approach with the work of Clark and Marshall (1981) mentioned above in which four types of information are proposed that speakers take into account when assessing mutual knowledge. Let us now take a look at one theory that has tried to address in more detail the problem of how we take context into account in interpreting utterances.

Sperber and Wilson (1986; Wilson, 1994; see also chapter 8) have recently proposed an alternative to the Gricean explanation of how a hearer recognizes a speaker's intended interpretation of an utterance. Their approach – **relevance theory** – rests squarely on Gricean foundations, but stresses that, if contextual assumptions affect the way an utterance is understood, then the most fundamental question to ask is how the hearer is able to select and use the set of contextual assumptions intended by the speaker. That is, while it has been long recognized that context makes a contribution to utterance understanding, Sperber and Wilson argue that the problem of how the hearer is able to identify the intended context out of all those possible has not received much attention by pragmaticians.

Consider the following scenario (from Wilson, 1994). You and your friend are keen tennis players. You have just recently begun playing with a new doubles partner and your friend asks what he is like. You reply, "He has much in common with John McEnroe."

For most readers, the intended interpretation of your utterance will be clear – you are saying that your new doubles partner is bad-tempered on court. But why should this be? Why is it not the case that your friend will interpret you as saying your new doubles partner is a gifted tennis player, has played on Center Court at Wimbledon, or is very rich – all things which are also characteristics of John McEnroe?

Sperber and Wilson argue that the reason we know the intended interpretation is that the new doubles partner is bad-tempered has nothing to do with maxims or rules, but rather has to do with a very general assumption about human cognition, namely that " human cognition is relevance-oriented: we pay attention to information that seems relevant to us" (Wilson, 1994). Applied to the communicative enterprise, this feature of human cognition means that we as hearers are equipped with a single, very general criterion for evaluating possible interpretations, namely we expect them to be *relevant.*

Now, one might ask, out of all of the possible interpretations that may be a little, somewhat, or very relevant, how do we know to choose the one that the speaker intended. To answer this question, Sperber and Wilson suggest that two factors influence how relevant an interpretation might be. First, the greater the *contextual effects* derived from some information (i.e., from an inference or interpretation), the more relevant that information will be. Contextual effects, the authors explain, are achieved when newly-presented information interacts with a context of existing assumptions in one of three ways: by strengthening a previously held assumption, by contradicting and eliminating a previously held assumption, or by combining with a previously held assumption to yield a logical implication derivable only from the new information and the context combined. Second, the authors propose that the more *effort* needed to derive an interpretation, the lower its relevance will be. According to the authors, some interpretations will require more mental effort (e.g., imagination and efforts of memory) than others to understand. For example, they would argue that remembering that McEnroe is ill-tempered on court is probably more readily available to someone than the fact that he is a Wimbledon champion. As such, the hearer will favor the interpretation requiring the least amount of cognitive effort provided it has a satisfactory set of contextual effects. Together, these two factors combine to make obvious the *optimally relevant interpretation*: the interpretation that gives the hearer enough contextual effects to be worth his or her attention without putting the hearer to any gratuitous processing effort (Wilson, 1994).

Functional Approach to Pragmatics

One approach to defining pragmatics has a focus quite different from the rest– the **functional approach**. Historically, this approach was based in philosophical arguments that language structure is not independent of the *uses* to which it is put and the notion of speech acts (we will look in more detail at the notion of speech acts later; e.g., Austin, 1962; Searle, 1969). We will be exploring this view further in sections to come, but, in general, its proponents have sought to define gross functional categories of language use such as the *imperative* (use of language to make requests for action), *declarative* (use of language to assert something), and *interrogative* (use of language to request information). As you will see, this approach has heavily influenced investigations of children's earliest pragmatic abilities.

Pragmatic Focus of this Chapter

The many varied definitions of pragmatics that exist have motivated different research strategies and examinations of pragmatic skills in children and adults. Depending on the definition that is used, different topics are of interest. The discussion of the development of communicative competence in children in this chapter will not be guided by one particular definition of pragmatics, but it will be narrowed to several topics within the general domain of pragmatics. First, we will focus on those aspects of pragmatics that could apply to any language and are not specific to a particular language or culture. For example, Grice's maxim *be informative* could be assumed to be an aspect of pragmatic understanding that will apply to communication in any language. In contrast, politeness markers are a feature of language use that vary with different languages and different cultures (though the phenomenon of politeness is probably universal). It is pragmatic skills of the first type that will be the focus of discussion here. Second, we will focus specifically on the development of two pragmatic abilities in children: (1) how children first begin to communicate intentionally and infer the intent behind speakers' utterances, and (2) how children are able to make inferences about what knowledge is or is not shared by their communicative partners. Both of these abilities are basic to the ability to communicate competently with other people.

Inferences about Communicative Intent

A Definition of Communication Must Take Intention into Account

Consider this definition of *communication*: communication is the transmission of information between a sender and receiver. Does this definition seem an adequate one to you? If it does, how would you categorize instances in which another person yawns or sneezes while you are talking to them? You might infer from these events that your interlocuter is bored or coming down with a cold, but would you want to say that he or she communicated that information to you? Probably not. Indeed, this is the problem with the definition of communication above – it fails to distinguish between instances in which information is *involuntarily or unconsciously* conveyed (e.g., yawn, sneeze) and instances in which information is *intentionally* conveyed. To avoid this problem, it is necessary to include intention in a definition of communication.

Grice (1968) proposed a definition of communication that included intention and went one step further. He argued that the sender must not only intend to get across a message, but he or she should also intend that the receiver recognize that intention. According to Grice, *successful communication* can only be said to have taken place if the sender's communicative intention has become mutually known to both the sender and the receiver. Whether or not the receiver then subsequently acts as the sender intended him or her to is another question, but the receiver must at least have recognized the sender's communicative intention. So, for example, in the scenario, Lucy's utterance *Mom, have you seen the tray?* can be said to have been an instance of successful communication because her mother recognized not only that she was trying to communicate something, but that her intention was to specifically request information about the location of the tray. Thus her mother does not even answer yes (which could be a relevant response to the question in another context) but proceeds directly to provide information about the location of the tray.

SAQ 9.3 How would Sperber and Wilson (1986) adapt Grice's definition of successful communication, above, in line with their relevance theory of communication?

Keeping the above example of an instance of successful communication in mind, consider the communicative behaviors of much younger children. By 12 months of age, a child may request your help by reaching out towards an object out of reach. Would you consider this nonverbal behavior to be an instance of intentional communication?

When do Children Begin to Communicate Intentionally?

The age at which children begin to communicate intentionally has been the focus of intense debate. By the time children are about 8 months old they are beginning to make requests for objects that often take the form of the child staring at the object, reaching for it, and accompanying this with a grunting noise or *protoword* vocalizations such as *ma* or *da* (Bruner, Roy, and Ratner, 1982). Because these early communications on the part of children begin long before you can ask them about what they are trying to do, researchers have had to turn to indirect means in order to decide whether and when children are communicating intentionally. Generally, researchers have turned to two other behavioral indices to determine if children are communicating intentionally. First, a child's communicative behavior is claimed to be intentional if the child *persists* with the behavior when it is not responded to by an adult and only stops the behavior once the adult has responded (Bates, Benigni, Bretherton, Camaioni, and Volterra, 1979). Second, if a child accompanies his or her gesture and/or vocalization with an *alternating glance* between an adult and the object being requested, this is considered to be an even more decisive sign that children are communicating intentionally. This alternating glance is regarded as indicating an awareness that communication can be used to get a person's attention and to get that person to achieve a goal (Bates et al., 1979). As children's communications are not often accompanied by such glances before about 9–12 months of age (although the exact age is a topic of considerable debate), it is only after this age that most children are thought to be communicating intentionally.

What is the Nature of Children's First Intentional Communications?

Speech acts

Apart from the question of when children begin to communicate intentionally, researchers have also been interested in the nature of these first intentional communications. This approach has been widely influenced by the work on speech acts and functional approaches to pragmatics (e.g., Austin, 1962; Searle, 1969).

In his book, *How to do things with words,* Austin (1962) launched his theory of speech acts based on his initial observation that certain declarative sentences such as *I apologize* and *I give you my word* are not used just to *say* things (i.e., to describe a state of affairs) but to *do* things (e.g., apologize, promise). Indeed, Austin argued that every time speakers utter a sentence they are attempting to do something with the words. Austin used the term *illocutionary act* to refer to what the speaker is doing by uttering a certain utterance, although *speech act* is the term used more commonly now.

Searle (1976) extended Austin's theory of speech acts by distinguishing five basic types of action that can be performed in speaking by means of the following five types of utterance:

1 **representatives**, which commit the speaker to the truth of the expressed proposition (e.g., asserting, concluding).
2 **directives**, which are attempts by the speaker to get the addressee to do something (e.g., requesting, questioning, ordering, forbidding, permitting).
3 **commissives**, which commit the speaker to some future course of action (e.g., promising, threatening).
4 **expressives**, which express a psychological state (e.g., thanking, apologizing).
5 **declarations**, which effect immediate changes in the institutional state of affairs and which tend to rely on elaborate extra-linguistic institutions (e.g., declaring war, christening).

Functions of children's early utterances

In discussions of children's early language, attention has slowly been shifting from the lexicon and grammatical systems underlying the child's utterances to the functions that those utterances perform and the nature of the interactional context in which they occur. As we will see, researchers have tried to develop taxonomies for describing the early functions of children's first – prelinguistic – intentional utterances (Bates, 1976; Halliday, 1975). Because children who are prelinguistic cannot be questioned about their communicative intentions, these taxonomies represent the researchers' best guesses about the functions of children's utterances.

Bates (1976) distinguished between two types of early illocutionary acts: *protoimperatives* and *protodeclaratives*. She argued that these two acts are the nonverbal equivalents (i.e., have the same effect) as adult imperatives (e.g., *Give me that*) and declaratives (e.g., *That's a* . . .). For example, a child may request some milk by reaching towards the refrigerator or indicate an interesting sight by pointing to it. Indeed, there is consensus among researchers that requesting (imperatives) and asserting (declaratives) are among the

Devil's Advocate Box 9.1

Does constructing a typology of speech acts seem useful to you? Can you think of reasons why Searle's (1969) typology may not be definitive or exhaustive? Strawson (1964), for example, claimed that Austin (1962) was misled in his thinking about the nature of illocutionary acts by focusing mainly on utterances that are *institutionally-based* – that is, require a whole host of specific social conventions to link the words (e.g., marriage vows) to the institutional procedures. For example, the act of pronouncing two people as husband and wife is linked to a very specific set of procedures laid down in the law and as a result not everyone can perform this illocutionary act simply by uttering the relevant utterance. Strawson argued that such conventional and culture bound illocutions do not represent the *fundamental part* of human communication. But what alternatives might be possible? Strawson suggested that we should look to more general conventions affecting communication, such as Grice's (1975) notions of relevance and cooperation in order to gain a more accurate characterization of the major illocutionary acts.

Sperber and Wilson (1986) have also questioned the importance of speech acts to the study of pragmatics. They argue that many speech acts can be successfully performed without being identified as such either by the speaker or the hearer (e.g., asserting, denying, warning). In addition, they point out that while the identification of some speech acts such as telling, asking, and saying may be an essential part of the comprehension process, these acts are not institutionally-based the way promising and thanking are. Rather, these acts appear to be genuine, universal communicative categories. If this is the case, it may be valuable to consider the earliest illocutionary acts of children as they are acquiring language, as these may represent a category of basic and fundamental illocutionary acts.

earliest speech acts to emerge. Children's earliest prelinguistic requests consist of gestures used in combination with vocalizations such as grunts, *da*, or *ma* and emerge around 8 or 9 months of age. The first assertions generally consist of a gesture such as a point, plus an early term for *there* or *that* such as *da* around 12–14 months of age. Children may also use one or two primitive expressives such as greeting someone with *hi* from quite early on (Clark and Clark, 1977).

A more detailed and extensive description of the functions for which children use their earliest communicative means has been suggested by Halliday (1975). He proposed the following taxonomy of seven functions in which the first four functions emerge around the time a child is 12 months of age, and the last three emerge around 18 months of age:

1 The **instrumental** function: communicative means are used to satisfy needs and get both goods and services.
2 The **regulatory** function: communicative means are used to control the behavior of others.
3 The **interactional** function: communicative devices are used to establish interaction with other people. It includes greetings and names for people.
4 The **personal** function: communication is used by children to express their awareness of themselves, for example, by expressing their feelings, likes, dislikes, interests, and so on.
5 The **imaginative** function: communicative devices are used to create an imaginary environment such as in pretend play.
6 The **heuristic** function: communicative devices are used to find out about the world, for example by asking *why?* questions.
7 The **informative** function: language is used to convey information to other people about things not visible in the immediate environment. This function differs from all the others in that the message to be conveyed must be done with language and cannot be done nonverbally.

Even if there is disagreement among researchers as to how to best classify the functions of children's early prelinguistic communications, the fact that different functions exist is generally accepted. The process of acquiring language does, therefore, appear to rest on a prior ability to communicate a number of illocutionary acts prelinguistically. But Halliday's taxonomy is quite different from the taxonomy of speech acts we just saw proposed by Searle (1976). Levinson (1983) has suggested that descriptions of children's earliest uses of language may be much better served by the Gricean (1975) intentional view of speech acts than by the convention-based accounts of Austin (1962) and Searle (1969). That is, it may be more accurate for a theory of communication to build on the notions of communicative intention, utterance function, and interactive context than on the notion of speech acts (Levinson, 1983).

Children's More Sophisticated Understanding of Speech Acts

Our discussion so far has focused mainly on what children, as speakers, may be aiming to achieve. Let's turn now to consider in more detail how children become more sophisticated at actually using their language to achieve an intended effect, and also to consider their developing ability, as listeners, to perceive the intentions of another speaker.

Most of the research examining how children are using their language to perform certain acts has focused on their ability to make *requests.* In the scenario, Zara's request *More milk* represents a very typical request of a 2-year-old child. During the second year, children's reach and point gestures are combined with the names of desired objects and words such as *more, want,* and *gimme* to produce more sophisticated requests (Slobin, 1970). Requests are used by children at the two-word stage to get action (e.g., *Want gum*), request information (e.g., *Where doggie go?*), and to make refusals (e.g., *No more*). By the age of 3, children can produce requests with embedded imperatives, such as *Would you open this?* (Read and Cherry, 1978).

Children begin to use more and more indirect requests as they get older. Garvey (1975) found that 5-year-old children used twice as many indirect requests (e.g., *Why don't you tickle me?*) as 4-year-old children. Children's comprehension of indirect requests has been one area in which it has been possible to see the development of children's ability to understand a speaker's intention when it is not directly available from the linguistic structure of the utterance itself. Children as young as age 2 have been found to respond appropriately to such indirect requests as *Can you find me a car?* or *Are there any more sweets?* (Shatz, 1978), but at this age they tend to always treat such utterances as requests for action and not for information (Shatz and McCloskey, 1984). It is only by about 6 years of age that children reliably use the preceding context to interpret utterances that can be biased towards a literal meaning (e.g., a statement or a question) or an indirect request (Ackerman, 1978).

It has been difficult to determine what cognitive abilities might underlie children's more sophisticated requests. Levin and Rubin (1983) postulated that children's perspective-taking skill may be related to three requestive abilities: (1) the use of indirect requests, (2) the formulation of requests in such a way as to anticipate a successful outcome, and (3) the use of different strategies when faced with noncompliance. These authors measured children's perspective-taking skill by seeing how they could recount an eight-picture cartoon sequence to someone who had not seen it before. Contrary to their prediction, the authors found that the perspective-taking ability of preschoolers and grade-three children (as measured by the story task) was only weakly related to their requestive ability. Levin and Rubin concluded that it may not be very useful to simply assume that the production of various request forms necessitates perspective-taking skills. A more useful approach, they suggest, may be to carefully consider the content of children's messages and keep in mind that certain requestive skills may be attributable to perspective-taking while other requestive skills may be more influenced by cognitive abilities, such as monitoring the possible reasons for communicative failure.

As children get older, they not only begin to use new ways of expressing each speech act (e.g., direct versus indirect requests), but they also add new types of speech acts to their repertoires. For example, in addition to requests, children learn to carry out various other directives such as asking, ordering, forbidding, and permitting. Grimm (1975) found that 5-year-old children managed quite well to ask, order, or forbid a story character to do something, but had trouble permitting him to do something.

Speech acts other than directives have generally been little studied, although children's understanding of commissives, in particular the speech act of promising, has received some attention. To understand what a promise is, the child must understand that the speaker expresses an intention to perform some act in the future that will be of benefit to the listener, and furthermore that the speaker realizes that the expression of this intention is a commitment to future action. Astington (1988) found that it was not until 9 years of age that children could reliably distinguish between promises and predictions. Children younger than age 9 were found to be unable to make this distinction and would confuse promising, in which the speaker has control over the future event (e.g., *I'll bring your books back tomorrow*), with predictions (e.g., *It will be a nice day tomorrow*) in which the speaker has no control over the eventual outcome.

Knowledge Needed to Make Inferences About Speech Acts

Let us turn to consider now the kind (or kinds) of knowledge that might underlie the ability to derive the intention behind the utterances of speakers. When Searle (1969) first proposed his typology of speech acts, he proposed a set of necessary and sufficient conditions to accompany the definition of each type of speech act. However, since that time, it has been shown that many utterances are simply too variable and situation dependent in nature to make such definitions very useful (Levinson, 1979).

SAQ 9.4 Try to think of an example of an utterance that might have a variety of meanings (and therefore be classified as a different speech act in each case) depending on whether you said it to a peer, a teacher, a parent, a stranger on a bus, and so on.

Scripts

Cognitive psychologists and researchers in artificial intelligence have been particularly concerned with understanding the role of world knowledge in intelligent behavior and in trying to specify more precisely the nature of the knowledge that we are using, for example, to make inferences about the interpretation of utterances. One reason that such researchers have been interested in precisely specifying the knowledge used in making inferences is so that the process can be modeled by a computer. Minsky (1975) proposed that we have *frames* for what such things as rooms in general consist of and for what specific rooms consist of. The notions of *scripts* (Schank and Abelson, 1977) or *schemata* (Rumelhart and Ortony, 1977) are similarly concerned with the body of knowledge that all adults have about everyday activities, such as going to a restaurant or riding on a bus.

Our script knowledge has most often been investigated in the context of understanding stories. For example, Schank and Abelson (1977) gave subjects short stories such as the following to read:

> Terence got on a bus to go to work. He sat down. When the conductor came, he realized that he had left his money at home, so he had to walk to work.

They found that readers generally had little difficulty going beyond the material presented in the text to answer such questions as "Did the conductor give Terence a ticket?". The tricky problem was to explain how people were making these types of inferences so easily. The authors suggested that people have scripts stored in memory that describe the relevant persons (e.g., travellers, bus driver, conductor), objects (money for bus fare, ticket, seat on the bus, door), and events (waiting for the bus, getting on the bus, paying the fare) occurring during a bus journey and appeal to these to answer questions that require going beyond the text.

Within the context of discourse, on the other hand, the influence of script knowledge has been little studied in adults. However, there has been extensive research with children investigating the development of script-like knowledge. Nelson and her colleagues (Nelson, 1981; Nelson and Gruendel, 1979) found that preschool children readily talked about their experiential knowledge in script-like form and that the hypothesized scripts affected the way they interpreted and remembered stories and everyday events. Script knowledge has also been shown to help sustain peer interaction in young preschoolers. Nelson and Seidman (1984) found that when children engaged in pretend play with a theme familiar to each child (e.g., making dinner), children were consistently able to maintain longer play-centered dialogues than when they were simply engaged in verbal episodes with peers outside of a play situation. The authors attribute this finding to the fact that the shared script knowledge provided a context within which each partner could make contributions that were understandable to the other.

Speech event

Ethnographers who study how language is used in everyday situations in different cultures around the world have found the notion of a *speech event* to be a useful way of capturing more precisely the way language is used and how individuals are able to decipher the intention of a speaker's utterance (Bauman and Sherzer, 1974; Hymes, 1972). A speech event is defined by ethnographers as "a culturally recognized social activity in which language plays a specific, and often rather specialized role (like teaching in a classroom, participating in a church service)" (Levinson, 1983, p. 279; see also Hymes, 1972). Other people have similarly argued that the functions of utterances appear to depend heavily, and quite decisively, on the social situation (i.e., the speech event) in which they are used. For example, Levinson (1979) has pointed out that, in a classroom, one's interpretation of the utterance *What are you laughing at?* is quite different from the interpretation that one might have under other circumstances. Levinson (1983) notes that this view of speech events is compatible with Wittgenstein's (1958) older notion of a *language game.* Wittgenstein (1958) denied that what we can perform with language can be captured in a small set of speech acts and argued that there are an indefinite number of language games that humans engage in and so, correspondingly, there an indefinite number of speech acts.

At this point, our discussion of inferring communicative intent has progressed from describing how people use language to considering the types of background knowledge that may come into play when interpreting utterances. We will now concentrate for most of the remaining part of this chapter on the latter topic and examine the development of a number of pragmatic skills related to the ability to make inferences about the shared background knowledge of communicative partners. Interestingly, as you will see, children seem to acquire a new and more sophisticated understanding of how to assess other people's knowledge between the age of 3 and 4.

Inferences About Shared Knowledge

We will begin by looking at four pragmatic skills that are related to the task of planning what to say at the *global* level of the sentence itself: introducing a topic, considering given and new information, contributing relevant information, and recognizing and repairing misunderstandings. We will then examine two further skills that come into play at the level of planning the *constituents* of the sentence itself (e.g., once you've already decided what you want to say). As you read about the development of these pragmatic skills in children, consider how the understandings highlighted at the beginning of this chapter might play a role in accounting for these developments.

Initiating a Topic and the Problem of Joint Reference

Consider Zara's comment, "Gran send me bear." This comment reveals a very important feature of language: language can be used to inform others about events and objects that are not immediately present. The very force of language as a tool is the fact that it can be used to refer to absent entities. The fact that we can use language in this way has freed us from the need to experience everything firsthand. An event can be witnessed by only one person and related to many others via language. Much remains to be known about how children first recognize that language can be used for this purpose. What is known, however, is that children begin to use their language for this function from very early on (e.g., Halliday's informative function).

By the age of 17 months, children have progressed from the ability to request objects that are close by, or far away (but in view), to being able to request objects that are out of view (Bruner et al., 1982; Greenfield and Smith, 1976). Children at this age will also search for objects that are currently out of view when asked, for example, *Where's your teddy?* (Huttenlocher, 1974; Zukow, Reilly, and Greenfield, 1982).Around this time, children also begin to use such linguistic terms as *gone, allgone,* and *no more* to describe such events as finishing all the food in their bowl, dropping a toy off a high chair, and so on. The use of such absence marking terms as *gone* has been shown to be correlated with children's ability to track and retrieve objects that are moved out of sight (Gopnik, 1982; Gopnik and Meltzhoff, 1986).

The ability not only to request an absent object, but to make a comment about an object not in sight (such as Zara's comment above), represents a more sophisticated understanding and use of displaced reference. Children begin to produce such comments around 2 years of age (Sachs, 1983). Usually these first comments take place in the context of *conversational routines* and the absent object or person mentioned is a unique referent that requires no further specification (e.g., Daddy). For example, a child may repeatedly ask the whereabouts of someone such as asking *Where's Daddy?* or *Daddy work?*. The structure of the child's utterance is usually similar from time to time, but the exact content and linguistic forms are varied (Sachs, 1983). These conversational routines are probably used precisely because, early on, there is no easy way for parents and children to arrive at agreement about a referent (Sachs, 1983).

Children's own ability to talk about past events appears to progress from talking about events that have just occurred, to talking about events earlier in the day, and, finally around 30 months of age to talking about events both before the present day and future events beyond the present day (e.g., Eisenberg, 1985; Hudson, 1993; Miller and Sperry, 1988). Researchers have suggested two reasons why the ability to initiate communication about a non-present or abstract topic may take a while to develop. First, the nonverbal devices that children have relied upon to initiate topics about present objects cannot be used. Second, the grammatical elements that are used to initiate such abstract topics are late in developing. These include the use of the definite article *the* as a marker of old information, anaphoric pronouns (e.g., *that, he*) as a means of referring to previously mentioned objects and persons, relative clauses as a means of specifying referents more clearly, and tense markers as a means of indicating of time relations (Keenan and Schieffelin, 1976; McTear, 1985).

Providing New Information (Maxim of Quantity)

When people communicate with others, they assess what is old and new information for them (Clark and Marshall, 1981; Grice, 1975; Keenan and Schieffelin, 1976). The ability to assess another person's knowledge is integral to successful communication. Piaget (1926) proposed that before 7 or 8 years of age children tend to communicate without tailoring their message to what a listener knows or doesn't know (they communicate *egocentrically* as he referred to it). Although early empirical work supported Piaget's (1926) claim, these methodologies were not specifically geared to younger preschool children (e.g., Flavell, Botkin, Fry, Wright, and Jarvis, 1968).

Recent **theory of mind** research focusing on children's developing understanding of mental states (e.g., Astington, Harris, and Olson, 1988) has shown that, as young as age 2, children are beginning to take the knowledge of other people into account and tailor their communication accordingly. O'Neill (in press) has demonstrated that 2-year-old children show an ability to tailor their communication to their parent's knowledge state. In these studies, two-year-old children had to ask a parent for help in retrieving a toy. On each trial, a child was first introduced to a new toy that was then placed in one of two containers on a high shelf. The parent either witnessed these events along with the child, or did not, because she or he had left the room or had covered her of his eyes and ears. It was found that when asking for help in retrieving the toy, children significantly more often named the toy, named its location, and gestured to its location when a parent had not witnessed these events than when she or he had.

Menig-Peterson (1975) found that when conversing with an adult partner, 3-year-old children talked more and needed less prompting to talk about events that their partner had not experienced, than those which had been jointly experienced by the child and the adult partner. Perner and Leekam (1986) examined how 3-year-old children adapted the information they gave to a 4-year-old communicative partner regarding two possible actions that a toy could produce. In one condition, the 4-year-old partner was absent while the 3-year-old watched both actions (fully ignorant partner), whereas in the other condition the partner witnessed one of the two actions together with the 3-year-old (partially ignorant partner). When later asked about the toy's actions by their fully ignorant partner, older 3-year-old children mentioned both actions and, when asked by their partially ignorant partner, they mentioned only the action which was new to their partner. Younger 3-year-old children correctly mentioned only the new action in the partially ignorant condition. However, in the fully ignorant condition, they tended to be underinformative, mentioning only one of the two actions even though further questioning revealed that they had not forgotten the other action. Three-year-old children have also been shown to display appropriate use of the articles *a* and *the*, a skill that requires children to take into account information that may be old or new for a communicative partner. Emslie and Stevenson

(1981) found that, when narrating a story, 3-year-old children used *a* to introduce a new referent, and *the* for subsequent mention, in a manner similar to adults.

Coherence (Maxim of Relation)

Coherence is concerned with the relationships between ideas expressed in utterances. The notion of coherence is implied in Grice's (1975) maxim of relation (and in Sperber and Wilson's (1986) development of his idea into the single overarching principle of relevance) whereby speakers aim to make their contributions relevant to the ongoing conversation. That is, in a coherent discourse, each person's contribution to the conversation relates to the ones that precede and follow it, so that the whole discourse or conversation is about something (Reinhart, 1981).

Bloom, Rocissano, and Hood (1976) looked at the ability of children, from 2 to 3 years of age, to make relevant contributions to a conversation. When a child's utterance shared the same topic as the preceding adult utterance and added information to it, it was classified as *contingent* (as opposed to non-contingent if it was off-topic). A number of different ways were noted in which children provided contingent utterances. At first, around 2 years of age, children tended to respond to an adult's utterance with a lexical item or phrase that either added to or replaced information in the preceding adult utterance. For example:

(1) *Adult*: Put this on with a pin
Child: sharp

(2) *Adult*: You gonna make a house again? (playing with blocks)
Child: tunnel.

At the next stage, between 2 and 3 years of age, children were found to repeat part of the adult's previous utterance, while at the same time adding or replacing certain information in the adult's original utterance, for example:

(3) *Adult*: What did I draw?
Child: draw a boy.

(4) *Adult*: Can I read it?
Child: I wanta read it.

Children were also found at this time to be able to both reword the adult utterance and add to it, or reword and repeat part of the adult utterance while also adding to it as in the following two examples:

(5) *Adult*: You do that one.
Child: now I do that one.

(6) *Adult*: What are you going to do? (child is setting up a doll's house)
Child: I'm gonna to make something in this.

Finally, by about 3 years of age children were able to produce contingent utterances in the form of an added sentence:

(7) *Adult*: Well, where's Gary?
Child: he doesn't come to the party.

Recognition of Misunderstanding and Children's Repairs

When, in the scenario, Zara changes her utterance *Mommy, -ops* to *Pops, pops* in response to her mother's query *Hm? What darling?*, this is an instance of Zara repairing an initial utterance. Children's repairs in miscommunication episodes have been argued to reveal the ability, even of prelinguistic infants, to recognize "a partner's capacity to *understand* a message" and attribute "an internal state of *knowing* and *comprehending*" to a communicative partner (Bretherton, McNew, and Beegly-Smith, 1981, p. 339; see also Bates et al., 1979; Bretherton and Bates, 1979).

In response to signals of miscomprehension by an adult, children show a tendency to repair their utterances from a very early age. Children usually first begin to repair their utterances around 12 months of age by repeating their original signal, or augmenting it gesturally or with vocal emphasis. By about 16 months of age, children substitute the original signal with a verbal or non-verbal communicative behavior of similar meaning, or substitute a new request for the old one (Golinkoff, 1986; Wilcox and Howse, 1982). Research (Wilcox and Webster, 1980; Anselmi, Tomasello, and Acunzo, 1986) has also shown that, at 17 months of age, children differentially adapt the amount of verbal material they repeat depending on whether an adult responds to their initial verbal request with a general query (e.g., What?), a specific query (e.g., You want what?), or treats the original request as a declarative statement (e.g., Yes, I see).

Findings such as these have generated considerable debate in the literature as to what they imply about the child's understanding of a communicative partner's mind. Golinkoff (1983, 1986, 1993) has argued that these findings suggest that preverbal infants have considerable communicative skill and treat their mothers, not as omniscient, but rather as communicative partners who need more information of a specific nature. Other researchers, however, have hesitated in attributing to children such a sophisticated understanding of a communicative partner in explaining these repair behaviors. Instead, it has been argued that children may have learned to make such repairs through observing the responses of adults to such queries (Anselmi et al., 1986; Shatz and O'Reilly, 1990; Wilcox and Webster, 1980). Overall, the key problem with this research has been that the factors influencing children's communications have not been adequately controlled for. To draw firm conclusions regarding the ability of children to take into account the mental state of a communicative partner, the partner's knowledge must be manipulated directly (e.g., O'Neill, in press).

In older children, research has focused not only on their ability to repair their utterances in response to listener feedback, but also on their ability to monitor their own comprehension as listeners and to appreciate that communication failures can be the fault of either the speaker or the listener. Space does not permit a review of this research, but the findings suggest that even into elementary-school age, children often fail to notice when they do not understand. Markman (1977) for example, found that first-grade children were confident that they would know how to play a game when the instructions contained glaring omissions. It was only when these children were given a chance to enact the game that they began to give any indication of recognizing message inadequacy. In studies investigating how children attribute blame for communicative failures, Robinson and Robinson (1976a, 1976b) have found that younger children, of around 5 years, commonly fall in the listener blaming category. Speaker-blamers become more common around 7 years, and by 11 years of age children were found to blame the speaker in each case when appropriate.

We will now turn to two other skills that come into play at the level of planning the constituents of a sentence. At this level, the concern is with designing a sentence that fits into the discourse and accurately conveys just the right piece of information (or, according to relevance theory, has the right contextual effects). The two skills we will look at are deixis and presupposition.

Deixis

Deictic terms are words that refer to places, times, or participants in a conversation from the speaker's point of view. For example, *I, you, here, there, this, that, yesterday,* and *tomorrow* are deictic terms. Evidence suggests that at a remarkably early age, children acquire a pragmatic ability to interpret and use deictic terms. According to Clark (1978), the appropriate use of the pronouns *I* and *you* and the locational terms *here, there, this,* and *that* is based on an understanding that such terms are mapped to the roles of speaker and hearer, and not self and other.

Charney (1979) has shown that children as young as 2 and 3 years of age are aware of the relevance of speaker roles with respect to deictic terms. In her study, 2- and 3-year-old children were seated on the floor with the experimenter and two toys (T1 and T2). In one condition (the same perspective condition), the child and the experimenter sat together and T1 was placed nearby, while T2 was placed further away. In a second condition (the neutral perspective condition), the child was equidistant from each toy but the experimenter was closer to T1. In the third condition (the opposite perspective condition), the child was near T1 and far from T2 while the experimenter was near T2 and far from T1. In each condition the child's understanding of the deictic term was probed by asking, for example,

> *See the airplane?*
> *See the train?*
> *Which one is here/there?*

Children's responses in the neutral condition were of particular interest, because if children were only operating from their own perspectives, then given no information from that perspective, they should choose randomly or show confusion. If, however, they were taking the role of the speaker into account, then they should answer according to the experimenter's perspective. The opposite perspective condition was also of interest because in this condition information from their own perspective would favor choosing the wrong toy. Charney found that even the younger children performed about as well in the neutral perspective as they did in the same perspective condition, with the opposite condition being the worst (about 72% correct however!). Thus, she found no evidence for a stage in which the self is used exclusively as the reference point.

Two-year-old children have also demonstrated quite a sophisticated understanding of the use of deictic pronouns, such as *I* and *you.* Tanz (1980) posed questions such as *Ask Tom if I have blue eyes* to 2-year-old children. Two-year-old children experienced little difficulty in formulating the relevant question to be posed to Tom, namely *Does she have blue eyes?* Generally it has been found that children acquire the pronouns *I, you,* and *it* during the second year of life (e.g., Chiat, 1986).

Presupposition

Presupposition is a term that has many definitions related to pragmatics. In its most global sense, it can refer to any aspect of conversational ability that relies on taking background knowledge into account. As such, it encompasses many of the topics we have already considered. For this discussion, however, we will adopt the definition used within linguistics. Presupposition is therefore restricted to referring to certain pragmatic inferences or assumptions that seem to be built into linguistic expressions and which can be isolated using specific linguistic tests (e.g., constancy under negation, which will be covered in more detail later in this section).

Presupposition has a long history, but suffice it to say here that linguists and philosophers (e.g., Frege, Strawson) noticed that the use of certain linguistic terms (called *presupposition-triggers*) seemed to give rise to a predictable set of inferences that were referred to as "presuppositions." In other words, certain words (in particular verbs) seemed to have inferences built into them. Here are two examples of the many constructions that have been isolated by linguists as sources of presuppositions, taken from Levinson (1983):

(1) *Factive verbs* (e.g. Strawson, 1950)
e.g., Martha *regrets* drinking John's home brew
(2) *Implicative verbs* (e.g. Karrtunen, 1971)
e.g., John *managed* to stop in time.

Can you guess what presuppositions are "triggered" by the verbs *regret* and *managed*? The answer, in example (1), is that the presupposition is the complement of the verb *regret*, namely that Martha drank John's home brew. It would not make sense to speak of Martha regretting that she drank the beer if in fact she had never drunk any. In example (2), the presupposition is that John tried to stop in time. Linguists have defined approximately 13 core phenomena that are considered presuppositional. Two more are listed here with the inferences presupposed in brackets (for detailed discussion see Levinson, 1983):

(3) *Definite descriptions* (e.g. Strawson, 1950)
e.g., John saw *the* man with two heads (There exists a man with two heads)
(4) *Change of state verbs* (e.g. Sellars, 1954)
e.g., John *stopped* beating his wife (John had been beating his wife)

Many of these constructions have been investigated in children. The focus has been on understanding how children come to understand and express the particular presuppositions carried by various words and linguistic devices (for a detailed review of this research see DeHart and Maratsos, 1984). Investigations of this kind have shown that the answer to this question is by no means straightforward. Let's now take a look at some research that has focused specifically on children's understanding of the factive verb *know*.

According to Kiparsky and Kiparsky (1970) certain verbs such as *know*, *regret*, and *remember* presuppose the truth of a complement clause while other verbs such as *think* and *believe* do not. To illustrate, consider the following sentences:

(1) John knows that it is raining
(2) John thinks that it is raining
(3) It is raining

Sentence (1) presupposes the truth of sentence (3) while sentence (2) does not. Verbs that presuppose the truth of the complement, such as *know*, are called factives, while verbs that do not are called nonfactives. Kiparsky and Kiparsky (1970) provided a number of tests for differentiating factive from nonfactive verbs. One such test is constancy under negation. This test is based on the observation that the truth of the complement of a factive verb, but not a nonfactive verb, remains constant when the verb is negated. Thus, sentence (4) below also implies the truth of sentence (3).

(4) John does not know that it is raining

A number of studies have investigated children's understanding of presupposition by examining their recognition of this property of factive verbs. As you can probably anticipate, the understanding of this property of factive verbs is not an early development. For example, in one study (Scoville and Gordon, 1980) kindergartners, second, fifth, and eighth graders, and adults were told such things as "Dr Fact doesn't think/know the ball is red." Then they were asked "Is the ball red?". It was found that only the adults reliably recognized that negation of factive predicates (e.g., doesn't know) did not affect the truth of the sentence complements (e.g., the ball is red). The kindergartners and second graders were as likely to assume that negation of factive predicates did affect the truth of the sentence complements as to assume that it did not, and the eighth graders were only performing marginally better.

Other researchers have argued, however, that tasks such as that of Scoville and Gordon (1980) are inordinately difficult for children and underestimate what children actually know about factive verbs. These researchers (e.g., Abbeduto and Rosenberg, 1985) have argued that while the tests proposed by Kiparsky and Kiparsky (1970) were logical ones, their definition of presupposition is not strictly logical. Presupposition, they argue, is also used to refer to the *beliefs of the subject* of a mental verb (i.e., the person who thinks, knows, believes, and so on). According to this view, a term such as *know* can be differentiated from one, such as *think* in that *know* implies that the subject has unambiguous evidence, either perceptual or inferential, as to the truth of the complement, while *think* implies the subject does not have such evidence. Thus, an alternative approach to studying children's understanding of presupposition is to see if children can select the appropriate mental term (e.g., *know* versus *think*) to describe a person's mental orientation (either their own or that of some story character) towards a particular statement or event. The findings have varied somewhat, but in general they have shown that when presupposition is defined in this way, an understanding is seen to emerge around 4 to 5 years of age.

For example, in two studies (Johnson and Wellman, 1980; Miscione, Marvin, O'Brien, and Greenberg, 1978) it was investigated whether children recognize the mental nature of the referents of mental terms, in particular *know, remember,* and *guess.* In these tasks, the subject, or the character in the story, hid or found an object in one of several possible locations. Two conditions were varied – whether or not the person saw where the object was initially hidden, and the person's actual performance (i.e., whether or not the judgment about the location of the object proved to be correct). In this way, scenarios could be set up in which the subject or character knew or guessed the location of the object. Under these conditions, 4-year-old children have typically been found to judge mental states strictly with respect to performance, ignoring knowledge basis. Miscione et al. (1978) found that many children younger than 5-and-a-half years of age stated that they knew the location of the object when their performance was correct, even when they lacked any knowledge basis (i.e., they had actually just guessed correctly). Similarly, Johnson and Wellman (1980) found that 4-year-old children completely confused conditions of *knowing, remembering,* and *guessing,* and that 5-year-old children were just beginning to differentiate these mental terms. Further studies have suggested that in fact 4-year-old children may be able to differentiate *knowing* from *thinking* and *guessing,* but are still rather poor at determining what constitutes good evidence for knowing something (e.g., Abbeduto and Rosenberg, 1985; Moore, Bryant, and Furrow, 1989; Moore and Davidge, 1989).

How Is the Information We Use to Make Inferences About Shared Knowledge Represented?

Encyclopedia and diary

Clark and Marshall (1981) have argued that the proposed *scripts, frames,* and *schematas* discussed earlier leave unanswered how this knowledge may be compartmentalized in a way that allows us to access relevant background knowledge quickly and effectively when engaged in the activity of conversing with others. Consider all the instances in conversation with others where we use or interpret a definite reference. For example, suppose that in the scenario, Lucy had said to her mother *You know the guy I met in the bookstore?* It must be explained how Lucy's mother is able to use the knowledge she has in order to interpret Lucy's act of definite reference (the guy I met in the bookstore). Or as Clark and Marshall argue, we need to consider "how this knowledge might be compartmentalized according to what information is mutually known by a community or by two individuals, as would be required for definite reference" (1981, p. 54). Clark and Marshall have proposed that in addition to an *encyclopedia* that we consult for references that require knowledge based on community membership and indirect coherence, we also have a *diary* which is a personal log in which we keep track of everything significant we do or experience. Thus, for example, for Lucy's mother to interpret the utterance *The guy I met in the bookstore,* she must consult her diary for an entry that gives evidence of the physical, linguistic, or indirect copresence of him, Lucy, and herself. At present, not much is known about how this diary might be organized, but the important point made by Clark and Marshall (1981) is that to handle a definite reference we cannot simply have a *dictionary* that links a reference (the guy I met in the bookstore) with a referent (Mark) because we would need to generate these lists ad infinitum taking into account such things as those names that we know another person knows, and those names that we know another person knows we know, and so on. Very quickly, this list would become unmanageably large. The notion of a diary presents us with a much more manageable log of this type of information.

Cognitive psychology and artificial intelligence have made important contributions which have helped to clearly specify and make explicit what background knowledge must be represented in order for people to be able to make the kinds of inferences about utterance meaning that they do make. McTear and Conti-Ramsden (1992) have emphasized that a better understanding of the types of knowledge involved may be useful in understanding some of the difficulties faced by language-impaired children who have pragmatic difficulties that may be influenced by background knowledge that is missing or inappropriately used. This is the topic of the next and last section of this chapter.

Pragmatic Impairment

Autism and Semantic–Pragmatic Disorder

Communicative impairment is a hallmark feature of autism. This impairment manifests itself particularly as a problem with two-way communication and the pragmatic use of language. Comprehensive research reviews have shown that individuals with autism are not specifically impaired in the development of phonology or syntax (e.g., Fay and Schuler, 1980; Tager-Flusberg, 1981). Individuals with autism may learn and use a narrower range of grammatical structures such as word order, past tense, and negation, but the development is not deviant in form. In contrast, the development of pragmatic aspects of language is deviant among individuals with autism. Baron-Cohen (1988) reviewed the studies exploring the pragmatic features of autistic communication. Such features included turn taking, foregrounding and backgrounding of old and new information, interruption of the speaker at inappropriate moments, and faulty use of eye gaze during conversation. Individuals with autism were found to be impaired on every pragmatic feature studied. Even prelinguistically, children with autism have been shown to differ significantly in how they use the nonverbal communicative means at their disposal. Several studies have shown that although autistic children will use gestures and nonverbal behaviors such as eye gaze to instrumentally request something from an adult, these communicative means are not used to establish or regulate joint attention. For example, children with autism are unlikely to use any of these means, such as showing a toy or pointing out an event, to coordinate attention between themselves and a communicative partner in order to share an awareness of an object or event (e.g., Curcio, 1978; Loveland and Landry, 1986; Mundy, Sigman, Ungerer, and Sherman, 1986).

This schism between an ability to acquire the syntactical and grammatical components of a language as opposed to the interpersonal use of language is also seen in some children who do not display many of the other features of autism. These children have been described as having a semantic–pragmatic disorder of language. A case study of one such child provides a very compelling example of how this disorder affects this child's interactions with others. In this exchange, the father and the child (John, age 3 years, 3 months) are looking at a book (Blank, Gessner, and Esposito, 1979, p. 346):

Father: That's Pat's house. What's everyone doing at Pat's house?
John: knock, knock, knock. (Knocking on door in book)
Father: Come in!
John: Nobody's home.
Father: Nobody's home? Well, isn't Pat home? (Pat is evident in the picture)
John: Come back later.
Father: O.K., Let's go to Pat's new house.
John: Pat's old house (Looking at book)

SAQ 9.5 What features of the above dialogue between John and his father indicate John's difficulty with the pragmatics of language?

Explaining Pragmatic Ability and Disability

What is abundantly clear from research to date is that a single ability does not underlie pragmatic development and that many diverse kinds of knowledge are involved. Indeed, the field of pragmatics to date is only at the beginning stages of being able to present, in a clear and specific way, the kinds of knowledge underlying pragmatic development. At a very global level, there are three kinds of knowledge that are generally assumed to play a role and interact in pragmatic development: linguistic knowledge, cognitive knowledge, and social knowledge (see McTear and Conti-Ramsden, 1992, for an extensive discussion of these factors underlying pragmatic impairment).

Level of linguistic ability

Although some aspects of pragmatic ability appear to be relatively independent of linguistic ability, as, for example, the ability to get another person's attention, to determine what is old and new information, and to order the events in a narrative, linguistic skills nevertheless play a large role in the expression of children's pragmatic abilities. Initiating a conversation, for example, would seem to require some degree of linguistic ability. If you are trying to introduce an absent referent as a topic of conversation, linguistic skills are paramount and may range from simply being able to use the name of an object, to being able to use a relative clause (e.g., The guy who I met in the bookstore) to achieve mutual reference. Similarly, distinguishing between old and new information will rely not only on children being able to recognize the need to do so, but on a mastery of the necessary linguistic constructions such as *a* and *the*. Indeed, grammatical ability has been shown to affect the production of informative messages. Feagans and Short (1984) found that children's oral language deficits co-occurred with difficulties in producing narratives, due to a lack of syntactic structures and vocabulary inflexibility. Syntactic difficulties also appeared to affect the children's ability to paraphrase in their responses to clarification requests, while limited vocabulary – as well as word retrieval problems – affected the ability of language impaired children to introduce new referents to their listeners (Liles, 1985).

Level of cognitive ability

Background and world knowledge is crucial to the ability of people to make sense of what is said in conversation by relating it to previous experience and by making predictions based on knowledge about objects and persons in the everyday world. As we have seen, this knowledge has often been described in terms of scripts. In one study, McTear (1989) has examined the effect of deficient world knowledge on a child's use of language. In this case study, the child was asked to describe what was going on in a picture that showed a child locked out of his house and emptying his pockets to find the key. McTear found that the subject was unable to make the inference that the child had lost his key and was unable to enter the house. Instead, the subject persisted in describing superficial details of the picture, such as the fact that the boy's pencil was lying on the ground. The subject, therefore, seemed to have a problem using world knowledge to infer the information that was implicit and not explicitly represented in the picture. This problem was all the more striking given that although the subject's descriptions were not the predicted ones, they were generally reasonable (e.g., he described the boy as having lost his dinner money).

What might have accounted for this child's deficient world knowledge is another question. McTear points out that there are a number of places where things could go wrong. There may be problems with the initial perception of the event, which itself is constrained by world knowledge. Limited world experience may therefore be part of the problem or, perhaps, processing problems as well might affect the ability to perform the required cognitive operations of categorization, inference, and so on.

Level of sociocognitive ability

Sociocognitive knowledge encompasses our ability to make social inferences about the actions, beliefs, and intentions of other persons. These inferences will be crucial to our understanding of the behavior of others and our ability to adapt messages to their needs. This type of knowledge encompasses the understandings about people highlighted on page 247. For example, we cannot tailor our communication to what another person knows if we do not understand that other people possess mental states, such as knowledge. As we have already seen, research on children's theory of mind has told us much about the development in children of an understanding of mental states such as desire and belief and their role in understanding and predicting behavior, both in themselves and others (for reviews of much of this research see Astington et al., 1988; Butterworth, Harris, and Frith, 1991; Perner, 1991). The basic finding of this research has been that children begin to acquire the ability to distinguish between the world (the way things are) and their representations of the world (the way they and others may believe/think/know the world to be) around 2 years of age when they begin to understand pretence and can distinguish between desires and the world. By 3 to 4 years of age, children come to have an understanding of belief and its role in explaining and predicting the behaviors of others. More complex social understanding, such as distinguishing between a lie and a joke, or understanding the role of beliefs and intentions when making judgments of moral responsibility and social commitment, develop later around the age of 8. Similarly, more complex uses of language such as irony, metaphor, and sarcasm develop late because they depend on more sophisticated belief attributions, such as knowing what someone else knows about another person's beliefs.

Currently, one theory receiving a lot of attention and empirical support is that pragmatic deficits of the kind seen in people with autism may be the result of an *impaired theory of mind* (Baron-Cohen, Leslie, and Frith, 1985; see Baron-Cohen, Tager-Flusberg, and Cohen, 1993, for a collection of research papers). That is, it has been hypothesized that people with autism may have a general inability to represent another person's representations and thereby to consider another person's mental states such as their desires or beliefs. Without an ability to consider and reflect on one's own mental states and the mental states of other people, many pragmatic aspects of language, such as adapting communication to the knowledge state of a communicative partner and making relevant and informative contributions, become difficult if not impossible.

Summing-up

This chapter has introduced you to the study of pragmatics: the study of language usage. We began by looking at how researchers have tried to define pragmatics as an area of study separate from semantics. The issues of meaning and context were problematic in this respect: what aspects of meaning are we concerned with in pragmatics (e.g., sentence vs. utterance meaning and encoded vs. inferred meaning), and what aspects of context play a role in the pragmatics of language.

We then took a look at how children are developing, from a very early age, the ability to make inferences about a speaker's communicative intent and inferences about background knowledge shared between themselves and a communicative partner. It was revealed how these early pragmatic skills depend on very basic knowledge about other people (e.g., theory of mind research). We have concluded this chapter by examining the nature of communication when pragmatic skills are impaired and possible ways of explaining these cases of pragmatic impairment.

Further Reading

For an introduction to the field of pragmatics the following two texts are recommended: S.C. Levinson (1983), *Pragmatics*, Cambridge: Cambridge University Press, and D. Blakemore (1992), *Understanding Utterances: An introduction to pragmatics*, Oxford: Blackwell. The collection of essays of Clark and his colleagues, H. Clark (1992), *Arenas of Language Use*, Chicago: University of Chicago Press, is useful for its in depth discussion of issues relating to the ways in which people coordinate the "common ground" of knowledge between them in discourse. Good texts dealing with many of the developments of children's pragmatic competence mentioned here are: (1) H. Clark and E. Clark (1977), *Psychology and Language: Introduction to Psycholinguistics*; (2) S.H. Foster (1990) *The Communicative Competence of Young Children*, New York: Longman; (3) D.J. Messer (1995) *The Development of Communication: from social interaction to language*, Chichester: John Wiley & Sons. (4) R.L. Schiefelbusch and J. Pickar (eds) (1984), *The Acquisition of Communicative Competence*, Baltimore: University Park Press; and (5) M. Shatz (1994) *A Toddler's Life*, Oxford: Blackwell. For a detailed reviews of pragmatic disability in children see (1) M. McTear and G. Conti-Ramsden (1992), *Pragmatic Disability in Children*, London: Whurr; and (2) B.R. Smith and E. Leinonen (1992), *Clinical Pragmatics*, London: Chapman and Hall.

CHAPTER 10

Learning and Memory

Outline

In this chapter we look at a number of important questions about learning and memory. We see that memory can manifest itself in a variety of ways, and that a good deal of evidence points to the existence of distinct memory modules: a short-term store, as well as separate stores within long-term memory. Much of the relevant evidence comes from analysis of the ways in which memory can break down as a result of damage to the brain. We briefly examine the central role played by the hippocampus in memory storage and retrieval, and consider the way in which learning occurs in the brain via the plasticity of the synaptic connections between neurons. We also look at various explanations of forgetting, as well as some of the ways in which conceptual knowledge might be represented. Finally, we review some of the ways in which researchers are attempting to understand learning and memory from a computational perspective.

Learning Objectives

After reading this chapter you should be able to:

- discuss the definition of learning and memory
- explain the basic characteristics of classical and instrumental conditioning
- describe the evidence for a distinction between short- and long-term memory
- evaluate the evidence for dividing long-term memory into sub-modules
- identify the main types of memory problem that can result from brain damage
- discuss the role of attention in learning
- explain the notion of transfer-appropriate processing
- discuss why forgetting occurs
- describe how memories and concepts are stored in the brain
- debate some of the ways in which learning and memory can be modeled computationally

Key Terms

- amnesia
- concepts
- declarative memory
- procedural memory
- transfer-appropriate processing
- working memory

Introduction

Our scenario contains a number of illustrations of the role of memory in everyday life. Here are some examples:

(a) *Dad*: She didn't work last night. She sat in front of the TV.
(b) *Zara*: Gran send me bear.
(c) *Mom*: . . . Where did she go last year? I've completely forgotten.
Dad: The Grand Canyon.
(d) *Lucy*: Mom, have you seen the tray?
Mom: I just saw it in the living room.
(e) *Lucy*: I'm eating at Jane's tonight, Mom – OK?

In each of these cases, one of the participants is remembering something from the past. In (a)–(c), events that occurred between a day and a year ago are remembered, while in (d) Mom is remembering something over a much shorter interval, namely where she saw the tray a few seconds previously. Case (e) is interesting because Lucy is "remembering" a future event, but this is not as paradoxical as it seems: she is remembering that at some earlier time (perhaps yesterday afternoon) she arranged to eat out tonight. In each case, an event can be remembered, and this implies that at an earlier time something was learned which can now be recalled. Thus Dad has learned and retained the fact that Gran went to the Grand Canyon last year for her holiday, and he is able to retrieve this fact in response to Mom's question.

Although the significance of memory in these examples is obvious, the scenario also contains many other less obvious examples of memory. In fact, every word spoken represents an example of memory, insofar as the meanings of words are stored in what is known as **semantic** or lexical memory. Mom would not know how to answer Lucy's question in (d) unless she knew the meaning of the word "tray." This meaning was probably learned when she was young and has been remembered ever since. Similarly, Lucy's ability to make coffee depends on information stored in memory, in this case a "procedure" or sequence of actions that need to be executed in a strict order. This form of memory is known as **procedural**.

In this chapter we consider how learning and memory are possible. We are particularly interested in the following questions:

- What are the different ways in which memory can manifest itself?
- Is the human memory system made up of a number of distinct modules, or is it a single system?
- In what ways can memory break down as a result of damage to the brain?
- Which parts of the brain are involved in memory and how does learning take place?
- How can we understand learning and memory from a computational perspective?

We begin by considering what is usually regarded as one of the simplest and most prototypical examples of learning, namely classical conditioning, famously discovered and investigated by the Russian physiologist Ivan Pavlov (1927). By looking at this form of learning, we see that memory obeys a number of fundamental laws.

Classical and Instrumental Conditioning

Pavlov's well-known conditioning experiments involved teaching dogs to salivate to a bell that signaled the imminent arrival of food. While studying digestive physiology, Pavlov discovered that a natural reaction (e.g., salivation) elicited by a motivationally-significant event (in this case food) could be conditioned to a stimulus such as a bell that predicted that event. The traditional terminology is that the motivationally-significant event is called the unconditioned stimulus (US), the natural reaction to that event is called the unconditioned response (UR), the predictive event is called the conditioned stimulus (CS), and the learned reaction to the CS is called the conditioned response (CR). In honor of its discoverer, **classical conditioning** is also known as Pavlovian conditioning.

Plainly, an animal that produces a CR in the presence of the CS must have learned something about the predictive relationship between the CS and US. For instance, the first time a dog hears a bell prior to its food, it will not salivate to the bell because it does not know that the bell means that food is about to arrive. But with repeated presentations of the bell–food relationship, salivation will increase as the dog learns about the relationship. Pavlov showed that conditioned responses could be remembered over very long intervals: presenting the bell after an interval of a week, for example, continues to elicit salivation.

What are the main properties of Pavlovian conditioning? First, conditioning depends on the sequence of events. If the CS and US are simultaneous, or if the CS follows rather than precedes the US, conditioning is likely to be very weak, although not necessarily absent (Matzel, Held, and Miller, 1988). Secondly, when the CS precedes the US, the strength of conditioning depends on the interval between these stimuli, generally being better the shorter the interval. If the bell precedes food by a few seconds, salivation to the bell will be much greater than if it precedes food by many minutes. Thirdly, the CR will be abolished or undergo **extinction** if after conditioning the CS is presented repeatedly without the US. Unsurprisingly, animals will stop making conditioned responses if the CS is no longer a reliable signal of the US.

Instrumental conditioning, first systematically investigated by Edward Thorndike around the turn of the century, differs from classical conditioning in a number of important respects. In an instrumental conditioning procedure,

Devil's Advocate Box 10.1 Is conditioning an unconscious process?

A very common belief about conditioning is that it can occur unconsciously, which is to say that a response may be conditioned to a CS without the subject being aware that this is occurring. But is this really the case? Certainly, conditioning can occur in all members of the animal kingdom, down to such lowly creatures as the sea-slug, *Hermissenda*, and it is hard to imagine (though perhaps not impossible) that such organisms are conscious.

In fact, evidence from adult humans suggests that far from being an unconscious process, conditioning is always accompanied by consciousness. For example, many experiments which attempt to mask the conditioning procedure have shown that CRs only develop in subjects who can say what the relationship between the CS and US is; those who are unaware of the relationship do not develop CRs (Lovibond, 1993).

One apparently convincing example of unconscious conditioning comes from so-called "biofeedback" experiments showing that it is possible to instrumentally condition autonomic responses such as heart rate. In such experiments, subjects try to move an indicator (say on a computer screen) in a certain direction. In fact, the indicator's position is determined by the subject's heart-rate, but even though subjects are usually unaware of their heart rate, they nevertheless succeed in moving the indicator in the desired direction (e.g., Fowler and Kimmel, 1962). Surely this is a compelling case of unconscious conditioning? The problem with such experiments is that although the experimenter arranges a relationship between changes in heart rate and reinforcement, from the subject's point of view a different response may be reinforced. For instance, deep breathing tends to increase heart rate, so if the subject breathes deeply, he or she will be reinforced. Crucially, evidence obtained by Roberts, Williams, Marlin, Farrell, and Imiolo (1984) suggests that subjects who show conditioning will be aware of such "incorrect" reinforcement contingencies, while those who are unaware of such contingencies will not show conditioning.

Another superficially convincing example of unconscious conditioning concerns phobias. A person who is afraid of spiders is quite likely to acknowledge that spiders are perfectly harmless, yet still show conditioned fear responses such as increased heart rate and anxiety in the presence of a spider. But again, it is unclear whether this really counts as an example of unconscious conditioning, because although the person may not regard spiders as dangerous, he or she is likely to see them as frightening. In that case, the conscious belief (*spiders are frightening*) and the conditioned fear response are closely coupled.

some reward or punishment is given to the organism when it makes a particular response. The best-known example studied in the laboratory is lever-pressing by rats. When the rat presses a lever mounted in the side of a so-called Skinner box, a pellet of food is presented. In this situation, the food is called a **reinforcer** (rather than a US) because it serves to reinforce the behavior that produced it, and the behavior itself is called an instrumental response (rather than a CR). In contrast, if the response produced an electric shock rather than food, the shock would be called a punisher because it would serve to punish and hence suppress the behavior that produced it.

Instrumental conditioning has many of the same features as Pavlovian conditioning. Thus an instrumental response will be made more vigorously the sooner the reinforcer follows it, and an organism will cease making an instrumental response if the reinforcer no longer occurs. The relationship between the instrumental response and the reinforcer (or punisher) can be described by the famous "Law of Effect," first proposed by Thorndike (1911). According to this Law, a response is more (or less) likely to occur if it produces a satisfying (or unpleasant) state of affairs. Thus in our scenario, Zara is likely to say "more milk" because in the past she has learned that this behavior has satisfying consequences, and she will say it louder and more frequently if it actually causes her parents to give her milk than if they ignore it.

This discussion of classical and instrumental conditioning provides an overview of some simple forms of learning, and gives a preliminary look at some of the main features of learning. Before considering more complex properties of memory, we first turn to some comments concerning the definition of learning and memory.

SAQ 10.1 Before reading the next section, try to think what the terms "learning" and "memory" actually mean. Can you produce rough definitions of them?

Definition of Learning and Memory

Before Pavlov's investigations of conditioning, the experimental and analytic study of human learning and memory had been initiated at the end of the last century by two pioneering researchers, the German psychologist Hermann Ebbinghaus (1885) and the French neurologist Théodule Ribot (1882). Ebbinghaus realized that it would be extremely difficult to uncover the basic laws of learning and memory in naturalistic settings, and so he set himself the highly artificial task of learning long lists of nonsense syllables (such as "wux"), reasoning that his memory for such lists would be uncontaminated by the meaningfulness of the to-be-remembered items. With these procedures, he was able to study under controlled conditions such fundamental properties of memory as the rate at which information is forgotten, the effect on later recall of extra list rehearsals, and so on. Ribot, in contrast, attempted to understand the principles of memory via the patterns of breakdown that occur following brain injury.

The terms **learning** and **memory** are often presented as different concepts, and the distinction between them is preserved in many ways in psychological research. Yet in reality they are different sides of the same coin. Memories are what are left behind as a result of learning, and we infer the existence of learning from the presence of memories. What exactly do we mean by the terms "learning" and "memory"? The definition of these apparently innocuous terms has been a topic of passionate debate by psychologists. In their enthusiasm to rid the subject of mentalistic concepts, advocates of **Behaviorism** (especially B. F. Skinner) tried to avoid using these terms as explanatory notions, instead arguing that the goal of psychology is to determine the laws that relate stimuli to responses. Accordingly, when we say that a dog in a laboratory Pavlovian conditioning experiment has learned and remembers something about the relationship between a bell and food, what we mean is just that a new behavior has been conditioned: the dog salivates to the bell, whereas previously it did not. On such a view, we should only use the terms "learning" and "memory" if there is some observable change in behavior, in which case the new behavior is the learning (and memory).

However, there are at least two obvious problems with this definition. The first is that learning may occur without any concomitant change in behavior: if a CS and US such as shock are presented to subjects administered with drugs that block muscular activity, conditioned responding may perfectly well occur to the conditioned stimulus when the paralytic drug has worn off (e.g., Solomon and Turner, 1962). Learning clearly occurs when the animals are paralyzed, even though no behavioral changes take place at that time. The second problem is that in many cases it can be established that organisms do much more than simply acquire new types of behavior. For instance, in a famous experiment, MacFarlane (1930) trained laboratory rats to run through a maze to obtain food, and found that when the maze was filled with water, the animals continued to take the right path to the food even though they now had to swim to reach it. Clearly, learning in this case does not merely involve the acquisition of a set of particular muscle activities conditioned to a set of stimuli, but instead involves acquiring knowledge of the spatial layout of the maze, with this knowledge capable of revealing itself in a variety of different ways.

A more cognitive view, therefore, is that memory is an abstract term that describes mental states which carry information, while learning describes a transition from one mental state to a second, in which the information is in some way different. Memories may perfectly well be formed without the development of any new behaviors and, furthermore, may manifest themselves in a variety of quite different behaviors. But, although this definition avoids the problems associated with characterizing memory in terms of behavior, it also has its shortcomings. For instance, how are we to distinguish between learning and forgetting? Forgetting, like learning, can be viewed as a change in stored information, except that in this case information is lost rather than gained. Our definition would plainly need to be supplemented by a proviso that learning involves the gain of information, but it is likely to be very difficult to specify what we mean by "information gain" without relying simply on behavior. We might find ourselves reverting to a behavioral definition of learning, which is precisely what we were trying to avoid.

Another problem with the cognitive definition is that it fails to deal satisfactorily with examples of what we might call "non-cognitive" learning. The cognitive definition refers to a transition from one *mental* state to another, and the reason for incorporating the restriction to mental states is to exclude examples like the following. Roediger (1993) reports that the average duration of labor for first-born babies is about 9.5 hours, while that for later-born babies is about 6.6 hours. Clearly, for second and third children the amount of time the mother spends in labor is much less than for first children. It seems strange to say that the female reproductive system is capable of "learning" and "remembering," so we would like to exclude this sort of case, despite the fact that the body has changed as a result of experience. The restriction to mental states excludes the labor case because the relevant changes take place in the body without any mental component. But then we seem to be committed to saying that all habits and skills, such as learning the piano or learning to ride a bicycle (which we *do* want to include as examples of learning) must be mental, and this seems unduly restrictive. Is it not likely that some aspects of learning a skill like riding a bicycle are really bodily rather than mental? Borderline cases like this probably illustrate the futility of trying to define the terms learning and memory.

Techniques for Studying Learning and Memory

Historically, human learning and memory have most often been studied using **recall** or **recognition** procedures. After learning a list of words or other materials, memory is tested with either a recall test, in which the subject attempts to generate as much of the studied information as possible, or with a recognition test, in which studied items have to be distinguished from non-studied ones. Our scenario includes examples of this sort of **episodic** (or **explicit**) memory, for instance when Dad remembers where Gran went on her holiday last year.

The almost ubiquitous use of recall and recognition tests serves to emphasize one of the many ways in which Ebbinghaus contributed to the study of memory. In addition to testing memory via such episodic tests, Ebbinghaus also used what we would now call procedural (or **implicit**) tests, in which the effects of memory are demonstrated indirectly via their influences on some ongoing task. An example is the use of "savings" tests: having learned a list of nonsense syllables, Ebbinghaus tested his memory at a later time not only by attempting to recall the list, but by making himself relearn it all over again. If the list is relearned in less time than it was originally learned, that is evidence of memory for the list persisting and aiding the relearning task. But this evidence is indirect in the sense that memory is assessed not via a conscious attempt on the part of the subject to recall the earlier presentation of the list, but instead via a detectable change in ongoing behavior. As mentioned earlier, another example of procedural memory occurs in the scenario when Lucy makes the coffee; again, such procedural knowledge is retrieved indirectly.

Ebbinghaus was long ahead of his time in recognizing the distinction between these two types of tests. In recent years, this distinction has been rediscovered and has fuelled an enormous amount of research, some of which will be described below.

Short- and Long-term Memory

If one considers how an engineer would construct a set of memory systems sufficient to deal with all the complexities of human life, in all probability the major distinction that would be drawn would be between **short-** and **long-term memory**. Almost all computers, for example, have a short-term store where information is maintained for brief amounts of time while it is relevant to the current task, and have a separate memory store for permanent information (e.g., a hard disk). In our scenario, the distinction between short- and long-term memory is apparent in the contrast between Mom remembering where she has just seen the tray and Dad remembering where Gran went on holiday last year. Clearly, the time-spans here are very different.

Figure 10.1 gives a schematic illustration of the way in which short- and long-term memory may be related. The system consists of a central processor which manipulates information, together with two short-term stores, one for auditory and the other for visual information (the basis for separating these will be discussed later), and, finally, a long-term store (LTS).

Given the common-sense appeal of the short- versus long-term memory distinction, it should perhaps come as no surprise that psychologists have devoted a great deal of effort to trying to see whether such a distinction exists in human memory. Here, we review some of the relevant evidence. Probably the most direct evidence for the existence of a separate short-term store (STS) comes from studies of so-called "digit span." For most people, there is a strict limit to the amount of information that can be held in memory at any one time. For instance, while it may be possible to keep hold of a 6- or 7-digit telephone number someone gives you, a 10-digit number is likely to be too long to remember without being written down. This suggests that the short-term store has room for about seven items at any one time, and no more (Miller, 1956; Shiffrin and Nosofsky, 1994). From a computational perspective, this obviously suggests the existence of a limited-capacity store in which new information displaces older information.

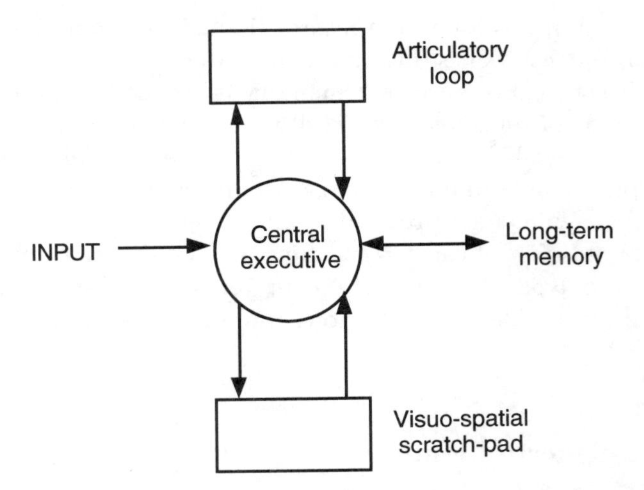

10.1 The working memory model of Baddeley and Hitch (1974). Information is processed first of all by a central executive system, which has access to two short-term 'slave' systems for verbal (the articulatory loop) and visuo-spatial (the visuo-spatial scratch pad) information. The central executive also controls the long-term memory system.

SAQ 10.2 If the short-term store has a fixed capacity, it should be impossible to increase one's digit span, yet in fact this can be done. Indeed, one person (see Chase and Ericsson, 1982) was able to increase his span to about 80 digits! How can this finding be reconciled with the notion of a fixed-capacity store?

Another type of evidence for distinguishing between short- and long-term memory comes from so-called free recall experiments in which subjects try to remember a list of items, such as words, and then remember them in any order they choose. Here is an experiment to try on yourself:

> First get a pen and paper. Then read slowly through the list of words printed below, taking a couple of seconds to try to remember each word. Once you've read a word, don't go back to it! (You might want to use a piece of paper to cover the words as you read them.) When you reach the end of the list, close the book and try to write down as many of the words on the list as you can, in any order you choose.
>
> scissors, lemon, snake, doorknob, dress,
>
> spoon, pineapple, basket, flower, accordion,
>
> mushroom, potato, glass, carrot, glove.

If you have behaved like subjects in laboratory versions of this experiment, you should have found it much easier to remember the first and last few words on the list than the ones in the middle. The solid line in figure 10.2 shows the results of just such a free recall experiment conducted by Glanzer and Cunitz (1966), except that subjects heard rather than read the words. The likelihood of recalling a word is plotted as a function of its position in the list. The figure shows that items appearing early in the list tend to be well recalled (the **primacy effect**) as do items appearing towards the end of the list (the **recency effect**). Items in the middle of the list are less well remembered.

Ignoring the primacy effect for the moment, it is clear that a likely explanation of the recency effect is in terms of a limited-capacity short-term memory store. Suppose that each item in the list is stored briefly in a short-term store prior to either being transferred to the long-term store or to being forgotten. As new words come in, earlier words are displaced from the STS. Plainly, a recency effect will emerge because the last items on the list can be recalled directly from the STS. Some of the earlier items, in contrast, will have been forgotten.

It turns out that various experimental manipulations can affect the recency and pre-recency parts of the serial–position curve differently, and this provides quite good evidence for the STS/LTS distinction. For example, the dotted lines in figure 10.2 show that when Glanzer and Cunitz required subjects to do 10 or 30 sec of some unrelated mental activity after the last item of the list, the

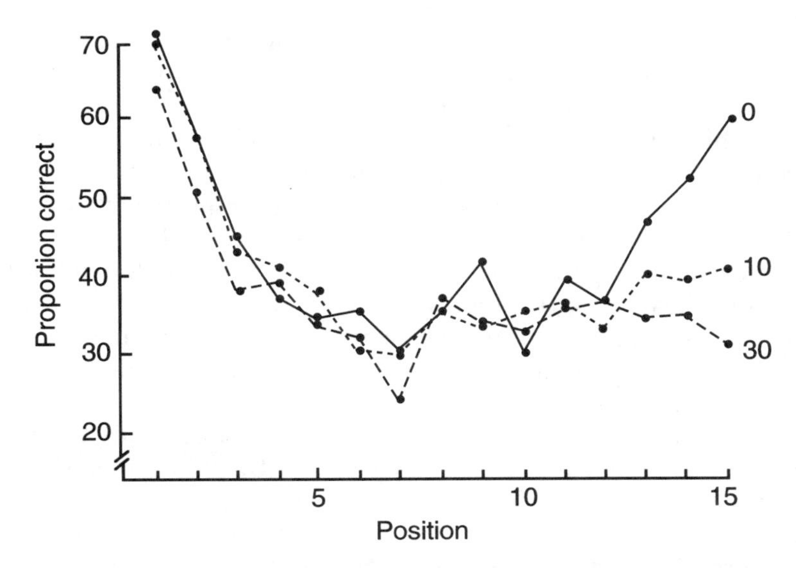

10.2 Serial position curves obtained by Glanzer and Cunitz (1966). Subjects heard a list of 15 words and then tried to recall them. When there was no delay after the last word (the 0 curve), the first and last words were better recalled than ones in the middle of the list (these are termed the 'primacy' and 'recency' effects, respectively). When there was a filled delay of 10 or 30 sec after the last item on the list (the curves labelled 10 and 30), the recency effect was abolished but recall of earlier words was unaffected. This is interpreted as evidence that the most recent items on the list are capable of being retained in a short-term memory store, but when a delay is interposed, these items are displaced from the store. Items not in the short-term store can only be recalled if they have been transferred to the long-term store.

recency effect disappeared, but recall of earlier items was unaffected. In another experiment, increasing the number of words in the list reduced performance over the pre-recency part of the curve, but left the recency effect intact. Thus two experimental variations yielded opposite effects on the pre-recency and recency parts of the curves, and this is consistent with the idea that different memory stores are involved in free recall.

It is important to note that the number of items which receive a boost over the recency part of the curve is pretty much the same whether the items are words, digits, or something else. This emphasizes the crucial fact that what gets maintained in the short-term store is a set of "chunks" of information, where a chunk can be a word, a digit, or whatever. This explains why some people can develop digit spans of much greater than seven or so items (see SAQ 10.2). The subject SF who could recall up to 80 digits was able to do so by breaking the list down into about seven or eight chunks, with each chunk in turn consisting of ten or so digits that represented a meaningful sequence to him. Thus, although you can increase your short-term memory capacity, there is a sense in which it is still constrained to consist of a small number of items; the items just get bigger.

What is the explanation of the primacy effect? In a clever experiment, Rundus (1971) required his subjects to rehearse aloud during the presentation of the word list. Figure 10.3 shows the relationship between the number of times a word was rehearsed and the probability of it being recalled. As the graph shows, these two are closely related for all items up to the last ones on the list, and the primacy effect can be explained simply by the fact that the first few words on the list get rehearsed more times on average than do items in the middle of the list, with extra rehearsals increasing the probability that a word is stored in the LTS. When the subject hears the first word, there is only one item to rehearse, but by the time the 10th word is presented, much less time can be spent rehearsing it if the previous words are not to be forgotten. The fact that rehearsal is unrelated to performance for the last items on the list confirms the hypothesis that these items are being recalled directly from the STS rather than from the LTS.

Rehearsal is an important mechanism whereby information undergoes **consolidation** in memory, but the simple STS/LTS model is misleading in implying that consolidation simply relies on the number of times a piece of information is rehearsed (and hence whether it is transferred to the LTS). Instead, consolidation also seems to depend on a sufficient passage of time after the learning episode to allow the memory to be more permanently registered by the brain, probably as a result of biochemical memory processes that continue for some prolonged period of time. Evidence for this comes from animal and clinical studies of **retrograde amnesia**. As a result of brain injury, drugs, electroconvulsive therapy (ECT), or a variety of other causes, memories stretching back in time may be disrupted. For instance, Squire, Slater, and Chace (1975) studied a group of patients who had undergone ECT, which is a treatment for depression in which electric shocks are administered to the brain. The patients were asked to recognize the titles of television series that had been broadcast in different years, and the results showed that memories going back about three years were temporarily disrupted by the ECT treatment, whereas memories for older programs were intact. This pattern of results suggests that the consolidation process can continue for as long as three years, but that after that time a memory is fully consolidated and hence immune to retrograde amnesia.

Retrograde amnesia is also seen in patients who have permanent amnesia as a result of brain injury or surgery. Marslen-Wilson and Teuber (1975) have found that patient HM's (see box 10.1) retrograde amnesia stretches back several years, again suggesting that consolidation may be a very long-term process.

Returning to the STS/LTS distinction, another important piece of evidence for the distinction comes from neuropsychology. Given the conceptual scheme shown in figure 10.1, it should in principle be possible to find people who as a result of brain injury have an impaired short-term store together with normal long-term memory, and such patients have indeed come to light. For example,

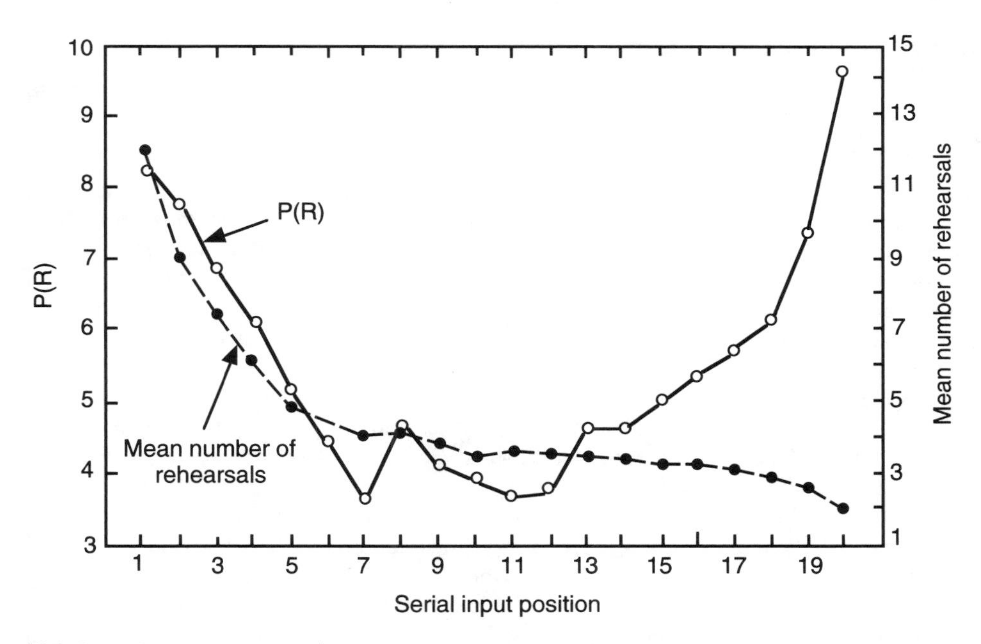

10.3 Data from Rundus (1971). Subjects took part in a serial free recall experiment but rehearsed items aloud. The solid curve (left-hand axis) shows the normal serial position curve with a primacy and recency effect, while the dotted curve (right-hand axis) shows that subjects rehearsed the early items more than any others. Taken together, the results suggest that the primacy effect in free recall is due to the fact that early items on the list are more frequently rehearsed than other items. However, the recency effect is not due to rehearsal: instead, the data are consistent with the idea that the last few items on the list are maintained in a short-term store.

the patient KF studied by Warrington and Shallice (1969) had a digit span of only one item, but had no long-term memory problems. This, of course, is consistent with a selective deficit to the STS.

Conversely, it should be possible to find other neurological patients whose STS is intact but whose long-term memory is impaired. HM (see box 10.1) provides an example of just this pattern of performance. Moreover, this pattern of deficits should lead to a predictable result in a serial free recall experiment: such amnesics should show normal recency effects, but should be poor in their pre-recency performance. Exactly this pattern of results has been observed (Baddeley and Warrington, 1970).

SAQ 10.3 Write down a list of pairs of tasks that you think you could comfortably perform simultaneously, and a second list of pairs of tasks that would be very difficult to perform in combination. Later in this chapter we will return to the issue of combining tasks.

Finally, there is a good deal of evidence to suggest that instead of being a single module, the STS should be divided up into a number of distinct modules as in the **working memory** system shown in figure 10.1. The evidence

Box 10.1 The case of HM

Perhaps the best-documented case history of the effects of brain damage on memory concerns the patient HM (see Corkin, 1984). At the age of 27, HM had most of his temporal lobes removed by the neurosurgeon William Scoville in an attempt to cure very severe epilepsy. In the operation (during which H.M. was awake!), the hippocampus, amygdala, and other adjacent structures were removed on both sides of the brain. While the operation was partly successful in its principal aim, a catastrophic side effect was to render HM highly amnesic.

Although HM's IQ is excellent (about 110) and his linguistic abilities unimpaired, his amnesia is severe and extends both to events occurring prior to his operation (retrograde amnesia) and to those occurring subsequently (anterograde amnesia). For instance, he remembers very little autobiographical information about his life prior to the operation, and subsequently has been unable to learn such basic things as the name of the current US President, despite regularly watching TV. He does not know where he lives or what his last meal was, and when he is asked what year it is, can give an answer as much as 40 years out. When he is repeatedly tested by a psychologist, HM usually has no recollection of ever having seen the person before.

But HM's amnesia is not global. For a start, his short-term memory is normal, with a digit span of about six items. Furthermore, he shows near-normal learning and retention of a number of motor and perceptual (or, more broadly, procedural) skills. For instance, he can learn how to solve a paper-and-pencil maze, and shows a decrease in solution time each time he does it. Indeed, for many procedural tasks, HM's learning and retention is within the normal range.

for this is complex, but comes mainly from dual-task experiments. These are memory experiments in which subjects are required to do two things at once, for example, learning a list of words and doing some mental arithmetic. It will come as no surprise that memory is normally impaired when a second task is performed concurrently, but Baddeley and his colleagues have found that there are combinations of tasks which, generally, can be performed together without interfering to any great degree with one another. For instance, visual and verbal short-term memory tasks can often be performed with very little interference. For this reason, Baddeley and Hitch (1974) proposed that STS is made up of at least three sub-modules, as shown in figure 10.1: an articulatory loop, which is like a tape-loop that can circulate and rehearse a few seconds worth of verbal material; a visuo-spatial "scratch-pad," used for visual and spatial tasks, such as remembering the layout of a set of objects; and a central executive, the part of the system that coordinates processing and directs attention. On this scheme, patient KF has a deficit to his articulatory loop. Because verbal and visual short-term memory are dealt with in different systems, there is the possibility of doing two tasks at once with only minimal interference.

SAQ 10.4 Try to specify the computational characteristics of the short-term store. How exactly does this store operate?

Procedural and Declarative Memory

We have seen that there is a good deal of evidence for separating short- and long-term memory, but in what other ways is the memory system modular? A considerable amount of research has examined the question of whether long-term memory consists of distinct subsystems. We have already seen a number of terms used to describe different types of memory (episodic, semantic, procedural), and in this section we will consider whether these distinctions reflect something about the underlying nature of memory. Specifically, we will consider some of the evidence for a well-known distinction, first proposed by Cohen and Squire (1980), between procedural and **declarative** memory. The relationship between procedural, declarative, episodic, and semantic memory is shown in figure 10.4. This shows that semantic and episodic memory are subdivisions of declarative memory.

Selective Deficits of Declarative Memory in Human Amnesia

The distinction between procedural and declarative memory has its origins in the pattern of memory breakdown observed in the classic **anterograde memory** syndrome, first described by Korsakoff (1889). As a result of brain injury or

Devil's Advocate Box 10.2 Linking memory to language

A slightly different perspective is cast on short-term memory when looked at from the perspective of word processing and speech production. In expanded models of the kind presented in figure 6.7, there are a number of input modules which process information at the pre-lexical and lexical levels in both visual and auditory modalities. Similar modules have been postulated for object processing and separate ones for face processing. In addition, to handle a variety of phenomena in speech production there are output modules at the lexical and post-lexical levels all with specified functions. It is believed that, when stimuli are processed or output is produced, the specific activity in each module continues for a while. In the case of the input lexicons, this activity is detectable a day or more later and is the origin of the priming effects discussed in chapter 6.

One interesting phenomenon that is nicely accounted for by this approach is the fact that there is typically a large difference in serial recall following auditory versus visual presentation of stimuli. While the model in figure 10.1 has no obvious means of explaining such an effect, it can be attributed to phonological information concerning the final item which is left behind in the pre-lexical processes in the auditory speech channel (see Morton, 1970). Also, the source of the strong recency effect in serial recall is put down in part to the output buffers, which are necessary if we are to explain a variety of phenomena in speech.

From this perspective, then, the question is not whether short-term memory is different from long-term memory, but, rather, how many sources of short-term information are available, and how do they all contribute to performance in different tasks? Which approach is correct? Well, it turns out that each approach has its advantages and can account for some of the data. Neither will serve on its own, and instead we need to work towards a unified model which will maintain the advantages of both.

disease, people can become densely amnesic. Within a few minutes of being given a list of words to remember, HM is unable to recall them. Another group of patients, suffering from Korsakoff's Syndrome (often brought on by prolonged alcoholism), experience similar memory problems.

In the 1960s however, it began to emerge that patients like HM do not have

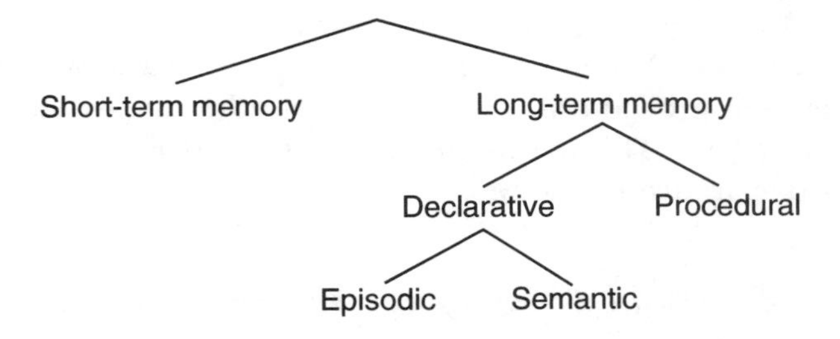

10.4 Relationships between different proposed memory systems. According to this taxonomy, procedural and declarative memory are subcomponents of long-term memory, with the declarative system being further divisible into a semantic and an episodic system.

a uniformly impaired memory, since they appear to be able to learn and remember what we might call "procedures" (Corkin, 1984). For instance, Warrington and Weiskrantz (1968) showed their amnesic patients a series of highly fragmented pictures such as that reproduced in figure 10.5. The patients were shown the most fragmented picture and asked to try to identify the object. If they could not do so, the next most fragmented picture was shown, and so on until the object was identified. Over the next two days, the procedure was repeated, and the point at which the object was identified was noted. Warrington and Weiskrantz found that both the normal control subjects and the patients took less time to identify the object on its second and third presentation than on the first one, indicating that some learning must have been occurring. Crucially, the patients showed just as much **priming** as their normal control subjects.

It is now known that amnesics who otherwise have acute difficulty learning and remembering new information can perform very well on a whole range of procedural learning tasks. Examples include not only priming effects such as that described above, but also conditioning and motor skills (Squire, 1987). For instance, amnesics are just as good as normal subjects at acquiring the skill of reading mirror-reversed text in which all of the letters are printed backwards.

What is the important distinction between those tasks that amnesics can perform and those they cannot? This is an important question because its answer may tell us something profound about the modular organization of memory (see chapter 3 for background on modularity). In their study, Graf, Shimamura, and Squire (1985) presented words either visually or auditorily to normal and amnesic patients, prior to a free recall and a so-called word stem completion test. In this latter test, the first few letters of a word are provided, and the subject is asked to complete the stem with the first word that comes to mind; no reference is made to the previous presentation (hence this is another indirect or implicit test of memory). The typical finding from such tests is that seeing a word on the study list increases the likelihood that subjects will later complete the stem with that word. This is an example of priming similar to that seen in the Warrington and Weiskrantz (1968) study discussed above.

Figure 10.6 shows that although the amnesic patients freely recalled fewer words than normals, there was no difference between words that had been presented visually and auditorily. However, on the word completion test, the converse pattern was found: amnesics and normals performed identically, but both groups completed more words that had been presented visually than auditorily.

Why does this interesting pattern of results occur? Ignoring the auditory/visual effect which, we will return to later, Graf et al.'s data show a dissociation between performance of normals and amnesics on a declarative, but not a procedural memory task. Declarative memory refers to memory for facts (semantic memory, e.g., *Angel Falls is in Venezuela*) as well as memory for past events (episodic memory, e.g., *where Gran went for her holiday last year*) that is to say, any memory that can be stated as a proposition. In contrast, procedural memory is non-factual; as we have seen, Pavlovian conditioning and the ability to ride a bicycle would be good examples. According to this theory, different brain areas are involved in these different forms of memory, and hence selective brain damage may affect only one particular sort of memory. In amnesia, it is declarative memory that is affected. An example of the converse,

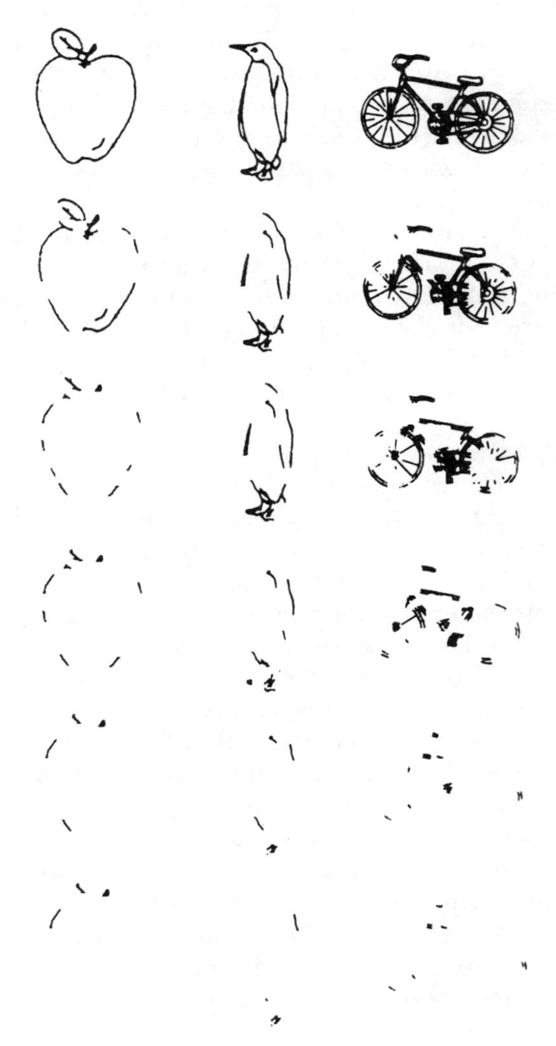

10.5 Fragmented pictures of the sort used by Warrington and Weiskrantz (1968). In a 'priming' experiment, subjects are shown the most fragmented version of a picture and asked to try to identify it. If they are unable to do so, the next most fragmented picture is shown, and so on until the object is identified. This procedure is repeated at a later time and the point at which the object is identified is noted. Warrington and Weiskrantz found that both normal subjects and amnesic patients took less time to identify objects on their second presentation than on their first one, indicating that some learning ('priming') must have been occurring.

namely selective impairment of procedural memory, may be Huntington's disease, which involves damage to structures in the brain called the basal ganglia. There is some evidence that patients with this disease have normal declarative memory, whilst being impaired in priming tasks and at learning new skills (Martone, Butters, Payne, Becker, and Sax, 1984).

Problems for the Procedural/Declarative Distinction

There are a number of problems with the idea that human memory is divided along the procedural/declarative dimension. Probably the main worry concerns data suggesting that amnesics are more impaired in some forms of declarative memory than others. As noted above, declarative memory covers both memory for past episodes and memory for plain facts, and although amnesics certainly have great difficulty learning both sorts of information, their episodic memory is far more seriously impaired than their semantic memory. Evidence for this comes from studies in which amnesics and normal control subjects are taught facts (e.g., *Angel Falls is in Venezuela*) and then tested for their recollection of them. If the controls are tested after a longer retention interval than the amnesics, it is possible to arrange for the fact memory of the two groups to be approximately equal. In spite of this, the amnesics will,

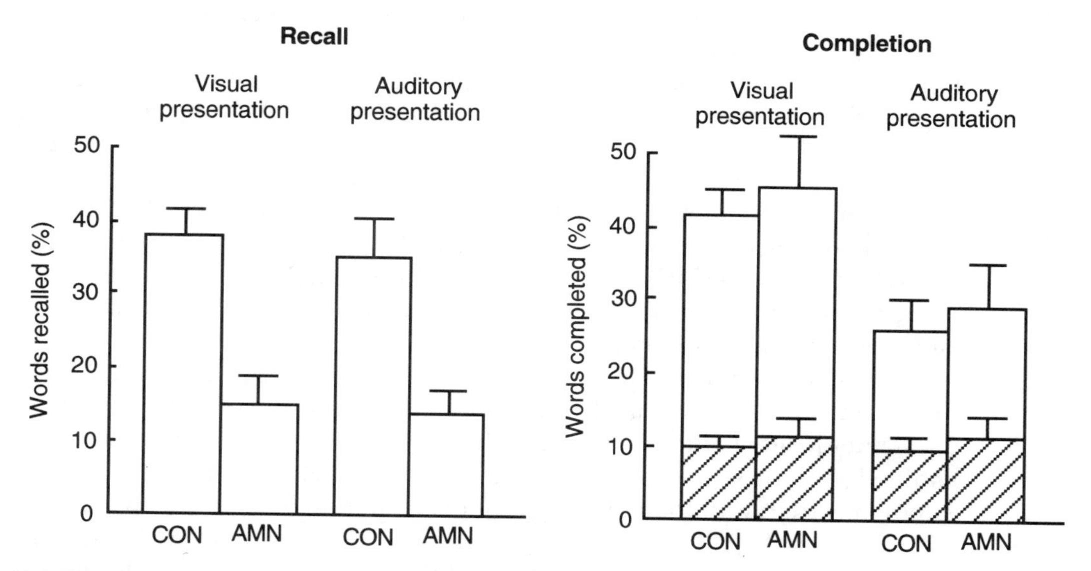

10.6 Data from an experiment by Graf, Shimamura, and Squire (1985). Amnesic patients (AMN) and normal subjects (CON) either read or heard words, and then either tried to freely recall them or to complete fragments of those words. The left-hand panel shows that amnesic patients are much worse than normals at free recall, but whether the words were read or heard makes no difference. In contrast, the right-hand panel shows that amnesics and normals showed equivalent levels of priming in the fragment completion test, but that more fragments of read than of heard words were completed. The shaded bars indicate the probability that a word fragment would be completed if it had not been studied in the first stage of the experiment. The fact that the levels of completion for studied words are much higher than this shows that priming is occurring.

nevertheless, be far worse than the normals at episodic recall, that is remembering where they learned the facts (Shimamura and Squire, 1987). This finding suggests that it is mistaken to think that amnesics are impaired equally at all forms of declarative memory. But at the same time, it would be erroneous to imagine that it is only their episodic memory that is impaired: amnesics also have great difficulty acquiring new semantic knowledge (Tulving, Schacter, McLachlan, and Moscovitch, 1988).

Another piece of evidence against the procedural/declarative theory comes from normal as opposed to memory-impaired subjects. A natural prediction of the theory is that it should be possible to observe cases of procedural learning in the absence of accompanying declarative knowledge, yet it has proven very difficult to find convincing examples of this. For instance, if conditioning is a prime instance of procedural learning, then it should be possible to perform experiments in which people acquire conditioned responses as a result of pairings of a CS and US, but where they are unable to report declaratively the nature of the CS–US relationship. Many experiments have been conducted to detect such a dissociation, but success has not been evident (see devil's advocate box 10.1).

We can illustrate the way in which procedural and declarative knowledge seem to "march in step" by considering an experiment by Perruchet and Amorim (1992). They required subjects to perform a reaction time (RT) task in which a light appeared in one of four locations (A–D) and the subject had to press as fast as possible a key corresponding to that location, then the light appeared in another location, and so on. Unknown to the subjects, however, was the fact that the lights appeared in a consistent sequence (DBCACBDCBA) which repeated over and over again. After many trials with this sequence, Perruchet and Amorim switched the subjects to a random set of trials, and found that reaction times increased. This suggests that the subjects had procedural knowledge of the sequence: when the stimuli appeared in a regular sequence, subjects were able to predict where the next stimulus would appear, and hence respond very quickly.

If procedural and declarative knowledge can be teased apart in normal subjects in the same way as they can in amnesia, then despite having procedural knowledge, perhaps subjects in this experiment did not have declarative knowledge of the sequence? This possibility was evaluated in the following way. After many trials with the repeating sequence, subjects were asked just to try to tap out on the keys the order in which the stimuli had occurred. Thus, if the subject remembered the sub-sequence DBCA . . . , he or she should just tap out that sub-sequence, and we can take this as a test of declarative knowledge despite the fact that it is nonverbal. Contrary to the prediction, it appeared that subjects had excellent declarative knowledge of the sequence, and were able to reproduce many parts of it. This close correspondence between procedural and declarative knowledge occurs in numerous experimental tasks (see Shanks and St John, 1994). In sum, experiments on normal (as opposed to amnesic) subjects have failed to provide compelling support for the idea that separate procedural and declarative memory modules exist.

SAQ 10.5 Say whether each of the following is a procedural or declarative memory task, and in the case of declarative tasks, say whether it is semantic or episodic: knowing the meaning of a word; Mom's remembering where the tray is; conditioning; free recall; riding a bicycle; savings in relearning; completing a word fragment; "remembering" a future event; knowing the meaning of a word in a foreign language you are learning.

Transfer-appropriate Processing

The results discussed above provide something of a paradox. On the one hand, it seems that the pattern of memory difficulties seen in patients with amnesia is consistent with a distinction between procedural and declarative memory. On the other hand, obtaining a separation between these two forms of learning in non-brain-damaged people seems difficult. In light of this, the notion that the human long-term memory system is made up of a number of isolable subsystems has been resisted by a number of researchers. In this section we will examine a rather different perspective on memory.

From a computational point of view, it is natural to think of memory in terms of the storage of facts. But an alternative view is that memory is simply a record of the mental operations carried out on stimuli. What sort of mental operations are there? The most obvious distinction is probably between **data-driven** and **conceptually-driven** operations. Data-driven processes involve bottom-up analysis of perceptual features. Thus, for example, deciding how many vowels there are in the word WHITE is a data-driven process, and reading versus hearing a word will require different data-driven operations. In contrast, conceptually-driven processes are much more top-down and involve the extraction of meaning. Knowing that white is the opposite of black would be an example.

If memory is a record of mental operations, then it should be possible to show that retention is affected by the sort of operation carried out on the stimulus at study, and in fact it is easy to show that memory is indeed affected in this way. In a set of highly influential experiments on the so-called **depth of processing** effect, Craik and Tulving (1975) presented subjects with a series of words to remember. In the study phase, subjects made superficial judgments about some words (is it in capital letters?) and deeper judgments about others (would the word fill the blank in the following sentence . . . ?). The results were clear-cut in showing that subjects remembered the words for which they made semantic judgments much better than those for which they made perceptual judgments. This result is very hard to reconcile with the theory that memory is a simple record of the stimuli that are presented. Instead, it is much

Box 10.2 Doing two things at once

We saw earlier in the chapter that Baddeley and Hitch's working memory model allows for the possibility of two short-term memory tasks (e.g., one verbal and one visual) being performed concurrently with little interference between them. However, it is perhaps more often the case that doing two tasks at once – that is, dividing your attention between two things – can be very difficult to do. Although it is possible, say, to talk and drive a car at the same time, other combinations of tasks seem to be incompatible with one another. For instance, for most people it would be tremendously difficult to read a story and write to dictation at the same time. Why are some combinations of tasks so difficult to combine? And can we overcome the problem of combining incompatible tasks?

One explanation of why certain tasks may be hard to combine is that we have a central attentional system with a limited amount of capacity. In the working memory model, this would amount to the suggestion that the central executive can only do so much at any one time, and on such a theory, combining two tasks creates difficulties (insofar as they are attention-demanding) because only a limited amount of attention can be allocated to each by the central executive. Psychologists often talk of attention as a "mental resource" which is limited (like most resources) and so has to be rationed. This theory runs into difficulties, though, because it fails to explain why the similarity of the tasks is important. If the two tasks are very dissimilar, the chances are that they can be performed concurrently without too much difficulty, but if they are very similar it is likely to be hard to combine them. For example, if tasks involve the same input modality, internal representations, and output modality, they are likely to be hard to perform simultaneously. In all likelihood, the task combinations you produced in answer to SAQ 10.3 reflect this fact. This finding is problematic for the "central capacity" theory. If two tasks A and B are each attention-demanding to some degree, then that implies that more attentional capacity will be required when the tasks are combined. Since attentional capacity is limited, there will be a dual-task decrement in performance, but this decrement should be unaffected by the overall similarity of the tasks. All that matters, for this theory, is the degree to which a central resource is needed for each task.

Another key piece of data concerns the effects of practice on dual-task performance. Intuition suggests that as we have more practice at combining tasks, the extent to which they interfere with each other can be dramatically reduced. For instance, holding a conversation when you are a novice driver can be very difficult, but after a few years' driving practice these tasks can be comfortably combined. In the laboratory there have been several impressive demonstrations of the effects of practice on dual-task performance. Consider a study by Hirst, Spelke, Reaves, Caharack, and Neisser (1980). These researchers trained subjects over a period of several weeks simultaneously to read stories and write lists of words to dictation. With sufficient practice, subjects became able to perform these tasks in combination so well that their comprehension of the stories was as good when they were having to write to dictation as when they were not. Hence, after sufficient practice the task of writing to dictation became so well learned that it did not interfere with story comprehension.

Many researchers have interpreted such findings in terms of the notion of *automaticity*. On this view, initially advocated by Schneider and Shiffrin (1977), tasks can be divided into two sorts, those that are attention-demanding and those that are not. A task that demands attention is called "controlled" while one that requires no attentional resources is called automatic, and sufficient practice will allow an initially controlled task to eventually become automatic. Supporting evidence for the automaticity theory comes from numerous studies which have shown that the attentional demands of a task can be reduced with practice. For example, Brown and Carr (1989) conducted an experiment in which subjects' primary task was to generate a specific sequence of keypresses in response to a given letter, with the sequences being two, four, or six keypresses in length. The measure of interest was the time subjects took to generate the target sequence. Over the course of four hours of practice at this task, subjects unsurprisingly became much faster, but the key result concerned the attentional demands of the task. A secondary task that the subjects occasionally had to perform concurrently with the keypressing task required them to remember an 8-digit number while generating the sequence of button presses. Brown and Carr found that early in training, it took subjects on average 1151 msec longer to generate the target sequence in dual-task than in single-task conditions, but by the end of the experiment, the decrement was only 560 msec. This clearly suggests that the attentional demands of the keypressing task were dramatically reduced over the course of extended practice. Perhaps with even more practice, there would have been no "cost" at all to performing the two tasks at once.

According to Schneider and Shiffrin (1977) and other proponents of the theory, there are several ways (in addition to the extent to which they interfere with other tasks) in which automatic and controlled processes can be distinguished. Specifically, automatic processes are assumed to be very fast and to be unavoidable in the sense that they are always evoked provided that the eliciting stimulus is present; they cannot be inhibited by voluntary control. Although the evidence for these other aspects of automaticity is controversial (see Cheng, 1985), there is no doubt that the automatic/controlled distinction captures a deep truth about the way in which multiple concurrent tasks relate to one another.

more readily reconciled with the view that memory is a record of the mental operations carried out on those stimuli.

Perhaps the strongest claim about the importance of **encoding operations** – those mental operations carried out on a stimulus when it is presented – comes from experiments by Endel Tulving and his colleagues. It is, of course, a fairly obvious fact about memory that some "cues" are more effective than others at helping a lost memory to be recovered. For instance, most readers will be familiar with the annoying phenomenon of being unable to remember a person's surname. But hearing the person's first name may allow their surname to be remembered, in which case the first name has acted as a **retrieval cue**. A simple view of memory might be that any cue that is associated with the "target" item (e.g., a first name in the case of trying to remember the surname) will be effective in aiding retrieval. However, Tulving has conducted numerous experiments showing that things are more complicated than this. In fact, cues that were not present at encoding will be much less effective than those that were, illustrating the crucial importance for later retrieval of the cues that were present at the time encoding took place.

As an illustration, Tulving and Osler (1968) conducted an experiment in which the overlap between cues present at encoding and at retrieval was varied. Prior to the experiment, a set of to-be-remembered target words was chosen, and two words (A and B) weakly associated with each target were also selected. For the target HEALTH, for example, two weakly associated words are BODY and VIGOR. Each to-be-remembered word was accompanied by weak associate A or weak associate B at encoding, by both of these (AB), or by neither. At test, cues A, B, AB, or no cues were provided for retrieval, and subjects had to freely recall as many of the targets as possible. When a particular cue was present at both encoding and retrieval, performance was good. But a cue at retrieval which had not been present at encoding was completely ineffective. This led Tulving and Osler to propose that a retrieval cue will facilitate recall if and only if its relationship with the to-be-remembered item was processed at encoding, and this is known as the **encoding specificity principle**. There are reasons to question whether this principle is universally true, but it does clearly capture an important rule of thumb about encoding operations.

From this perspective, it is straightforward to derive what is now known as the **transfer-appropriate processing** (TAP) theory of memory. This theory states that performance on some memory test benefits from a prior learning episode to the extent that the mental operations needed to complete the test overlap with those required during the learning episode. When one learns a new motor skill such as hitting a tennis ball in a particular way (e.g., a forehand volley), it is much easier to carry out the new skill in the way that has been learned than in some new way (e.g., on the backhand). So it is with memory, according to the TAP theory: having analyzed a stimulus in a certain way, what is remembered is the analysis itself. If a later memory test requires this analysis to again be undertaken, then performance is good. If not, performance is impaired.

We can illustrate the application of the TAP theory by considering an experiment by Weldon and Roediger (1987). These researchers presented subjects in the first stage of their experiment with a mixed list of pictures and words. Thus, one study item might have been a picture of an elephant and another the word "GIRAFFE." Later, subjects were given one of two memory tests. In the word-fragment completion test, incomplete words like _I_A_F_ were shown to subjects with instructions to try to work out what the word could be. In the picture-fragment test, subjects tried to name degraded pictures like those shown in figure 10.5. For both of the tests, some items were ones included on the study list and others were novel.

Weldon and Roediger found that subjects were more successful at completing a word fragment if the item in question had been presented in the study list as a word rather than as a picture. In contrast, degraded pictures were better recognized if the item had been presented as a picture rather than as a word. Thus, if subjects saw a picture of an elephant and the word "GIRAFFE" on the study list, they completed the word fragment _I_A_F_ better than the fragment _L__H_N_. However, they identified a degraded picture of an elephant better than one of a giraffe.

How is this interesting pattern of results to be explained? According to the TAP theory, it comes about because of differences in the extent to which the mental operations required in the test recapitulate those demanded at study. Identifying a degraded picture at test clearly shares many mental operations involved in identifying the original studied picture, and working out a word fragment shares operations involved in reading a study word. In contrast, performance is poor if the memory test involves operations different from those carried out at study.

SAQ 10.6 Try to see if you can explain the results of Graf et al.'s experiment (figure 10.6) in terms of the TAP theory. As a hint, it might help you to decide roughly what sorts of mental operations are involved in (a) reading a word, (b) hearing a word, (c) freely recalling a word, and (d) completing a word fragment.

The theory of transfer-appropriate processing is a way of accounting for memory phenomena in processing rather than in modular terms. Instead of explaining dissociations between declarative and procedural tests in terms of different underlying brain structures, such dissociations are viewed from within a single memory system in which a variety of different operations may be performed on a stimulus. Memory is simply the record of these operations.

Forgetting

On the face of it, the idea that memories will be retrieved if the encoding context is recreated is, to say the least, counterintuitive because it seems not to allow for the ubiquitous phenomenon of **forgetting**. With the exception of very unusual people such as Luria's (1968) famous mnemonist, S, who could apparently remember everything that had ever happened to him given sufficient retrieval cues, we all regularly have the vexing experience of being unable to recall something we once knew. What, then, is the explanation of forgetting? Can we understand the processes that cause it?

It is important to note, as a preliminary point, that the term "forgetting" is ambiguous and has at least two distinct meanings. First, the term is used descriptively to refer to a person's behavior. At the behavioral level, "forgetting" refers to a person's inability to retrieve items of previously-learned information. Secondly, the term is used in an explanatory sense, to mean that information has been lost or erased from memory, and people usually attribute forgetting (in the behavioral sense) to forgetting (in the explanatory sense). But in fact it has been controversial as to whether forgetting in the latter sense does in fact occur.

Traditionally, it has been assumed that all forgetting (as behaviorally defined) from long-term memory is caused by **retroactive interference** (RI). The idea is simply that memory traces are interfered with by later things that we learn. Suppose a subject has learned an association between two events, A and B, and then learns a new and contradictory association between A and C. Retroactive interference refers to the fact that the subject's ability to remember the original A–B relationship is impaired as a result of having learned the new association, because in some way or other the A–C association has interfered with the A–B association.

Typically, the process of interference (and hence forgetting itself) has been attributed to one of two sources. The first (the changed-trace hypothesis) is that the A–C association may have "overwritten" the earlier A–B association and hence led to unlearning, with the original association literally lost or at least fragmented as a result of the later learning. But there is a second possibility, called retrieval failure: the interfering information may have made the original association difficult to retrieve, without actually having destroyed it. Such a view suggests that memories can always be uncovered if we try hard enough, and it should be obvious that advocates of the notion of so-called "recovered memories" (for instance, of child abuse; see Loftus, 1993) believe that forgetting is in general due to retrieval failure.

What is the evidence for and against each of these theories? Let us consider the results of some experiments conducted by Chandler (1993) which used the A–B, A–C design with the elements of the associations being forenames and surnames. In Chandler's first experiment, subjects initially read a series of A–B target names (e.g., Robert Harris) and were given an immediate cued recall test (Robert H–?) to ensure that they had learned them. Some of the targets were experimental items and some were control items: this refers to the fact that on a second list of names, which also had to be learned, there appeared A–C names (e.g., Robert Knight) that had the same first name as the earlier experimental names, but there were no items on the second list that had the same first name as the earlier control names: instead, there were a number of unrelated D–E names. Then, on a final memory test, subjects were given a mixed-up list of the first names and surnames that had appeared in the original list (the A and B elements), and were asked to match them up. Note that on this test, the potentially-interfering surnames (e.g., Knight) from the second list did not appear. Chandler found that subjects were able to correctly match 59 percent of the control names but only 46 percent of experimental names, representing a sizeable RI effect.

As an aside, it is important to note that this result disproves the possibility that forgetting is merely the result of a pure temporal decay process, whereby information is lost merely as a result of the passage of time. Since subjects are more likely to forget an A–B association when it is followed by an A–C than by a D–E item, despite the fact that the time interval between study and test is equated, it would not be sufficient to say that forgetting is due to simple trace decay. This finding cannot be explained on the notion of trace decay because the A–B association should have decayed equally whether it is followed by an A–C or by a D–E pairing.

Chandler's experiment obviously provides evidence of interference-induced forgetting, in that subjects were poorer at remembering names when they were followed by other similar names. But what is the basis of this RI effect? On the changed-trace account, the effect is attributed to the fact that the similar names (Robert Knight) led to the earlier names being unlearned or overwritten in memory, with the A–B association being permanently lost. In contrast, the retrieval failure account attributes the effect to the fact that the names from the second list have somehow blocked retrieval of the target names.

How can we discriminate these hypotheses? In a further experiment using the same general procedure, Chandler obtained a result that cannot be accounted for if the target words were really lost from memory. In this study, Chandler merely varied the delay interval between the two lists of names and the final memory test for different groups of subjects. In one case, the delay between the first list and the test was 5 min (during which the second list was learned), in a second condition the delay was 15 min, and in a final condition it was 30 min. Once again, experimental names on the first list were followed on the second list by similar names while the control names were not. For the groups tested after 15 or 30 min, the second list was presented immediately after the first list and was then followed by some filler tasks before the test took place.

A surprising finding emerged in the results, namely that the amount of RI actually declined as the retention interval was increased. At intervals of 5 and 15 min, the results of the first experiment were replicated, with RI of about 8 percent being obtained. However, by 30 min, there was no evidence whatsoever of RI. Such a result is impossible to explain on the changed-trace account, because the account has to attribute the RI obtained at the shorter retention intervals to genuine unlearning of the original experimental items, at least relative to the control items; but if the experimental items have to some extent been unlearned, they should still be harder to recall than the control items at the 30 min delay.

In sum, Chandler's results suggest that forgetting is not in the main due to real unlearning or fragmentation of memory traces. Instead, it seems to be due to the fact that later information blocks the retrieval of earlier information. Chandler's results suggest that this blocking process only occurs when the potentially-interfering information is active in memory. As the delay before the test is increased, the interfering information gets less active in memory and is less likely to block retrieval of the target trace. In a last experiment, this interpretation was confirmed. Chandler (1993) presented the final memory test 30 min after the original A–B list, but for different groups gave the second list either just after the original study list or just before the test. The outcome was that no RI occurred in the former case – replicating the result from the second experiment – but that RI did occur when the interfering list was learned just prior to the test. Such a result confirms the idea that blocking of the target items by subsequent items only occurs when the latter are active in memory. When the intervening items occur long before the test, sufficient time has elapsed for them to become inactive in memory and hence unable to block retrieval of target names. It appears that the behavioral phenomenon of forgetting is not in the main due to a genuine loss of stored memory traces.

From a computational point of view, the phenomenon of forgetting (in the behavioral sense) may seem at first sight to represent a profound limitation of the human cognitive system; after all, computers do not forget. But in fact it is possible to argue that forgetting is not at all a limitation but, instead, is quite a sensible feature of the memory system. Suppose we are interested in why forgetting occurs at the rate and in the precise manner that it does. If the memory system is limited in capacity, some loss of information over time may be necessary to allow new information to be stored. From a "rational" point of view (see chapter 1, "Strategies for Cognitive Science"), it turns out that the best thing to do is to forget as a decreasing power function of time, since that is the rate at which information in the environment becomes redundant. Anderson and Schooler (1991) demonstrated this, at least for one situation, by examining an electronic version of the New York Times. They found that if a certain name such as "Qaddafi" appeared in a headline in the newspaper on a given day, then the likelihood that it would appear on a later day was a decreasing power function of elapsed time. Having established that this is what the system should do, we can then ask whether humans or other organisms do indeed forget at this rate: the answer appears to be "yes" (Wixted and Ebbesen, 1991), indicating that memory is beautifully adapted to the nature of the environment.

How are Concepts Represented in Memory?

A major issue in the study of learning and memory, as well as in other branches of cognitive science, involves the character and properties of **conceptual knowledge**. As a result of exposure to a large number of chairs, for instance, a child acquires a concept of what a chair is, as well as the ability to recognize new chairs. The child knows, for example, that chairs are for sitting on and that they are man-made artifacts. Although many concepts have labels attached to them (the word "chair" is associated with the concept chair), there are many other concepts for which we probably do not have words. In this section we address a fundamental question about memory: how exactly are concepts represented? That is to say, in what form is conceptual knowledge stored in memory? In chapter 8 we discussed the conceptual basis of the

Box 10.3 Brain basis of memory

Studies aimed at locating which parts of the brain are most involved in memory have identified temporal lobe regions, particularly the **hippocampus** and structures related to it, as crucially important. The hippocampus is part of a circuit in which signals are transmitted from the subiculum (the "output" part of the hippocampus) to the mammillary bodies via a pathway called the fornix (see figure 10.7). From the mammillary bodies, signals are sent via the cingulate cortex back to the hippocampus. Evidence that these regions are involved in memory comes from detailed examination of brain lesions leading to amnesia. Although HM's surgery removed structures in addition to the hippocampus, evidence suggests that highly localized lesions to the hippocampus, fornix, or mammillary bodies can create profound amnesia (Squire, 1992).

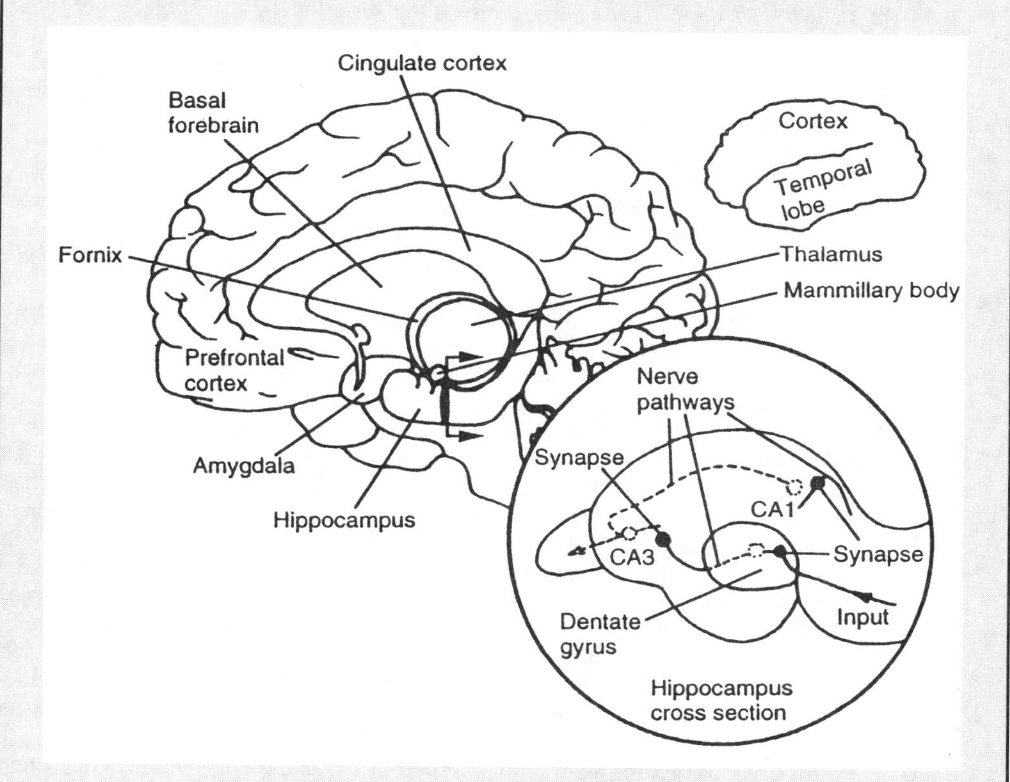

10.7 The principal brain areas involved in learning and memory. The hippocampus, which receives inputs from sensory cortex, is thought to be the storage site of episodic memories prior to their consolidation in the association cortices, and the main reason for thinking this is that lesions to the hippocampus cause profound anterograde amnesia. Patient H.M., who suffers from both profound anterograde and retrograde amnesia, had many temporal lobe structures in addition to the hippocampus removed, but later studies have shown that it is largely the hippocampus that is important. In the hippocampus itself, long-term synaptic changes (long-term potentiation) can be shown to occur as a result of stimulation.

The hippocampus, though, is not so much the site of memories as the system involved in consolidating them. It cannot be the site of declarative memories since many amnesics with damage to this region do not have extensive retrograde amnesia. Such patients can perfectly well remember things that happened to them prior to their brain damage, and thus their memory traces per se are intact. Instead, the hippocampus seems to control the formation of new memories.

So where are memories themselves stored? The answer seems to be that they are laid down in the cortex. Some evidence for this comes from pioneering experiments by Wilder Penfield (Penfield and Perot, 1963). Penfield was able to electrically stimulate areas of the cortex of patients who were awake while undergoing surgery for focal epilepsy, and he found that stimulation of the temporal lobes led most of his patients to experience very vivid and realistic images. These often involved hearing music or somebody's voice, such as when one person said "I hear someone talking . . . I think it was about a restaurant or something" (p. 627), and they were often reported by the subjects as feeling very familiar, as they would if they were memories. Penfield assumed that subjects were in fact recalling past events that had happened to them, and proposed that memories are stored at very precise cortical locations (as in a computer's memory) and can be elicited by presenting a small current through an electrode in contact with that location.

But Penfield's data have been criticized, for instance by Loftus and Loftus (1980), on the grounds that many of the reported experiences were fantasies and could not possibly be real memories. In fact, Penfield was unable to provide clear evidence that any of the reported experiences were true memories. Rather than being stored in precise locations, what is more likely is that memories are stored in a distributed fashion across the cortex, with many different parts of the cortex contributing to a given memory. What Penfield detected were probably not memories themselves, but rather combinations of the basic experiences that make up true memories.

While the study of the anatomical basis of learning and memory stretches back over a century, study of their neural basis has a rather more recent history and was given its greatest impetus in the post-war era, principally by Donald Hebb (1949) and Karl Lashley (1950). Modern brain research has established that memories are encoded in the brain via the plasticity of synaptic connections between neurons, with individual memories being stored in parallel across huge numbers of neurons organized into "macrocolumns" in the cerebral cortex (Squire, 1987). Each of the approximately 10^{11} neurons in the brain receives inputs from very many other neurons, and these inputs consist of neurotransmitter molecules that attach themselves on receptor sites on the dendrites of the neuron. When the input activation reaches a sufficient level, channels are opened which allow ions to be admitted into the neuron. These ions cause an electrical impulse to be generated, which then travels down the output pathway (the "axon") of the neuron, and which leads to neurotransmitter molecules being released, which can then act on other neurons.

When two connected neurons are excited at the same time, the synaptic connection between them may grow stronger, leading to facilitation between them. This is thought to be the basic way in which memories are stored. The process of synaptic plasticity has been extensively studied in the hippocampus, where so-called **long-term potentiation** (LTP) has been observed (see figure 10.7). As a result of sending a train of impulses down a microelectrode attached to a part of the hippocampus, increases in the strengths of connections are observed, such that a signal from one neuron elicits a much stronger response from another neuron than would normally be the case. These changes can last for weeks or months. Although controversial, it is widely thought that LTP is the basic process whereby the brain stores information (see Morris, 1994).

meanings of words, but their representation in memory and how these representations are formed was not considered.

Since the pioneering work of Eleanor Rosch in the 1970s, an influential approach has been to say that categories of stimuli are represented by mental **prototypes** and that learning a concept involves abstracting the appropriate prototype. The concept CHAIR, for example, might be represented by a typical chair that has been mentally abstracted from our experience of a large number of actual chairs. On this account, responding to a new stimulus is a function of its similarity to the prototype. As test stimuli get closer to the prototype, they should, therefore, become easier to categorize, an effect that is readily demonstrated in the laboratory. For instance, Rosch, Simpson, and Miller (1976) asked subjects to categorise artificial stimuli such as random dot patterns. A pattern from one of four categories was presented on each trial and the subject made a classification decision, with corrective feedback for incorrect responses. For each category, the patterns were the category prototype plus one pattern at each of five levels of distortion. After learning the category assignments, subjects were instructed to continue classifying the patterns as rapidly as possible and their response times were recorded. Finally, subjects rated each of the patterns in terms of how typical it was of its category.

On the basis that typical items are closer to the category prototype, the prototype view predicts that differences should be observable in responding to the stimuli as a function of their distance from the prototype, and this is exactly what Rosch et al. observed. Items judged highly typical were classified more rapidly in the test stage than ones judged less typical.

Perhaps the most compelling reason to believe that abstraction of the prototype underlies categorization is the abundant evidence that the prototype stimulus itself will be classified accurately and rapidly, even when it has never been presented in the training stage of an experiment. For instance, Homa, Sterling, and Trepel (1981) trained subjects to classify geometrical patterns into three categories which varied in size. Three prototype patterns were defined, and training patterns were constructed by highly distorting these prototypes. In each block of the study phase, subjects saw 20 different patterns from category A, 10 from category B, and 5 from category C. One of these patterns was presented with corrective feedback on each trial, and subjects continued until they had achieved two errorless blocks. In the transfer phase, which occurred either immediately or after 1 week, the original training patterns plus the unseen prototypes were presented for classification.

Homa et al. found that subjects were in some cases more likely to correctly classify the prototype, which they had never seen, than any of the specific training instances, and this was particularly the case when the category contained a large number of instances (e.g., 20). Figure 10.8 shows that the benefit for the prototype over the original training items was enhanced when a long interval (one week) intervened between training and testing. Here, the prototype was correctly classified on 96 percent of trials, while the original

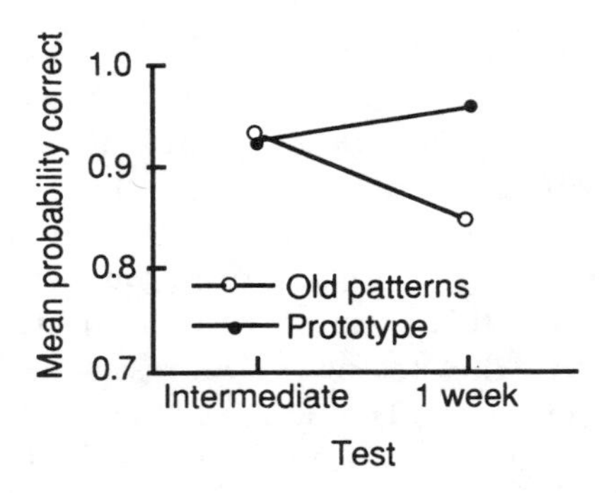

10.8 Results from Homa, Sterling, and Trepel's (1981) experiment. The figure shows the mean probability of correct classification responses for the original training patterns and the novel prototypes. Subjects were trained to classify geometrical patterns into three categories. Either immediately or after one week, the training items and prototypes were presented as test stimuli. When tested after a week, subjects were more accurate in classifying the prototype than the original training stimuli. Also, the old training items were more susceptible to forgetting than the prototypes.

patterns were only classified correctly on 85 percent of trials. Such results seem to imply that the prototype, at least in some instances, is mentally represented.

A further interesting result emerged from the experiment. When classification performance was tested after a delay of a week, considerable forgetting was evident for the original training items: for the 20-item category, performance fell by about 10 percent. In itself this result is not surprising, but as figure 10.8 illustrates, no such forgetting occurred with regard to the prototype patterns. For these, performance if anything slightly improved across the delay. Thus prototype classification may continue to be highly accurate even when memory for the training instances has deteriorated, a result that is consistent with the notion that it is the abstracted prototype that is mediating classification.

However, there are some reasons for doubting the adequacy of prototype abstraction as the basis of concept learning. Perhaps the strongest motivation comes from experiments showing that information retained about specific training items influences classification. According to the logic of prototype theories, if the prototype is the only representation in memory that plays a role in the classification process, then specific training items that may have been studied should not affect classification performance. However, such **exemplar** effects can be demonstrated in a number of ways. The simplest is to compare classification performance on old and new patterns equated for distance to the prototype. If concepts are represented in terms of abstracted prototypes, then in experiments such as that of Rosch, Simpson, and Miller (1976) we should expect the original training items to be classified no better than new test items equidistant from the prototype. Instead, numerous experiments have found that old items are responded to better or faster than new items even when equated for similarity to the prototype.

Similarly, in Homa, Sterling, and Trepel's (1981) experiment where subjects were trained to classify simple geometrical shapes into three categories of different sizes, the original patterns were classified better than new ones equidistant from the prototype. Subjects were tested on new patterns that were distorted from the prototype as much as the original patterns, but which systematically varied in terms of their similarity to specific old patterns. Increased similarity to an old training item went with increased classification accuracy. If information about the specific training items is discarded in the formation of the prototype, it is difficult to see why such a result should emerge.

How can we go about accounting for these results in terms of computational systems? A number of investigators (e.g., Gluck and Bower, 1988; McClelland and Rumelhart, 1985) have proposed that concepts are represented in memory via the weights in a connectionist network (see chapter 2, "Connectionism and Parallel Distributed Processing, for background"). In a simple **parallel distributed processing** (PDP) model of classification learning, such as that shown in figure 2.3, the stimulus is encoded as a series of perceptual elements or micro-features that are presented to the input layer of the network. Each category is represented by a unit in the output layer, and the input and output units are connected, via so-called "hidden units," by modifiable weights. The task of the network is to activate the "chair" output unit whenever a chair is presented at the input layer, to activate the "table" output unit whenever a table is presented at the input layer, and so on.

How can this be achieved? The answer is to use a learning algorithm such as the "delta" rule. On each learning trial, a stimulus is given to the network as input, and activation spreads via the connections to the output units, activating some of the category units and not others. Early on in training, the correct output will probably not be selected, but the weights can be modified to ensure that next time such a stimulus is presented, the category is a little bit more likely to be selected. After sufficient training, a network can be extremely good at classifying diverse objects into their correct categories, with its conceptual knowledge being distributed across a large number of parallel weights between highly interconnected units. Networks like this are exciting because they store conceptual knowledge by appealing only to processes known to operate in the brain, namely the transmission of excitatory and inhibitory signals between units. Box 10.3 gives an overview of some of the basic brain processes and mechanisms involved in memory.

Before ending this chapter, it should be acknowledged that the sorts of theories of conceptual representation we have been discussing in this section are almost certainly incomplete, for at least two reasons. First, the theories assume that concepts are represented entirely in terms of perceptual features or combinations of features: on a prototype theory, for example, the concept DOG is represented in terms of typical values on each of a number of dimensions coding perceptual attributes such as size, colour, shape, and so on. But surely our concepts incorporate information about non-perceptual properties, such as functional or causal attributes? Surely the concept CHAIR includes specification of the fact that chairs are things that can be sat on? Moreover, it is quite possible that innate, non-perceptual attributes to do with such things as object permanence figure in many of our concepts. For a theory of the possible innate components of concepts, see chapter 8, "The Meaning of Words".

The second reason why these theories are likely to be inadequate derives from the fact that they assume that categorization and conceptual structure go hand-in-hand with **similarity**. Whether one thinks of concepts as being represented by prototypes or by the weights in a PDP network, it will always be the case (according to such theories) that an object is judged to be a member of a category to the extent to which it is perceptually similar to previous members of the category. It is assumed on a prototype theory, for example, that the relationships between objects represented in memory are largely fixed: an ostrich is further away in psychological space than a robin from the prototypical bird, and this is an immutable fact about our conceptual knowledge. But although similarity often does play a crucial role in category decisions, people also appear to be able to use concepts in a far more flexible way which is largely unconstrained by similarity. Indeed, there is abundant evidence (see Medin, 1989) that how close (or similar) something is to the prototype of its category can be highly plastic and context-dependent. For example, in the context of "milking," a cow is a very typical animal and a horse an atypical one, but in the context of "riding" it is the other way around. The key point is that the degree of similarity between two objects – such as a category member and the category prototype – is not a fixed relationship but can vary depending on the circumstances.

As a concrete example, consider a study by Rips (1989) in which subjects were told about a bird that became transformed (by exposure to toxic substances) into something that looked like an insect. When asked to say what the creature was, subjects were quite happy to say that it was a bird, despite the fact that they judged it to be perceptually more like an insect. Thus a category decision was made that was quite at variance with the object's raw similarity to previous category members. How are we to explain such effects? It appears that we must concede that in some circumstances concepts figure more like theories than like prototypes. Because people have a complex theory about biology and inheritance, they know that a bird is still a bird even when its external form is radically altered. In other words, our concept of a bird is not a simple representation (such as a prototype) but is instead an extensive theory, and the relationship between the concepts bird and animal is determined by their roles in that theory.

Summing-up

In this chapter we have looked at a number of important questions about learning and memory. We have seen that memory can manifest itself in a variety of ways, in episodic retrieval as well as on semantic and procedural tests. A good deal of evidence points to the existence of distinct memory modules: a short-term store, as well as separate stores within long-term memory. Much of the relevant evidence comes from analyses of the ways in which memory can break down as a result of damage to the brain. The amnesic syndrome represents a fascinating window onto memory processes.

We have also discussed briefly the central role played by the hippocampus in memory storage and retrieval, and have considered the way in which learning occurs in the brain via the plasticity of the synaptic connections between neurons. We have considered some of the ways in which researchers are attempting to understand learning and memory from a computational perspective involving the transmission of excitatory and inhibitory activation in artificial neural network models. Finally, we have looked at the various ways in which concepts might be represented in the memory system.

Further Reading

An excellent review of learning and memory from a cognitive science perspective is provided in *Learning and Memory*, New York: Norton (1991) by B. Schwartz and D. Reisberg. For more on the neuropsychology and brain basis of memory, and on the procedural/declarative distinction, see *Memory and Brain*, Oxford: Oxford University Press (1987), by L.R. Squire. For a good overview of current research on concepts, see the many excellent chapters in *Concepts and Conceptual Development: Ecological and Intellectual Factors in Categorization*, Cambridge: Cambridge University Press (1987), edited by U. Neisser. The application of connectionist models to learning and memory is reviewed in *Connectionism and Psychology: A Psychological Perspective on New Connectionist Research*, London: Harvester Wheatsheaf (1991) by P.T. Quinlan.

CHAPTER

11 How We Solve Problems

Outline

We start from the position that a central function of thought is to model reality so that we can act in the world. Initially we use the theory of mental models to examine the solving of certain simple spatial problems and then go on to consider more complex syllogistic problems. A mental model of how something works can be used analogically to help solve a problem in a different area. We describe and compare different computational models of this process. In the final section we consider the differences between experts and novices and how we can view the process of becoming an expert in computational terms.

Learning Objectives

After reading this chapter you should be able to:

- define the notion of a mental model
- describe a mental model account of spatial deductions and syllogistic reasoning
- define the nature of analogical problem solving
- compare the mapping procedures of different models of analogy
- describe a model of expertise
- question the adequacy of existing approaches to expertise
- identify a number of theoretical questions requiring solution

Key Terms

- analogical problem solving
- inference
- means-end analysis
- mental model
- problem space
- syllogism

Introduction

How shall a problem be defined? Here, we restrict the sense of problem solving to circumstances where some immediate goal is thwarted and some means must be found to reach the goal (see Duncker, 1945). There are examples in our scenario: Mom tries to open the parcel but fails initially to do so; Lucy tries to make toast but discovers that the toaster is broken.

Much problem solving involves reasoning a way to a solution. Given this information and that piece of information what can be concluded? Many everyday decisions involve *inferences* that go beyond the given information. Consider questions such as: which degree course should you do? Which job should you apply for? In reaching a decision you have to infer what you will enjoy most or which may yield the best prospects. Everyday problem solving and reasoning enables people to make decisions so that they can achieve their goals.

If thought and problem solving are in the service of needs, some of which can be met by acting in the world, there must be a relationship between thought, perception, and action. We suppose that a central function of mind is to construct models of reality so as to enable action in the world. After reviewing the notion of a **mental model**, we consider problems where the role of knowledge is relatively minimal before considering cases where beliefs are important. Following that we consider the process of problem solving when special knowledge is provided by way of an analogy. Such problem solving can also be viewed as a form of modeling. The final section examines expert problem solving.

Mental Models

Normally individuals have no reason to doubt their percepts: it makes sense to act on the basis of what they perceive. A shadow in an alleyway may signal an assault. From the other side of the street they may decide it was a trick of the lighting. Such a view proposes a close correspondence between perception and belief. People act as if their perceptions were true though they may later revise that belief. Likewise, on hearing a warning such as "There's someone in the shadows!" they may envisage a situation in which there is a person in the shadows and cross the street before checking. In this case, the warning is treated as if it were a perceptual datum. It is treated as if it were a true proposition about the world – this belief can still be revised (Gilbert, 1991). This position is consistent with a critical proposal regarding human thought, namely that it constructs models of reality – an idea that we introduced in chapter 1 and used in chapter 8. The basic idea is that a mental model has a similar relation-structure to the situation that it represents.

SAQ 11.1 Imagine Lucy and her two friends. Lucy is sitting to the left of Mary and Mary is sitting immediately to the left of Jane. Is Lucy necessarily to the left of Jane?

Each entity (object or person) in the world is represented by a corresponding token in the model; the properties of entities are represented by properties of tokens in the model and relations among entities are represented by relations among tokens (Johnson-Laird, 1983, 1993). Although a model may be constructed on the basis of language (as in the case of the warning), its structure corresponds not to the language used but to the situation described. The following sections apply the concept of a mental model to a number of simple situations and show how operations on the model allow problems to be solved.

2D Spatial Deductions

In constructing a mental model, we make simplifying assumptions. Take the problem described in SAQ 11.1 "Lucy is sitting to the left of Mary and Mary is sitting to the left of Jane. Is Lucy necessarily sitting to the left of Jane?" In order to construct a model, individuals have to understand the meaning of the terms and make an assumption such as: the three are sitting in a line. In this case, the conclusion is true. However, an alternative model of the situation is one in which the friends are sitting around a round table. In this case, Jane could be sitting opposite Lucy. This illustrates a crucial point: a conclusion reached depends upon the nature of the model or models constructed. In particular, it depends on what is explicitly represented and so, at least initially, on what is asserted. In order to determine whether or not a provisional conclusion necessarily holds, individuals have to determine whether or not there is some alternative model which is consistent with the statement of the problem. The theory of mental models, developed by Johnson-Laird and colleagues (e.g., Johnson-Laird and Byrne, 1991) predicts that a problem is more difficult if more than one model has to be considered. Additional models impose a load on limited working memory.

Imagine some utensils on a breakfast table. You are presented with the following description (see Byrne and Johnson-Laird, 1989) and are asked to write down the relation between two objects, which has not been stated directly. If you would like to experience the problem, cover the rest of the page, read the description, and imagine looking down on the layout.

Problem 1:
The fork is on the right of the spoon.
The cup is on the left of the spoon.
The knife is in front of the cup.
The plate is in front of the fork.
What is the relation between the plate and the knife?

We assume an interpretive device that captures the relationship expressed in each of the statements (premises) in terms of a three-dimensional coordinate system (see chapter 8, "The Meaning of Words"). The meaning of the term "on the right of" is a set of parameters that gives a location in this space. This information is used to construct a model of the array (see Johnson-Laird and Byrne, 1991, p. 173). The overall model of the state of affairs described can be depicted in a spatial diagram:

cup	spoon	fork
knife		plate

A procedure locates the target items (knife and plate) in the array and returns a description, which expressed in English, supports the conclusion "The plate is on the right of the knife." Suppose, we changed the second statement in Problem 1 to "The fork is on the right of the cup." This would yield a second problem:

Problem 2:
The fork is on the right of the spoon.
The fork is on the right of the cup.
The knife is in front of the cup.
The plate is in front of the fork,
What is the relation between the plate and the knife?

SAQ 11.2 Draw a spatial diagram for Problem 2. Is it the same as that for Problem 1? If it is, could it be different?

This change is consistent with the model of Problem 1, but now the relationship between spoon and cup is unspecified. There is an alternative spatial model:

spoon	cup	fork
	knife	plate

In order to draw the correct conclusion in this case, individuals have to consider more than one model. According to the theory, this problem should be more difficult to solve.

How else might this problem be solved? An alternative possibility is that such problems are solved by using mental rules of inference (Hagert, 1984; Ohlsson, 1984). Such an account requires a scheme (a postulate) that specifies the logical properties of a relation such as "on the right of." In this case the postulate specifies that the relation, "on the right of," is transitive:

> For any object x, any object y, and any object z, if x is on the right of y, and y is on the right of z, then x is on the right of z

Let us redraw Problem 1 detailing the explicit relations; the direction of the arrow indicates the nature of the relationship specified.

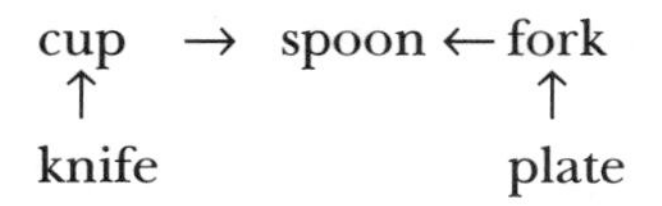

The only connection between plate and knife is via fork and cup. In order to apply the postulate, this relationship has to be inferred first. This requires two steps: first the relationship between spoon and cup has to be converted to "spoon on the right of cup." Next, the postulate can be applied leading to the transitive inference that the fork is on the right of the cup. Given that the knife is front of the cup and plate is in front of the fork, it necessarily follows that the plate is to the right of the knife. The crucial point is that two inferential steps are required to establish the relationship between fork and cup.

What of Problem 2? Here the relationship between fork and cup is given: "The fork is on the right of the cup." Hence the inference chain is shorter. Assume that the length of derivation correlates with mental difficulty: which problem does the rule theory predict will be easier? On the **rule theory**, problem 2 should be easier than problem 1. In contrast, model theory predicts the reverse result because problem 2 requires two models whereas problem 1 requires only one model. Byrne and Johnson-Laird (1989) found 70 percent correct solutions to one-model problems and 46 percent correct solutions to two-model problems. This counts as a strike against rule-based accounts. We consider next the mental model approach to some more complicated problems.

Syllogisms

Imagine a room with a number of people and then consider the following problem:

> Some of the students are women
> All of the women are drivers
> What follows?

Such a problem is known as a syllogism (see box 11.1 for additional background). With such a syllogism nearly everyone concludes: some students are drivers. That is, they generate an informative conclusion which could not have been inferred from either of the two statements (or premises) alone. Model theory supposes that individuals understand the premises and construct a model of the situation to which they refer. There are various ideas about the form of these models (e.g., Euler circles, where sets of individuals are represented by circles that can overlap; see Stenning and Oberlander, 1995). We adopt the notational scheme of Johnson-Laird (1993) since this representation has been used in other domains such as conditional inference and discourse.

The model for the first statement ("Some of the students are women") represents a situation in which each student ("s" in the model) is a woman ("w" in the model).

```
s        w
s        w
.    .   .
```

Each line refers to an individual member of the set, in this case an individual who is a student and a woman. The number of individuals represented is arbitrary – in this model we have represented just two and so there are just two lines. The room may, of course, contain different kinds of people. This possibility is indicated by the presence of three dots. The idea is that a mental model represents as little as possible explicitly in order to reduce the load on working memory. The three dots indicate other states of affairs, not made explicit as yet.

The second statement is "All of the women are drivers." This statement means that there are no women who are not drivers. We need a way to capture this meaning in the model. The tokens corresponding to women have to be exhaustively represented as being drivers (denoted by "d" in the model). This exhaustive relation is signalled by placing square brackets around each of the tokens for women and pairing each with the token for driver in the following way:

```
[w]   d
[w]   d
.    .    .
```

In short, the square brackets in the model of the second statement show that there are no women who are not drivers. Even if we flesh out what is represented implicitly, there will be no women who are not drivers. On the other hand, there could be drivers who are not women because drivers are not exhaustively represented. There is no square bracket around the "d's". The next step is to combine the two models by adding the model of the second statement to that of the first, which produces the following integrated model:

```
s    [[w] d]
s    [[w] d]
.    .    .
```

The second model builds on the tokens represented in the first model.

Box 11.1 The nature of syllogisms

Syllogisms involve reasoning about category relations and typically consist of two statements, called premises, that are to be treated as true. For example,

All psychologists are wine drinkers
All Italians are psychologists

One premise relates the subject of the conclusion (Italians) to a middle term (psychologists). The other premise relates the middle term to a predicate (wine drinkers). The task is to state a conclusion, if any, between subject and predicate terms. In this syllogism we can infer: All Italians are wine drinkers. The various types of relations between subject, predicate, and middle terms involve set inclusion, set overlap, and set exclusion, expressed by the words all, some, none, and some not. The number of possible argument forms is large since each of the two premises can involve any one of the four logical relations. Further there are four basic ways, known as figures, in which the subject (S), predicate (P), and middle (M) terms can be permuted:

Figure 1	Figure 2	Figure 3	Figure 4
M – P S – M	P – M S – M	M – P M – S	P – M M – S

Valid conclusions are invariably derived for some argument forms such as: All Italians are psychologists; All psychologists are wine drinkers; What follows? Answer: All Italians are wine drinkers. For other argument forms such as: Some Italians are not wine drinkers; All Italians are psychologists; What follows? rarely yield the correct conclusion: Some psychologists are not wine drinkers. Not all syllogisms have valid conclusions, so, for example, a syllogism of the type: All of the A are B and Some of the B are C has no valid conclusion. Despite over 60 years of research there is, as yet, no generally accepted account of the various results, though arguably model theory provides the most complete account.

Whenever a "w" token occurs a "d" token is added to it. Furthermore, since "w" is exhaustively represented with respect to "d," an integrated representation of these two statements must continue to ensure this. This state of affairs is indicated using square brackets around women drivers: of the students already represented, all of them are women drivers. The three dots continue to signal other states of affairs not yet made explicit.

What other kinds of people could exist in the room that are consistent with the premises? Thinking about these and representing them mentally fleshes out the implicit model. There could be male students and some of these could be drivers so there needs to be a model in which "s" and "d" co-occur. There could be other individuals who are not drivers – so "s" can occur alone. There could also be some drivers who are not students, indicated by "d" alone. Hence, a full set of models for the state of affairs consistent with the syllogism is:

s	[[w] d]
s	[[w] d]
s	
s	d
	d

Given this representation, in which there is no longer any implicit model, one conclusion is: "Some students are drivers," which is the conclusion drawn initially. In this case a valid conclusion could be reached without fleshing out the implicit model. We contrast this problem with Problem 4 where such fleshing out is required.

SAQ 11.3 Before reading further, what conclusion could you draw from a syllogism with the following premises: All the psychologists are Italian and None of the psychologists is a smoker? Try your hand at representing it in mental model terms. Health warning – this syllogism is difficult!

Problem 4
All of the women are students.
None of the women is a driver.

The first two lines in the model below establish that there are no women who are not students. This is why the set of women is exhausted with respect to the set of students. However, there can be students who are not female so the set of students is not represented exhaustively. In addition, the set of women and the set of drivers are exhausted with respect to one another because none of the women is a driver: this is the meaning of the square brackets around the tokens for women students and the tokens for drivers.

```
[[w]       s]
[[w]  s]
           [d]
           [d]
 .     .    .
```

This model supports a conclusion:

None of the students is a driver

However, this conclusion is invalid because it is refuted by an alternative model of the state of affairs described by the statements in which there is a male student who is a driver. The initial explicit model can be changed in a way consistent with the meaning of the two statements:

```
[[w]       s]
[[w]  s]
      s    [d]
           [d]
 .     .    .
```

This supports a conclusion such as "Some of the drivers are not students". However, this conclusion is also invalid because all the male students might be drivers, as in the following model:

```
[[w]  s]
[[w]  s]
      s    [d]
      s    [d]
 .     .    .
```

Does anything follow? Yes: "Some students are not drivers." This problem is much more difficult than the single-model problem. Indeed the evidence is that one-model problems are consistently easier than multiple-model problems (see Johnson-Laird, 1993). Why? One possibility is that we stop when we have a seemingly valid conclusion – we are cognitive satisficers. A second possibility is that there are limits to our ability to keep these alternative models in mind because of constraints on our working memory.

What might govern our decision to stop? The believability of the conclusion generated or to be evaluated may be one factor. The data of Evans, Barston, and Pollard (1983) support this view. In their study, undergraduates were asked to determine whether or not a conclusion necessarily followed from two statements. An example follows:

No addictive things are inexpensive
Some cigarettes are inexpensive
Therefore, some addictive things are not cigarettes

The conclusion is believable but invalid. If the terms in the statements are rearranged, an invalid and unbelievable conclusion can be derived:

No cigarettes are inexpensive
Some addictive things are inexpensive
Therefore, some cigarettes are not addictive.

In both cases, reordering the terms in the conclusions yields a valid conclusion. Evans et al. found that individuals generally accepted more valid than invalid conclusions. However, they were also influenced by believability. They accepted many more believable than unbelievable conclusions. But in addition, the effects of believability were much greater for invalid syllogisms: their sample accepted 71 percent of invalid but believable conclusions.

The apparent willingness to accept believable but invalid conclusions is consistent with model theory, but does it mean that human beings are irrational? Our basic supposition is that cognitive processes of reasoning are adaptive, that is, they evolved to further our practical goals (see Evans, Over, and Manktelow, 1993). Is it adaptive to reason only selectively? The data show that only conclusions which are unbelievable are likely to be examined closely. Presumably, it is reasonable to maintain our beliefs, unless there is good reason to do otherwise: revision is effortful and time-consuming. Hence, believable conclusions may fail to elicit the search for an alternative model. In contrast, a conclusion which contradicts prior beliefs should be scrutinized and, if possible refuted, since it threatens the coherence of our beliefs.

What is the evidence that constraints on working memory are also important in affecting whether individuals generate correct conclusions? Few studies have addressed this issue directly. However, existing studies suggest that the central executive component in working memory (see chapter 10) is particularly important (Gilhooly et al., 1993).

We might still ask, though, whether individuals construct mental models in the way proposed. Research requiring individuals to think aloud whilst solving syllogisms (Ford, 1994) suggests that students may represent relations in a spatial manner (close to Euler circles) or may remain close to the verbal form. The representations of untutored individuals are not well researched except that we know from the work of Luria (1971) and Cole and Scribner (1974) that rural peasants will often refuse to respond to arbitrary syllogisms or to syllogisms whose content is outside their experience. This outcome is not inconsistent with a mental model theory because this presumes that individuals reason by building mental representations of situations. Devil's advocate box 11.1 presents a radically different view of reasoning based on rules.

Analogical Problem Solving

In seeking to solve a problem individuals can use a solution to a problem they already know. This is problem solving by analogy. They create a model of the new problem based on a model of a known problem – technically, they are engaged in second-order modeling (see Holland, Holyoak, Nisbett, and Thagard, 1989). Individuals may use analogies to solve problems where they lack adequate technical knowledge. In the area of electricity, for instance, individuals seem to have one of two broad kinds of model. According to the flowing-fluid model, electricity flows along wires as water flows through pipes. According to the moving-crowd model electricity moves along wires as a crowd moves through a passage. These analogies differ in their effectiveness for different kinds of problems. Consider these two problems:

> *The battery problem*
> What happens to the current flowing in a wire when two batteries that produce the same amount of current are linked (a) in series (one after the other) (b) in parallel (side by side)?
>
> *The resistor problem*
> What happens to the current when two identical resistors are placed (a) in series or (b) in parallel?

SAQ 11.4 What does the moving-crowd model predict about current flow when two resistors are in parallel? Treat the resistor as analogous to a turnstile.

The flowing-fluid model relates rate of flow to current and pressure to voltage. Water pressure depends on the difference in height between reservoir and user. If the height is doubled there will be twice the pressure and twice the flow. This model correctly predicts that if two batteries are put in series (analogous to doubling the height of water) there will be twice the current, whereas if the two batteries are connected in parallel (water remains at the same height) there will be no difference. However, the analogy is less valid when it comes to considering the effects of resistors in the circuit. According to the analogy any obstacle should impede flow, so two resistors should reduce the current regardless of whether they are arranged in series or in parallel.

In contrast, the moving-crowd model correctly predicts that two resistors in parallel produce a greater current than a single resistor. The resistor functions like a turnstile. Two such stiles in parallel allow more people through than just one. However, this analogy suggests no clear conception of the nature of batteries. Gentner and Gentner (1983) found that individuals adhering to the flowing-fluid model solved the battery problem better than the resistor prob-

Devil's Advocate Box 11.1 Reasoning as the application of formal rules of inference

Mental model theory presumes that in reasoning, individuals construct a mental model of the situation and manipulate it. An alternative view is that the basic reasoning process is based on the manipulation of purely arbitrary symbols that lack meaning (e.g., Braine and O'Brien, 1991; Rips, 1994). The mind contains a content-free mental logic. Such a view accounts for our ability to reason about either familiar or more abstract material. On this view, the process of comprehension extracts the logical form of a statement rather than constructing a model of a situation. Formal rules of inference are then applied, a conclusion derived, and then this is translated back into the content of the problem. There is no complete theory of syllogistic inference based on formal rules, though there is some evidence to suggest that subjects can apply simple, superficial rules in order to generate conclusions such as matching the logical form of the premises (see Gilhooly et al., 1993). However, in this case, they are not really reasoning at all. We will look at a rule view in the context of conditional inference.

The comprehension process extracts the same logical form, "If p then q," for the statement "If Lucy works hard, she gets tired" and for a statement with a completely different content, such as "If Lucy is in New York, then Peter is in Moscow." Suppose individuals have to derive conclusions from two premises. The comprehension process would extract the corresponding logical forms and the reasoning process would proceed by matching the logical form to a set of elementary rules of inference. One such rule is termed modus ponens and has the structure given below:

Natural Language Statements	*Extracted Logical Form*
Premise 1 If Lucy works hard, then she gets tired	If p then q
Premise 2 Lucy works hard	p
What follows?	?

The premises match those of the modus ponens rule and so the conclusion "q" is derived. This conclusion must then be translated back into the content of the problem to yield a valid conclusion: "She gets tired." An identical process occurs with the other argument.

How does such a proposal explain the fact that some inferences are more difficult for us to make than others? Let us look at the psychologically more difficult inference known as modus tollens:

Premise 1 If Lucy works hard, then she gets tired.
Premise 2* Lucy isn't tired.
What follows?

In this case, some of us conclude that nothing follows. Rule theory explains this by supposing that only some inference patterns are mentally represented. The inference pattern corresponding to modus tollens is not mentally represented, that is, there is no mental rule corresponding to:

Premise 1 If p then q
Premise 2* not-q
Therefore not-p

How then do some of us draw the correct inference, "Lucy isn't working hard." The idea is that we do so indirectly. As before, we extract the logical form of the premises:

Natural Language Statements	*Extracted Logical Form*
Premise 1 If Lucy works hard, then she gets tired	If p then q
Premise 2 Lucy isn't tired	not q
What follows?	?

Since there is no matching inferential rule, other rules are employed. One is a supposition-creating rule. In this case, we can suppose premise 2, namely p. We can then apply the modus ponens rule because premises 1 (If p then q) and premise 2 (p) match the modus ponens rule. We can then conclude q. But application of a further rule for conjoining two assertions gives rise to a contradiction: q (by modus ponens) and not-q (asserted as a premise). By another rule (Reductio ad Absurdum) any premise (bar premise 1, which cannot be changed) that leads to a contradiction can be negated. We therefore conclude not-p. This proposition can then be converted back into natural language and leads to the conclusion: "Lucy isn't working hard."

More steps are involved in reaching a conclusion for this inference and so reasoners will make errors. For a rule theorist, modus tollens always require more steps than modus ponens. In general, mental difficulty is a function of the number of steps in the mental derivation. Such a view contrasts with that of the theory of mental models where difficulty is a function of the number of explicit models that have to be constructed. Modus ponens requires just one explicit model whereas modus tollens requires three for the conditional interpretation. It is conceivable that human reasoning involves both rules and models, but such hybrid accounts only make sense when simpler one-mechanism accounts have been found wanting.

lem, whereas those with the moving-crowd model showed the reverse pattern. This study shows that a person's model of a phenomenon affects their problem solving. What factors affect their ability to utilize analogy? Consider the following problem first of all:

The radiation problem (Duncker, 1945)
A doctor wants to destroy a tumour inside a patient's body without

> damaging the surrounding healthy tissue. There is a device for delivering rays which can destroy the tumour, but at the intensity needed these rays also destroy the surrounding healthy tissue. What should the doctor do? If you are unfamiliar with the problem, why not try to solve it?

Gick and Holyoak (1980, 1983) wrote an analogy to this problem which described how a General attacked a fortress. The roads to the fortress were mined to prevent the movement of large groups of men, and the fortress was impregnable to the attack of a single small group. The General divided his army into a number of small groups, one to each road and captured the fortress by having these groups converge on the fortress at the same time. This is known as the convergence solution. Given an explicit hint to use the (source) story as a means of solving the (target) radiation problem, a substantial number of subjects proposed that relatively weak rays could be focused on the tumour from a number of different angles. In this way, the tumour could be destroyed, leaving the healthy surrounding flesh intact. This convergence solution maps properties of the source analogy onto properties of the target domain. This solution is one of a number of possible solutions, for instance, the rays could have been directed down a tube to the tumour, but it has the merit of being relatively non-invasive.

Surprisingly, the number of individuals spontaneously noticing and using the analogy was quite small. However, once prompted to use it, individuals were relatively good at mapping it. What factors affect these processes of noticing and mapping? The spontaneous retrieval of an analogy presumably depends on its similarity with the target problem. Holyoak and Koh (1987) distinguished different types of similarity: surface and structural. Suppose that the radiation problem stated that only one ray machine was available – the convergence solution used by the General to attack the fortress would then be precluded since this feature violates the structure of the solution plan. In contrast, the precise way in which the General attacked the fortress (by dividing his army) and the use of divided rays to destroy the tumour are surface differences. Holyoak and Koh showed that both differences affected the likelihood of spontaneously noticing and mapping an analogy. In contrast, once a hint was given to use a prior analogy, structural differences were relatively more important.

This difference between noticing and mapping carries practical implications. It suggests that individuals may be unable to solve a problem because of an inability to access a suitable analogy. It is perhaps for this reason that the discovery of profound analogies, which underlie some innovations in science and art (see Hesse, 1966) are relatively rare. Indeed, there may be no effective and practical procedure for the discovery of a suitable analogy, unless one can severely constrain the number of potential sources (cf. Johnson-Laird, 1989) or one can find some more abstract relation that connect different domains. The focus of the next section is on computational models of the mapping process.

Three Computational Models

How can analogical problem solving be defined at the task level (Marr, 1982)? Gentner (1983; 1989) supposed that areas of knowledge, termed domains, have an inherent structure of relations. An analogy is a mapping of knowledge from one domain to another such that a system of relations obtaining in the source domain is conveyed to the target domain. As you may have noticed, this view is compatible with the broader concept of thought as a means to model reality. An analogy focuses on the relational commonalities between two domains independently of the objects in which those relations are embedded. Thus, objects in the two domains are placed in correspondence by virtue of their roles in a common relational structure. Gentner proposed that only relations, not properties, are transferred. Just as groups of men converged on the fortress, sets of rays could converge on the site of the tumour. In contrast, the fact that armies are composed of men and that men have two legs is not transferred.

The Structure Mapping Engine (SME)

Which relations are transferred? In her structure-mapping theory of analogy, Gentner argued that coherent sets of relations are mapped. This preference is termed the systematicity principle. So, for instance, in the radiation problem, individuals map the relational structure which corresponds to the General dividing his forces so that they converge and attack the fortress at the same time. This is mapped onto the medical domain so that divided rays converge on the tumour at the same time.

Gentner (1989) and Falkenhainer, Forbus, and Gentner (1989) applied the theory to a particular analogy and described an algorithm (the structure-mapping engine or SME) for establishing the relational mapping. This algorithm is not intended to model how individuals actually achieve an analogical mapping but demonstrates that relational structure can be computed, and provides a way to state what is achieved by an analogy. Consider the following problem:

> *The heat-flow problem*
> There is a hot cup of coffee with a silver bar sticking out to which an ice-cube is attached. The ice-cube is melting. Why?

Assume that someone does not understand that the cause of the ice-cube melting is the difference in temperature between the coffee and the ice-cube, though they know that heat flows from the hot coffee to the ice. We can diagram their current understanding as in figure 11.1.

We could use an analogous situation to help them gain that understanding. Individuals are shown a picture in which water flows from a large beaker along a pipe to a small jar or vial. Assume that the person knows that a difference in water pressure between the beaker and the vial causes water to flow between them. Figure 11.2 diagrams this understanding.

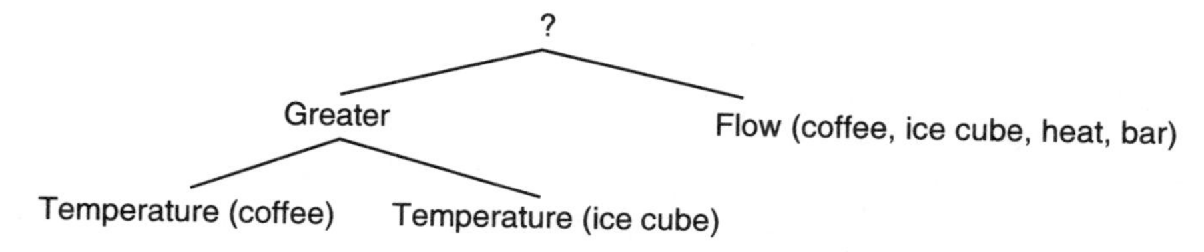

11.1 Diagram of initial understanding of heat-flow problem.

SAQ 11.5 Which relational structure will be selected by the systematicity principle?

The program first identifies identical relations between the source and target domain. Here, it identifies the relation "greater" and entertains two possible mappings: one in which "pressure" is matched with "temperature" and one in which "diameter" is matched with "temperature." Where a relation is matched, then the objects associated with it are matched appropriately. Thus, pressure in the beaker is matched to the temperature of the coffee. Next, it selects those matches consistent with the systematicity principle. It selects the pressure and temperature match. Why? The relation "greater" in the context of pressure has greater connectivity than the relation "greater" associated with the diameter of the vessels because the former is part of a causal structure linking difference in pressure and water flow. The diameter relation is eventually discarded. Such systematic matches sanction candidate inferences. Thus, the pressure-difference causal chain sanctions the inference that the temperature difference causes heat flow. The effect of the mapping is to create a causal pathway between temperature difference and heat flow. In other words, figure 11.1 is transformed into figure 11.3.

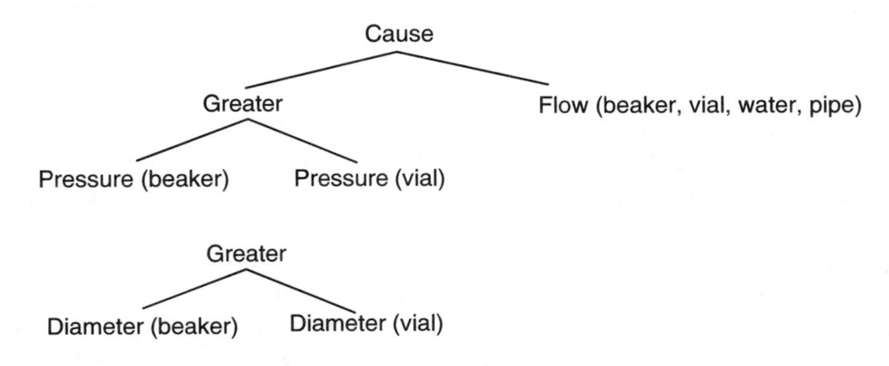

11.2 Diagram of understanding of pressure and water-flow.

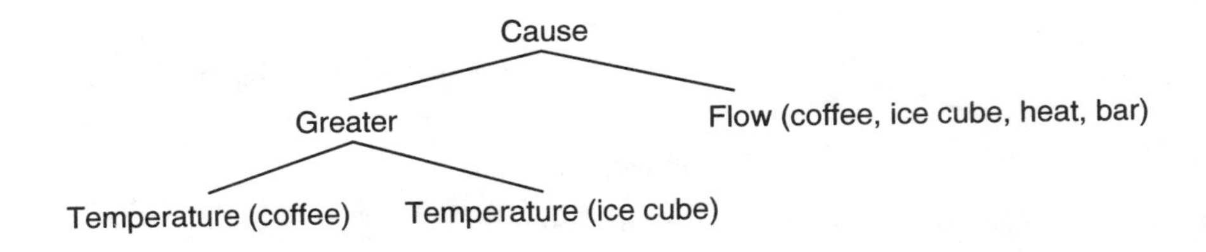

11.3 Diagram of final understanding of heat-flow problem.

Some general properties of SME are worth commenting on. First, as originally conceived (e.g. Falkenhainer et al., 1989) the program constructed all possible matches and selected the best on the basis of systematicity. Second, it is domain-general since it operates on relations rather than seeking specific kinds of content. Third, the mappings from a given source domain will depend upon the target domain. Gentner (1989) illustrates this. Instead of two objects differing in temperature suppose there were two objects differing in specific heat; a marble (with relatively high specific heat) and a ball bearing (with relatively low specific heat). Which has more heat capacity? If they have the same mass they will differ in heat capacity. In this case, the natural analogy concerns the diameter of the two vessels. The beaker has a greater diameter than the vial and can hold more water. The analogy, therefore, is that just as the container with greater diameter holds more water (for a given level of water) so the object with the greater heat capacity holds more heat (for a given temperature). Fourth, the mapping process is independent of the systems goals and purposes. These goals and purposes influence what is represented in working memory prior to the mapping process and can check the outcome of the process: they do not however, affect its operation directly. Evidence for the importance of pragmatic factors was provided by Keane (1990): individuals typically used an analogy which had been shown to work in a prior story and ignored one that had failed previously, although it could solve the current problem. The next two models allow pragmatic factors to affect the mapping process directly.

The Analogical Constraint Mapping Engine (ACME)

The structure-mapping engine operates in a serial fashion. Holyoak and Thagard (1989) created a parallel implementation of the critical structural features and also allowed pragmatic factors (the goals and purposes of the analogical process) to affect the mapping process directly. ACME uses the technique of parallel constraint satisfaction on propositional representations. It first establishes a network of nodes. Each node is a match between a relation (water-flow, heat-flow) or an object (beaker, coffee) in the source and a target domain. Excitatory links connect matching nodes and inhibitory links connect nodes representing alternative matches to the same relation or object. In

ACME, nodes that match semantically are given extra activation from a semantic unit outside the network. Likewise, nodes that are pragmatically important are given extra activation from a pragmatic unit. So, given the goal to explain heat-flow: pressure-difference rather than the diameter-difference between the beaker and vial is critical. When the network is constructed it is run until the nodes settle into a stable state. The optimal match between the two domains is reflected by those nodes whose activation exceed a given threshold. The program, therefore, effectively produces just a single optimal solution. It has been run successfully on a number of different kinds of analogies: for example, it produces an appropriate mapping for the radiation problem and the heat-flow problem.

Holyoak and Thagard's demonstration is interesting from a number of points of view. It shows that an important thinking skill can be modeled in a parallel fashion – though this parallelism is still simulated on a serial machine – and it shows one way in which pragmatic factors can be integrated into the process of mapping.

How plausible are the algorithms proposed by SME and ACME? Both of them characterize what needs to be computed in order to achieve an analogical mapping. However, in order to develop an adequate cognitive model we need to consider further constraints. Human thinking is constrained by an inability to hold more than a few things in mind at the same time. This limitation is recognized in the theory of mental models. Here it suggests that people may not hold multiple alternative mappings in mind at the same time. Instead they may discover a mapping in incremental fashion.

The Incremental Analogy Machine (IAM)

This program, developed by Keane (e.g., Keane, Ledgeway, and Duff, 1994), also generates a single optimal mapping, like ACME, but builds this mapping incrementally in a serial fashion. The algorithm first finds a best guess or seed group by finding the part of the source domain with the highest degree of structure. In the water-flow/heat-flow analogy it selects the causal structure identified in figure 11.2. Before mapping the seed-group, a specific relation is identified which links many objects (e.g., water-flow). This relation is matched to heat-flow in the target domain and so permits a set of mappings (eg; pressure-difference is matched to temperature difference,and pressure in the beaker is mapped to temperature of the coffee).

At the next step, those relations in the seed-group for which there is no one-to-one mapping are transferred to the target domain as candidate inferences. In this case, the difference in temperature is treated as the cause of heat flow. The program then evaluates the mapping according to an evaluation function (e.g., if more than half of the relations in the seed-group have been matched, then accept as optimal). If successful, the mapping process stops. In short, the program maps just one of the possible groups in the source domain and expands this mapping incrementally. In the event of any failure, it considers alternatives serially and undoes previous mappings. Since IAM chooses an initial seed-group for mapping, the properties of the initial seed-group should be an important determiner of the ease of solving an analogy. Consider the attribute-mapping problem.

The attribute-mapping problem (after Holyoak and Thagard, 1989; see also Keane et al., 1994)
The task is to decide which things in List A correspond to which things in List B (ignoring the meanings of words) and to write these down next to the appropriate names and attributes, presented below list A.

List A	*List B*
Steve is smart	Fido is hungry
Bill is tall	Blackie is friendly
Bill is smart	Blackie is frisky
Tom is tall	Rover is hungry
Tom is timid	Rover is friendly

Steve
Bill
Tom
smart
tall
timid

The task is difficult since there are a lot of possible matches: "smart" might match "hungry" or "friendly" or "frisky." What is the unique mapping?

SAQ 11.6 How could this problem be used to test the incremental mapping notion?

If you tried the problem, you may have noticed that there is a character in each list mentioned only once (the singleton). If these are paired (that is, Steve = Fido), then smart = hungry and in consequence, Bill = Rover, tall = friendly, Tom = Blackie, and timid = frisky). The key to achieving the correct mapping is identifying the singleton since this effectively disambiguates the whole set of matches. If the singletons are matched first then the correct mapping should be achieved more easily than if other attributes are matched first. This can be tested by putting the singleton in list A at the bottom of the list. The experimental data confirmed this prediction: individuals solved the problem more quickly when the singleton was first, rather than last on the list (178 seconds vs. 363 seconds; Keane et al., 1994, Experiment 2B). For IAM, the effect of order on this problem arises because there are no structural constraints. The seed-group in List A is, therefore, the first group encountered (e.g., smart Steve). Furthermore, since the single element in this group can be matched with any of those in List B (e.g., hungry Fido or friendly Blackie), the only basis for a matching choice is the one that is encountered first in List B. If the singleton is encountered first then it gives rise to an unambiguous mapping.

Comparing the Three Models

How might these three models be compared? One measure is the number of alternative mappings that they compute to reach a correct solution (Keane et al., 1994). Holyoak and Thagard (1989) used this measure to assess the performance of ACME. Each cycle is effectively a possible mapping. Likewise, this measure can be determined for IAM by looking at the number of times the program backtracks (e.g., to find a new seed-group). The SME model produces all alternative mappings scored in terms of their structural goodness so this model can yield measures of problem complexity only. How might an index of number of cycles relate to human performance? Holyoak and Thagard (1989) showed that the number of cycles ACME goes through before it reaches a correct mapping reflects the frequency of correct solutions among subjects; the more cycles the lower the solution rate. We might also expect that the more mappings needed for the problem the longer the time required to reach a solution. For problems where correct solutions are anticipated, reaction time measures provide a test of different models.

Keane et al. (1994) used this measure to test the predictions of the three models on the attribute mapping problem. SME computes all 32 possible mappings. ACME requires eight mappings before it settles in a stable state and this number (as is also true for SME) does not vary as function of order. IAM requires just one mapping for the singleton-first condition and an average of 2.9 mappings for the singleton-last condition. Human subjects take twice as long on average so IAM predicts the direction of the difference but not the precise magnitude of the difference. Such data suggest the need to consider a serial feature to analogy models.

The aim of constructing a computational model is to predict detailed aspects of human performance. The comparative testing of these alternative models against empirical data is a very useful development. However, there are no agreed ground rules for performing such tests (see also chapter 2, "How does Simulation Assist Explanation?"). Furthermore, the focus to date has been on understanding the process of mapping. There are also models of the retrieval of analogy (e.g., Thagard, Holyoak, Nelson, and Gochfeld, 1990; Gentner and Forbus, 1991), but the importance of different stages in different situations is unclear. Novick and Holyoak (1991) showed that a major source of difficulty in transferring solutions from worked examples in mathematics to novel examples was the ease of adapting the example. The example in box 11.2 indicates the importance of exploring the implications of an analogy. It takes time to flesh out these implications.

Box 11.2 Inventing a new paintbrush (Schon, 1979)

A group of individuals were attempting to improve the performance of a paint brush, which was currently producing a gloppy flow of paint. At some point a person suggested considering a brush as a kind of pump. On this analogy the act of painting was analogous to pumping paint through the channels created by the bristles. This analogy reorganized the perception of a paintbrush: rather than the bristles themselves being critical, it was the spaces between them. It was not how paint stuck to the bristles that was critical but how it was pumped through these channels. The act of painting (bending the bristles) was construed as pumping a liquid rather than as spreading paint. The group went on to develop a theory of pumpoids of which brushes were exemplars. Applying an analogy alters the possibilities for action.

Expertise

The first part of this chapter described the process of exploring a mental model in order to determine the adequacy of a conclusion or in order to make an effective decision. The second part described the use of an existing model to solve a problem. The actual examples differed in the extent to which they required knowledge of an area. In either case, a state of the world can be represented in a mental model and actions can be applied to that state to achieve some desired state (e.g., the generation of a conclusion or a decision). This suggests a general way to describe problem solving: as a search through a space of states (a **problem space**) in which one state is transformed into another by some operation. On this view, a mental model is a state within a problem space and operations on the model are equivalent to traversing this space. We first describe a view of problem solving based on this notion, proposed by Newell and Simon (1972), and then consider efforts to account for the effects of expertise on problem solving. Consider a simple task: making toast.

IF	the goal is to make toast	Goal
	and there is sliced bread	precondition
	and there is a toaster	precondition
THEN	toast the bread	Action

This task comprises a goal and a set of subgoals together with the actions required to achieve them. The process of transforming an initial state into a goal state can be viewed as a set of IF . . . THEN **productions**. Each production rule consists of a condition (i.e., a goal), any preconditions and an action (see chapter 2, "Production Systems"). The production is considered only when the situation in the world, or in the system's working memory, matches the conditions. What happens if a precondition for an action is not met? We can establish it as a new goal which has to be met. Suppose, there was no sliced bread:

IF	the goal is sliced bread	Goal
	and there is bread	precondition
	and there is a knife	precondition
THEN	slice the bread	Action

Once this subgoal has been achieved, the conditions in the world allow the first production to apply. Lucy was faced with the situation in which the toaster did not work. She had, therefore, to establish a new subgoal: find alternative means to toast the bread. Once she had done so, she could tackle the main goal: making toast.

The example illustrates a view of problem solving developed by Newell and Simon and incorporated in their General Problem Solver. In simple cases, the complete set of choices can be readily diagrammed in a state–action tree in which possible states of the world are represented together with actions that lead from one state to another. For instance, in the 3-ring Tower of Hanoi problem described in box 11.3, there are 27 different states.

SAQ 11.7 What are the advantages and disadvantages of studying how individuals solve novel problems?

If we consider a game such as chess, the number of possible moves (states in the search space) to achieve a check mate is astronomical. A typical game might involve 60 moves and given an average of 30 alternative legal moves at each point, the total number of possible moves is 30^{60}. Clearly, no computer and certainly no human being could choose an action by exploring each possible path through the state–action tree. The problem is intractable. Newell and Simon (1972) proposed some general rules-of-thumb or heuristic methods for reducing the size of the search space: **means-end analysis** being the most important.

The means-end heuristics involves breaking a problem down into subproblems (see the toast example above). GPS first finds a production rule in which the goal state is mentioned in the condition of the rule. It tries then to find a production which will eliminate the difference between the initial state and the goal state. This becomes a subgoal. GPS compares this subgoal with its current state. If they match the action is carried out so a new state is created. If they do not match then a further subgoal is constructed with the goal of making the conditions identical. These subgoals are stored on a stack (like rings on a peg: technically, a push-down stack; see chapter 2, "Push-down Automata").

As soon as a production is found that matches the conditions of the subgoal, the specified action is carried out and the subgoal is removed from the stack and the next one is retrieved and so on. In the case of the 3 ring Tower of

Hanoi problem, the problem could be subdivided in an interesting way into two 2-ring problems and a 1-ring problem. The two 2-ring problems in turn can be divided into three 1-ring problems. This is known as the goal recursion strategy. In the case of a conflict between productions (a situation which arises whenever the state of the world is compatible with more than one production), they proposed a number of conflict-resolution strategies. For example, in the event of conflict, select the production with the most specific conditions.

Newell and Simon proposed that productions are stored in long-term memory and their use is triggered by the current focus of a person's attention. Limitations of short-term or working memory restrict the conditions that can be considered and the number of productions that can be fired at any one time. They also assumed that the contents of short-term memory are open to conscious report. This assumption guides the use of thinking-aloud protocols in which individuals say what is in their mind as they solve problems. Overall, the account so far suggests a conscious, serial, and deliberative view of problem solving, though, as we shall see, their view does not preclude unconscious and parallel processes of thinking. Indeed, it allows perceptual tests on the environment to replace plans for action.

Studies using problems such as the Tower of Hanoi, which do not require specialist knowledge for their solution, suggested that GPS provided a reasonable description of problem solving with these materials, especially after some experience. Individuals adopt a representation of the problem derived from the problem description. In other words, they form a mental model of it and identify a set of operators. Modified versions of it have been used to explore the effects of brain damage on problem solving and such studies have shown that patients with damage to the front part of the cortex have difficulty establishing sub goals (see chapter 12, "Processes Enabling the Pursuit of Goals"). Normal individuals also show impaired performance when the number of subgoals to be held in memory increases (Egan and Greeno, 1974) and when they have to select a correct move that appears to increase the distance between their current state and the goal state (Atwood and Polson, 1976). The ease of solving a problem also depends upon the content of a problem (Hayes and Simon, 1977). For instance, alternative versions of the 3-ring Tower of Hanoi problem describe a situation in which three monsters (small, medium, and large) are holding three globes of different sizes. In the "move" version, the task is to move the globes so that the size of globe is proportionate to the size of monster under three constraints: only one globe could be moved at time, if a monster is holding two globes only the larger one can be moved, and a globe cannot be transferred to a monster holding a larger globe. In the "change" version, the goal has to be achieved by shrinking or expanding the monsters under analogous constraints. The "change" problem is more difficult either because the rules are more difficult to apply or because they are less easy to learn. Individuals also find it more difficult to solve a second problem if it involved a different representation from the first ("move" versus "change").

Box 11.3 The 3-ring Tower of Hanoi problem

There are three vertical pegs. On the first peg are three rings of increasing size. Your task is to transfer these pegs, one at a time, to the middle peg, so that they are in the same order. You must never place a larger ring on a smaller ring. How many moves are needed to reach a solution?

In the state–action diagram below, the actions of moving a ring from one peg to another are omitted for each of the displayed states. The diagram effectively represents different ways of solving the problem. A person's solution path corresponds to a sequence of states.

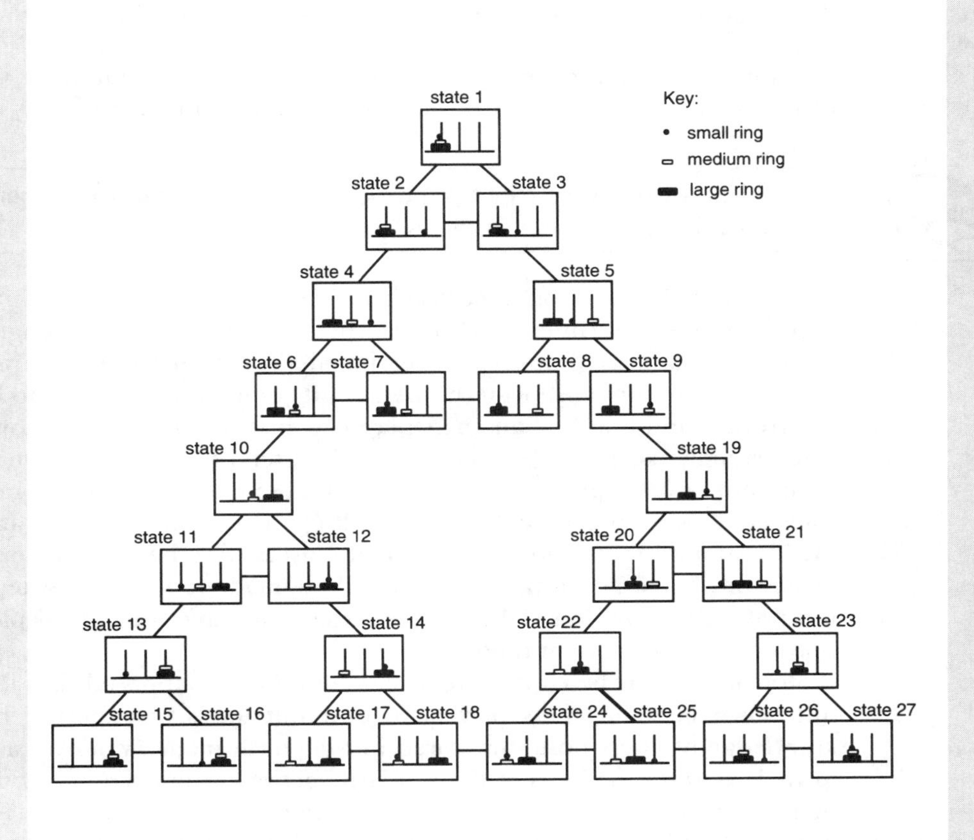

11.4 State-action diagram for the 3-ring Tower of Hanoi problem.

The use of a goal stack is not the only way to solve the Tower of Hanoi problem. Vera and Simon (1993) describe an alternative scheme in which the system is guided entirely by perceptual productions that test visible features of the external array of disks and pegs. The system notices first that the largest disk has not been placed on the goal peg. If there is a smaller disk on top of it, the system moves that disk. Otherwise it notices the largest disk that is impeding the move and sets the goal of moving this blocking disk to the other peg. This system is an example of situated symbolic problem solving. There is no memory representation of the situation nor is there a goal stack. The perceptual tests yield descriptions of the disks in functional terms (e.g., the largest blocking disk) and hence denote affordances. Individuals new to the problem often adopt variants of this perceptual strategy. With experience they move towards a recursive strategy. That is, they construct plans to move whole pyramids of disks (Anzai and Simon, 1979). Others, number the disks and remember a generative rule: for example, disk 1 is moved on every other step. The concept of a production is a general one: a system of such productions might provide a basis for a unified theory of cognition (see chapter 2, "Architectures" and chapter 3) and so be extended to account for the effects of expertise. Before reading on, please read and think about SAQ 11.8.

SAQ 11.8 In what ways might you expect novices to differ from experts in their problem solving?

The impetus for work on expertise came from the studies of de Groot (1965; first published in Dutch, 1946) on chess. Computer chess programs may consider thousands of possible developments (paths in the search tree). Given limitations on human short-term memory it comes as no surprise to learn that de Groot found that human chess players considered only a few moves ahead. Interestingly, excellent players (i.e., grandmasters) followed up no more moves than tournament players. They did, however, follow-up better moves and assessed these more quickly. In extending this research, Chase and Simon (1973) showed that more highly skilled players were better at reconstructing briefly presented board positions when these made sense in chess terms. They did not differ from less skilled players when the same number of pieces were not in a sensible configuration.

How might this be explained? Chase and Simon supposed that the critical difference between chess players of different levels of expertise lay in the knowledge of chess positions stored in long-term memory. Moves are chosen in their view on the basis of the similarity between the current position (state of the world) and the set of stored positions. Searching ahead is relatively unimportant by comparison. Chase and Simon proposed that a grandmaster might have 50,000 "chunks" of information representing important positions. Such chunks are derived by studying games of other players and developing a repertoire of tactics. Simon and Gilmartin (1973) modeled such performance

Devil's Advocate Box 11.2

Holding (1985) calls Chase and Simon's proposal the recognition-association method and argues that the evidence for it is only indirect. The skills used to represent a briefly presented board position (i.e., identification of chunks) may not be the same as those used to select a move. In fact, when more time is allowed for the analysis of random board positions, expert–novice differences do emerge (Holding and Reynolds, 1982; Saariluoma, 1985). But can differences in search skill account for differences in chess performance? Charness (1989) reported a case of a player who moved from being a good tournament player to international master status over the course of nine years with no change in search skill.

by writing production rules in which the condition identifies a specific configuration of chess pieces. If the configuration matches the condition, then the action part is carried out. For novice players, the conditions merely identified a specific piece. Hence, the contrast between novice and expert can be represented by domain-specific productions. Since these are automatically triggered, expert problem solving may involve parallel processes of thinking. Conscious deliberative thinking is restricted to the evaluation of possible actions.

Anderson (1983) offered the first substantial attempt to develop an account of both novice and expert problem solving, making use of the production system concept. In his theory, the Adaptive Control of Thought, ACT* (pronounced act-star), Anderson supposed that novices solve problems in the same way as GPS using domain-general methods. Experts, on the other hand, have domain-specific productions that fire automatically when their conditions are met. Anderson's proposal will be briefly described to provide a basis for discussing cognitive science issues of expertise.

Anderson's theory supposes three major components: (long-term) declarative memory, (long-term) production memory, and working memory. Declarative memory represents facts while production memory represents actions in the form of production rules. Working memory represents the state of the world and can store factual information in declarative memory and retrieve it as necessary. If the information in working memory matches the conditions on a production then the relevant action can be taken.

Learning consists of three stages: a declarative stage, a procedural stage, and a tuned procedures stage. Anderson supposes that information about a new domain is provided in declarative form and general problem-solving methods are used to produce the appropriate action. For instance, learning how to drive might consist of information supplied by an instructor, such as how to start the car and pull out into the road. Each fact is initially retrieved from declarative memory, stored in working memory, and used to work out a sequence of actions. Successful sequences are compiled into productions. That is, declarative knowledge is transformed into procedures for action allowing specific actions to be retrieved rather than having to be worked out. Further, productions that have the same goal and frequently occur together can be combined to produce a single macro-production. In the final stage, these productions are strengthened (through successful use); generalized (by replacing specific instances by a relevant variable); and discriminated (by blocking the use of a production unless certain conditions are met). Essentially, then, experts differ from novices in their use of productions. In a more recent version of the program (ACT-R; Anderson, 1993), the principles for activating productions and for resolving conflicts have been guided by a rational analysis of cognition, which essentially seeks to maximize the achievement of goals while minimizing the costs of computation (see also chapter 2, "Production Systems with Spreading Activation").

A second proposal, Soar (see chapter 2 and chapter 12, "Soar," for further details) that also aimed for a unified account was proposed by Newell (e.g., Newell, 1990; Laird, Newell, and Rosenbloom, 1987). It too considers the actual nature of the environment in which problem solving occurs. The theory eliminated the distinction between procedural and factual memory: all of long-term memory was a production system. Working memory contains a hierarchy of goals such that when a goal is selected, a corresponding problem space is created. The satisfaction of the goal requires a search in the problem space. Like ACT*, Soar learns. In Soar this occurs as a consequence of an impasse or block. Suppose, for example, there was no operator to transform one state to another, (e.g., the toaster was broken) a new subgoal would be created ("find another operator"). If Soar succeeded in removing the difficulty, then it stores a "chunk" of information in memory in the form of a production which states what action to take given a similar situation in future; for instance, given a broken toaster, make toast in a frying pan.

More recently, Soar has been applied to certain practical problems such as learning how to use an automated teller machine (Vera, Lewis, and Lerch, 1993). In using such machines, individuals just represent relevant aspects of the interaction that allow them to take advantage of information in the environment. It follows that people familiar with a task may not be able to recall all the steps (Payne, 1991) and that the complexity of the task can be reduced by providing cues to relevant actions. The model developed, met these assumptions and combined background knowledge of how to achieve certain actions (e.g., push a button) with relevant cues from the ATM itself.

Undoubtedly, there is a shift from novice to expert. Voss and Post (1988), for instance, found that experts in chemistry could not transfer their expertise to other domains such as problems in political science. For such problems they operated like novices: they described the causes of problems at a very concrete and specific level whereas domain experts used more abstract categories. Experts see and represent domains in a different way to novices. When asked to sort a large set of physics problems, novices grouped them according to the objects mentioned (e.g., planes and rotations) whereas experts grouped them according to the physical principles. Similar findings have been demonstrated in the area of radiology (Lesgold, 1988). It follows that experts should remember different aspects of a problem compared to novices. Indeed, computing novices appear to remember the programming code whereas experts remember the structure of the goals (Adelson, 1984); novices remember the propositions of a computing text, whereas experts construct a mental model of it (Kintsch, Welsch, Schmalhofer, and Zimny, 1990).

The notion that expertise is the result of the automatic firing of compiled or chunked productions (habits) captures the idea that experts perceive patterns rather than isolated elements (but cf. Holyoak, 1991). Indeed, for some kinds of expertise this may be the critical property. However, it suggests a rigidity in performance. But flexibility, as well as speed is a feature of expertise in some areas. In fact, experts, in contrast to novices, may acquire an understanding that allows them to be flexible in their use of pre-established routines. Box 11.4 describes an example of such flexibility.

The kind of change involved is captured by the representational redescription hypothesis (Karmiloff-Smith, 1992; see chapter 3, "Progressive Modularization and going beyond Modularity"). Essentially this proposes that individuals are able to redescribe their own stored knowledge. Knowledge embedded in special purpose procedures can become data available to other parts of the system. Knowledge can become explicit rather than being merely implicit. Such a view suggests an important role for hybrid computational systems that combine the benefits of connectionist and symbolic models (see Clark and Karmiloff-Smith, 1993; see also chapter 2, "Hybrid Models").

Experts in certain domains may well develop greater flexibility, but there is also something rather curious about the idea that experts never have difficulty. One becomes an expert by working at the limits of one's competence. Expert writers work longer and worry more about the same writing tasks than novices (Scardamalia and Berieter, 1991). Nor are scientific discoveries the result of automatic problem solving. Gruber's (1981) analysis of Darwin's work show the time consuming nature of the process; likewise Tweney's analysis of the diaries of Faraday reveal the extensive process of conjecture and refutation (Tweney, 1985). And this is surely a further important feature of some forms of

Box 11.4 Reading X-rays

Lesgold and colleagues compared the performance of expert and novice radiologists; each was presented with the chest X-ray of a healthy person who had a lobe of her lung removed 20 years before. This operation had caused the heart to shift position so that it appeared to be wider than normal. Most participants initially diagnosed congestive heart failure on the grounds of an enlarged heart. Experts, however, revised their opinion when informed of the prior operation and were assured of the person's current physical health. Novices, by contrast, insisted that the person was dying.

expertise: the ability to articulate and account for a certain judgment or decision. In the context of science and policy-making it is also the capacity to express arguments in support of a particular position which is crucial. Such considerations stress the need to look at expertise and problem solving in a social and technical context in which joint problem solving is the norm (see also D'Andrade, 1989). Expertise also brings with it the capacity to explore a problem space or a model of a problem: i.e. to play. Current models of expertise emphasize a kind of technical mastery but seem to omit any account of the pleasure in exploration and feelings to do with the quality of a solution to a problem.

Summing-up

Thought, we have supposed, is in the service of need and action in the world. Much problem solving is tied to specific practical issues and is situated in a rich environment. The notion that a critical function of thought is to model situations in the world naturally accommodates this fact. Adaptiveness implies limiting the complexity of the representations we construct to what is necessary for action. Such a view accords with limitations on working memory and with the supposition that our models make explicit as little as possible. The bulk of research discussed on mental model theory involved verbal problems of a relatively simple nature. Outstanding questions remain: what leads to the fleshing out of the model? Clearly knowledge of the various possibilities will be one factor. Indeed we explored this possibility in the section on analogical problem solving. Here we considered models with greater internal complexity. A model from one domain can be used to flesh out a model in another domain. Three different computational models of the mapping process were discussed. Comparative tests of these models are now possible and these will lead to further changes. The kind of criteria for comparative testing, and the relationship between the kinds of indices extracted from such computational models and measures of human performance remain to be clarified. Processes associated with the discovery of suitable analogies and their adaptation to specific cases remain to be explored computationally (although, see Clement, 1991). Analogical problem solving is one way to deploy one's expertise. Other aspects of expertise were considered in the third section. Current models account for the speed of performance and can accommodate the kinds of conceptual shifts that occur. Whereas current research in this area emphasizes the domain-specific nature of expertise, issues to do with flexibility also need to be addressed. In addition, we need to recognize that cognition is situated in a social sense: the explanation and justification of actions is also a vital human skill.

Further Reading

Human Problem-solving, Englewood Cliffs, N.J.: Prentice Hall by A. Newell and H.A. Simon (1972) is a difficult but classic text on the problem-space approach. For a wide-ranging coverage of research on problem solving *Thinking and Reasoning*, Oxford: Blackwell by A. Garnham and J. Oakhill (1994) is very readable. For detailed coverage of research on syllogisms and conditional inference *Human Reasoning: The Psychology of Deduction*, Hillsdale, NJ: Lawrence Erlbaum Associates by J.St.B.T. Evans, S.E. Newstead and R.M.J Byrne (1993) cannot be bettered. *Human and Machine Thinking*, Hillsdale, NJ: Lawrence Erlbaum Associates by P.N. Johnson-Laird (1993) provides an informative and detailed exposition of the computational issues, especially on the theory of mental models.

Exercises

1. Consider the following problem: three men (Mr Large, Mr Medium, and Mr Small) want to cross a river. They find a small boat which will only hold a weight of 200 lbs or less. Mr Large weighs 200 lbs; Mr Medium 120 lbs, and Mr Small 80 lbs. Assuming they have oars but no rope, how can they all get across? Draw a state–action diagram for this problem. What is the minimum number of journeys?
2. How distinct are the processes of understanding a problem and solving a problem?
3. Is a general theory of human expertise possible?
4. Construct the initial mental model representation of the conditional "If p then q" and of the biconditional "If and only if p then q." Does the modus tollens inference (see devil's advocate box 11.1) require the same number of explicit mental models in both cases? (See Johnson-Laird, Byrne, and Schaeken, 1992, Experiment 3 for the answer!)

CHAPTER 12 The Control of Thought and Action

Outline

One way to analyze behavior is as a sequence of actions, but how are those actions sequenced? How do individuals control their behavior so that appropriate actions are performed in the appropriate order and at the appropriate times? This is the topic of the current chapter. We begin with a lengthy discussion of the empirical side of action, focusing on data from neuropsychological investigations. The modeling of neuropsychological phenomena in computational terms is an increasingly important area of cognitive science, and the second half of the chapter adopts this approach. We look at two leading computational accounts of the control of action: the SAS/CS model of Norman and Shallice (1980, 1986), and Soar (Laird, Newell and Rosenbloom, 1987; Newell, 1990), providing an overview of each system and a discussion of the mechanisms by which they account for the data. Although each system deals very well with certain aspects of the control of thought and action, the challenge for cognitive science is to develop a single principled account for all aspects of the domain.

Learning Objectives

After reading this chapter you should be able to:

- identify and classify common errors in the control of action
- outline the main forms of neuropsychological deficit in the control of thought and action
- interpret the neuropsychological findings in terms of current theories of the control of complex behavior
- explain why different processing systems are thought to control action under routine and non-routine conditions
- compare and contrast two theories of the control of action and how they account for the common classes of error

Key Terms

- cognitive neuropsychology
- computational modeling
- controlled and automatic behavior
- disorders of thought and action

Introduction: the Organization of Action

It is not hard to appreciate why Mom burnt the toast. She was happily attending to the task of cooking the toast, and probably expecting the toast to pop-up when it was ready, when she was interrupted by a far more interesting task – opening a parcel from Gran. In similar circumstances we might all forget to monitor the toast, and not realize that it hadn't popped-up when it should have. But how are actions controlled? How is it that people can often interrupt, and then resume, long action sequences, such as making the toast? Why is it that sometimes things go wrong?

General Properties of Action

Before beginning in earnest, there are a number of general properties of actions that we can isolate without having much of a theory about how action is controlled. Perhaps the most obvious of these is that actions are sequenced. Actions occur through time. Although some actions may be performed simultaneously, many actions must be performed in sequence, simply because of the physical constraints involved. When Lucy is making the coffee, for example, she cannot pour coffee and milk into a coffee mug and add sugar all at the same time (unless she has three hands or a special coffee-making device). A different sort of physical constraint operates when she is adding the coffee granules from the coffee packet. Before she can dip the spoon into the packet she must open the packet and pick up the spoon. Dipping the spoon into the packet has preconditions (that the packet be open and that the spoon be held) which must be satisfied before the action can be performed. In this case, having more hands or special tools wouldn't help Lucy. The actions must still be sequenced.

A second observation is that action sequences can be interrupted. Making toast involves putting bread in the toaster, switching the toaster on, waiting a few minutes until the toast pops up, switching the toaster off, and removing the toast. Mom was able to interrupt this action sequence and switch to the action sequence for opening the parcel with no difficulty. If not burning the toast had been sufficiently important to Mom, she would, also, have been able to resume the toast preparation action sequence before the toast was burnt. On the other hand, if she was also required to prepare the coffee, feed Zara, answer the phone, and collect the parcel from the postman at the same time as making the toast, it is quite likely that something would go wrong. Thus, while action sequences can be interrupted or even interleaved if necessary, there is a limit to the amount of interleaving that we can maintain.

Thirdly, objects have appropriate actions. The appropriate thing to do with a glove is to put it on your hand. The appropriate thing to do with a pen is to write with it. The appropriate thing to do with a mug of coffee is to drink it. These are not the only things you can do with these objects – you could try to balance them all on your head – but if given any one of these objects and told to use it, most people would perform an action appropriate to the object. The notion of an appropriate action is relevant because, as discussed below, when attention is diverted, or in persons suffering certain forms of brain injury, it appears that actions appropriate to objects may be triggered without intentional control.

Object-appropriate actions are, however, not independent of context. Given a pair of shoes, an appropriate action is to put them on, but this is not appropriate if you are about to get into bed. Thus, our final property of actions is that they have an appropriate context. Actions can be performed in and out of this context, but when performed out of context they appear, at best, eccentric.

Action Lapses in Everyday Life

Actions don't always quite go as planned. Consider the apparently simple task of preparing a mug of coffee. It is not uncommon, after the coffee is made, to accidently start to put the coffee mug (instead of the milk container) in the fridge. Errors of this sort are particularly common when we are distracted, or when we are attending to something else (see chapter 10: box 10.4). The fact that we can attend to one thing while doing another suggests that the non-attended action is to some extent **automatic**, but lack of attention often leads to errors. These unintended erroneous actions are known as **action lapses**. Such lapses are not uncommon in people who are functioning perfectly normally, but in certain neurological patients they can be virtually incapacitating.

Reason (1979, 1984, 1990, 1993) and Norman (1981) have developed classifications of everyday slips, lapses, and minor blunders. Reason's full taxonomy consists of many types, some of the more common of which are:

- **Object substitution errors**, where an appropriate action is performed with the wrong object. The above example of the coffee mug in the fridge illustrates this form of error.
- **Place substitution errors**, where the correct objects for the action are employed, but some spatial location is incorrect. Pouring milk into the coffee percolator, instead of the coffee mug, is an example of such an error.
- Anticipation or **omission errors**, where actions are performed out of sequence. This class is exemplified by pouring boiling water into a coffee percolator without first putting the filter and coffee granules in place.
- Capture errors or **utilization behavior**, where some element of the current situation captures behavior and triggers an unintended, but appropriate, action sequence. Preparing coffee, when distracted and only moments before one intended to cook some toast, exemplifies this error class.
- **Perseveration errors**, where some action or sequence of actions is re-

peated, even when the goal of those actions has been achieved. Repeatedly spooning coffee granules into the percolator filter (and not stopping after the usual quantity) is an example of this form of error.

It is not always easy to classify an error as being a lapse of one of these types. Often it is necessary to know what action was actually intended. Some people, for example, might intentionally put milk in the coffee percolator: if everyone present has milk in their coffee, it may be an efficient way of making white coffee. To be sure that such an event corresponds to a place substitution error requires that we know what action was actually intended. While asking individuals can usually tell us this – people know what their intentions are – it may, if a person's outward behavior is all that we have available, be necessary to make assumptions about their intentions in order to correctly classify any errors in their actions.

SAQ 12.1 Normally people become aware of our action lapses when something goes wrong. They then usually correct their action and think nothing of it. But action lapses are surprisingly common. Indeed, it is generally only by taking note of such lapses that one realizes how common they are. Try to record and classify any of your own lapses that you notice over the next few days. If possible, score yourself on Reason's questionnaire of everyday minor lapses (Reason, 1993).

The Neuropsychological Data Base

The error patterns shown by intact individuals are not the only source of data for a theory of the control of action. Indeed, if anything, the problems exhibited by patients suffering various forms of neurological damage are more critical to theories of the domain, for these problems are often strikingly counterintuitive. But how can **cognitive neuropsychology** – the study of impaired and intact cognitive processes in brain-damaged individuals – tell us anything about the normal functioning of the brain? Box 12.1 provides the answer.

The problems that arise in the organization of complex behavior within brain-damaged individuals may be classified into four levels. These levels support the existence of cognitive structures, **schemas**, for the control of action. At the lowest level is the ability to make object-appropriate actions (e.g., to drink when presented with a mug of coffee). Building on this is the ability to schedule such actions into simple behavioral routines (e.g., to prepare a mug of coffee). Higher still is the level of voluntary control (e.g., to decide not to make a mug of coffee when presented with the necessary ingredients). Finally, there is the level where action and cognition meet, the level where conscious deliberate control is required over actions (e.g., to form a shopping plan). Brain damage can result in deficits at each of these levels.

Box 12.1 How does observing neurological patients help us understand the functional architecture of the brain?

Neuropsychology is the study of the effects of brain dysfunction on people's cognitive abilities. This brain dysfunction might have been caused by many different types of disease or trauma – for instance brain tumors, strokes (where blood vessels in the brain have been damaged), or head injury (e.g., road traffic accidents, gunshot wounds, etc.). But why should studying people who are unlucky enough to suffer these sorts of problems be of theoretical interest to the cognitive scientist? After all, at first sight it might seem odd to study people with damaged brains in order to learn about how an undamaged brain works.

Well, the answer is that we can learn a great deal about how the brain's cognitive processing mechanisms are organized (i.e., its functional architecture; see chapter 3 for background on **modularity**) by observing how the processes break down. The primary method used in cognitive neuropsychology is the investigation of **dissociations** of function. A dissociation between two abilities is where a patient can perform one task as well as they always could, but where their performance on some other task is severely impaired. Where this pattern occurs, it suggests that the processes underlying performance on the two tasks are not shared – if both tasks tapped the same processes (or set of processes) then performance on the two tasks would always be either impaired or spared in any one patient.

Neurological patients may show many such dissociations, with some more surprising than others. For example, two of the more common syndromes are classical **amnesia** (primarily an inability to remember new information – see chapter 10) and **agnosia** (a deficit in the visual recognition of items). Individual patients may show either the amnesic syndrome or agnosia singly, or occasionally they may be seen together. Such a dissociation – where either problem can appear independently of the other – is known as a double dissociation and suggests in this case that the brain systems which support visual recognition share little in common with those which enable us to learn new information. This, perhaps, seems hardly surprising since the tasks are so obviously very different. But what about the ability to recognize objects contrasted with the ability to recognize human faces? Since recognizing objects and recognizing faces might be similar, it is perhaps surprising to find that patients who have even very severe problems with the visual recognition of objects may, nevertheless, be able to recognize familiar faces. Moreover, where some patients are completely unable to recognize people by their faces (known as "prosopagnosia") they may, nevertheless, be able to recognize objects almost as well as they ever could. Even finer grain distinctions about the visual recognition system can be made. One patient, a farmer, reported by McNeil and Warrington (1993) found that following his brain damage, he was completely unable to identify his own family by their faces, but he could still recognize his sheeps' faces! From this we can learn that there is something specific about the way the brain processes information as regards human face recognition.

Two other common methods of deriving data from brain-damaged patients (apart from studying dissociations) are qualitative error analysis and the experimental manipulation of processing stages. Error analysis refers to the examination of patients' mistakes, which are either of a type not made by normal subjects, or where the errors are seen in normals, but appear in patients with abnormal frequency (see chapter 6 for examples). In the experimental manipulation of processing stages, some aspect of an experimental paradigm is systematically altered in such a way as to test a given theory.

These methods are not exclusive to neuropsychology (indeed the latter is the normal operational method in traditional experimental psychology, and a number of examples are given in chapter 10).

Data gleaned from all these three types of methodologies enables inferences to be made about the functional architecture of the brain, and examples of each will be used in the following sections where we outline neuropsychological evidence for the different stages and levels of processing involved in the control of complex behavior.

Level 1: The Ability to Make Object-Appropriate Actions

Above, we introduced the idea that objects have appropriate actions. Neuropsychology provides particularly strong evidence that this is the case since the ability to make object-appropriate actions can be selectively impaired in some patients. One of the earliest demonstrations of this type of impairment was reported by Pick (1905). He described cases who, when trying to light a candle with some matches, made mistakes such as attempting to strike the match on the candle (an object substitution error) or failing to blow out the match (an omission error). This type of disorder in patients – one specific to object-appropriate actions – is known as ideational **apraxia** (see McCarthy and Warrington, 1990, and Heilman and Rothi, 1985, for further discussion of this phenomenon) and most commentators agree that it provides good evidence for the existence of object-specific action "engrams." These are stored representations of actions which are accessed quite automatically in unimpaired individuals – one does not have to "think" what to do with objects as familiar as, say, a pen. However, neuropsychologists have not yet entirely resolved the debate as to whether the impairment in ideational apraxia lies with access to, or storage of, action-independent semantic representations of object knowledge, or if it is a disorder which is restricted only to actions.

SAQ 12.2 You should be able to find at least six examples of object-appropriate behavior in the scenario. What are they? Find two examples in the scenario of people using objects in non-routine ways.

Level 2: The Ability to Schedule Actions into Behavioral Routines

The ability to deactivate actions

Given that one is able to make an appropriate action, one must also be able to stop performing it once it is completed. This ability may also be selectively impaired: some patients who are able to make an appropriate action nevertheless have problems in being unable to stop themselves from unnecessary repetition of it. This is known as perseveration, and one well-known example is shown in figure 12.1, where A. R. Luria asked his patient to draw a circle.

Perseveration appears in many forms, and the most straightforward taxonomy of these types is given by Sandson and Albert (1984). They suggest that there are three different forms of perseveration (Stuck-in-set perseveration, Recurrent perseveration, and Continuous perseveration), all of which are distinct in terms of clinical features, cognitive process, and neuroanatomy. Stuck-in-set perseveration refers to the inappropriate maintenance of a current category or framework, involves an underlying process deficit in high-level functioning, and is related neuroanatomically to frontal lobe dysfunction. Recurrent perseveration is characterized by the unintentional repetition of a previous response to a subsequent stimulus, involves the abnormal "post-facilitation of memory traces" (p. 715), and is caused by posterior left hemisphere brain damage. Continuous perseveration is the inappropriate repetition of behavior without interruption, involving a deficit in motor output, and reflecting damage deep within the brain.

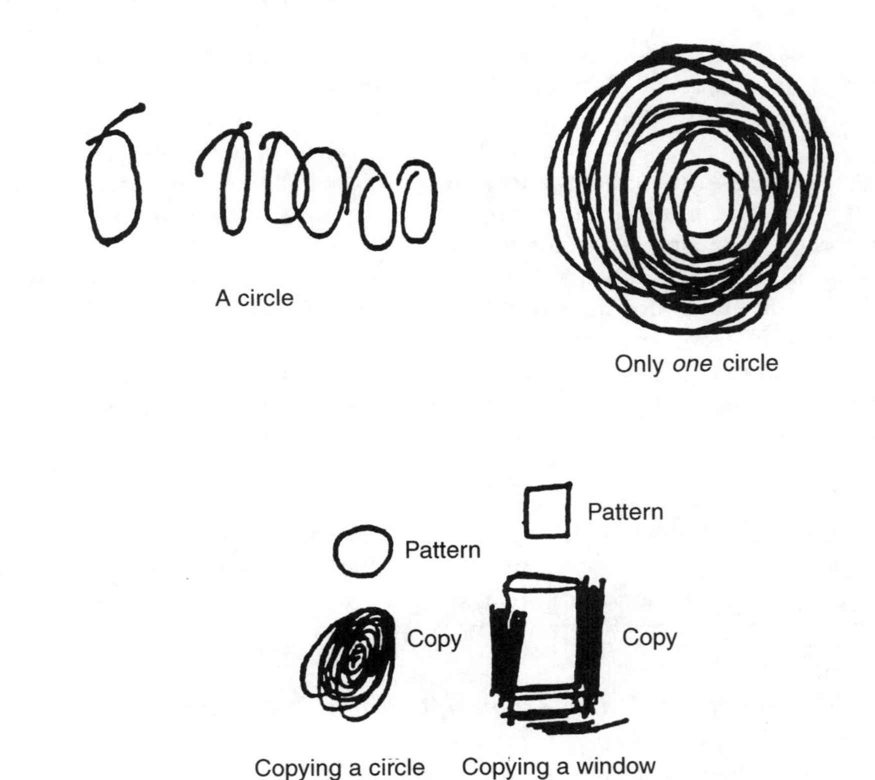

12.1 Examples of perseveration. The top left figure was the patient's response to the instruction 'draw a circle.' The patient was then reminded to draw only one circle, and responded with the drawing on the top right. The figures below these are the patient's attempts to copy the patterns indicated (adapted from Luria, 1973).

Perseveration may appear across a wide range of situations, however, making it difficult to decide whether the form shown in any particular task is conceptually distinct from another, or whether the differences merely reflect the constraints of the situation. Nevertheless, it is one of the more consistently observed forms of behavior disturbance following brain damage, and as such provides a real challenge to the cognitive scientist – any adequate account of the organization of behavior must be able to explain this apparently simple form of dysfunction.

The ability to schedule simple behavioral sequences
More complex behavior does not, of course, consist only of object-appropriate actions being randomly activated or deactivated. Most theorists agree that even the most basic level of purposive behavior involves some scheduling of these object- or situation-specific actions. For instance, if one is riding a bicycle towards a stop sign, one does not have to consciously consider ones actions at the level of detail of, for example, "Well, I'd better apply the brakes, remember to stop peddling, and when I'm at a standstill, then put my foot on the ground or else I'll lose balance." Each of these actions – which are appropriate to the different objects are normally activated (once overlearned) in the correct sequence.

The action disorganization syndrome (Schwartz, Reed, Montgomery, Palmer, and Mayer, 1991) provides evidence for such organization at a basic level. Schwartz et al. describe a patient who, following a subarachnoid hemorrhage, was generally able to make object-appropriate actions, but often the action made was incorrectly selected, or would not be appropriate to the actual object in use. For instance, he poured tomato juice onto waffles, and applied shaving cream to his toothbrush (both of which might be either object substitution errors or place substitution errors). Critically, at other times he would perform actions out of sequence (e.g., putting on shoes before tapered pants). Additionally, he would show some of the features already described, such as omitting individual components of a sequence and unnecessary repetition of tasks or task components (perseveration).

Schwartz et al.'s study is important, not only because it describes a fascinating case which provides data which can challenge theories of the organization of behavior, but also for the detailed empirical examination of the structure of action which it contains. They share a view of the basic organization of behavior as comprising hierarchies of temporally structured units (see, for example, Fuster, 1980), with the bottom level consisting of what they term A-1 events. These events serve objectives which are simple and immediate, such as making the contents of a closed container available by removing its cover. The immediately higher level they term A-2 scripts. These are clusters of A-1 acts which all serve to accomplish a given goal, such as making a cup of instant coffee. Of course at this point a new idea has been introduced: that purposive behavior involves the representation of goals (see, e.g., Duncan, 1986; MacKay, 1987; Shallice, 1988). Neuropsychology provides some evidence that this level of goal representation is separable from cognitive levels, which mediate **routine** behavior, because some patients may exhibit sequences of behavior which are internally coherent and where all the actions are carried out appropriately (unlike the action disorganization syndrome), but where the behavior itself is inappropriate to the patient's situation. Usually the behavior appears to be beyond voluntary control.

Level 3: Voluntary Control Over Behavioral Routines

In this section we outline three types of problems shown by some patients with neurological disease. These syndromes are evidence that "scripts" or sets of schemas can be triggered automatically when the cognitive processes, which usually govern their activation are damaged. At the most basic level, object-appropriate actions only are triggered by a stimulus in the environment. At the highest level, whole complex strings of behavior are activated. Studying these different patterns of behavior can tell us how behavior is organized and controlled by the brain.

Utilization Behavior

Lhermitte (1983) first reported an unusual behavioral sign, which he termed "Utilization Behavior," which he managed to elicit in patients with (usually) large lesions to the frontal lobes. The patient sits face-to-face with the examiner, and is not given any information about the test, and the examiner remains silent throughout. The examiner then proceeds by showing the patient a glass, which is placed within reach of one of the patient's hands. The patient with utilization behavior will automatically grasp it, and the examiner then holds a bottle of water towards the other hand, and the patient then too grasps this. After a short period of puzzlement, the patient invariably pours the water from the bottle into the glass and drinks it. The examination then continues with other objects, such as a knife, plate, and fruit; each time the examiner presents the objects, the patient grasps them and uses them without saying a word. If the patient is presented with three pairs of glasses, one at a time, the patient will put them all on, eventually wearing all three at once. Lhermitte reports that these patients will, after a short period of distraction, actually use the objects presented to them even if they have been directly instructed not to. This behavior was apparently never observed in 100 control subjects. Lhermitte concludes that for these patients "tactile, visual-tactile, or visual stimuli imply to the patient an order to grasp the objects presented to them and use them" (p. 252).

One possible interpretation of Lhermitte's finding might be that the patients, confused by the surprising behavior of the doctor, and knowing that doctors typically present objects to patients for some test purpose, simply thinks that the doctor expects the objects to be used. Indeed, this is precisely what Lhermitte's patients reported when questioned about the testing session.

This possibility is, however, challenged by the case reported by Shallice et al. (1989) who experimentally manipulated the conditions of Lhermitte's procedure to examine the effect of environmental variables upon utilization behavior. Shallice et al.'s patient, LE, showed utilization behavior not only when presented with objects, but also when he was actively engaged in other tasks and the situation was arranged to discourage it. Thus, LE was seated at a desk engaged in neuropsychological assessment, and on the far end of the desk was a tray on which various household objects had been placed. No reference was made to these objects by the examiner. Despite being engaged in the assessment procedures and a clinical interview with the neuropsychologist, patient LE, on a number of occasions leaned forward, selected an item (or items) from the tray, and used them. The behavior ranged from minor "toying" (e.g., a single action where an object was manipulated in a non-purposeful way, such as idly pushing a stop watch around the table top) to coherent activity, such as spontaneously, and without a word, picking up a pack of cards and dealing four cards to himself, and four cards to each of the other two examiners present. This type of behavior was termed "Incidental Utilization Behavior" and is evidence that the patient's behavior is not necessarily a consequence of Lhermitte's testing procedure, instead supporting the interpretation that the sight of the object is enough to trigger the action-appropriate behavioral sequences.

Imitation Behavior

Lhermitte (1986) later reported two further examples of stimulus-bound (i.e., environmentally-triggered in the absence of higher cognitive control) behavior. The first, "Imitation Behavior" refers to quite straightforward imitation of the experimenter's actions by the patient. The choice of behaviors that Lhermitte chose to elicit imitation is however quite novel, to say the least. Under similar conditions to the utilization behavior procedure, the experimenter made such gestures as thumbing his nose, military salutes, and chewing paper. Lhermitte found that a high proportion of patients with large lesions (which usually involved the frontal lobes) showed some degree of spontaneous imitation of the examiner, despite being explicitly instructed not to do so. Lhermitte reports that male patients "even imitated such socially unacceptable gestures as using a urinal, or urinating against a wall, in front of 20 or 30 people!" Unsurprisingly perhaps, none of the normal control subjects copied the experimenter, and most seem puzzled by his behavior or even laughed at him, thinking he was trying to be amusing. With imitation behavior, and unlike the cases of apraxia mentioned above, the patients' actions are perfectly formed, and the action sequences are carried out in the correct order.

Lhermitte (1986) argues that imitation behavior and utilization behavior are both manifestations of a basic disorder, and differ only in severity (p. 330). Interestingly, he maintains that imitation behavior is a voluntary act, rather than an "automatic or reflex response" and is thus distinguishable from echopraxia (Dromard, 1905), which is an automatic imitation of other people's gestures, apparently performed with the speed of a reflex action. However, the fact that Lhermitte's patients were apparently unable to stop themselves from copying the examiner despite having just been told not to (and being able to understand and remember this instruction), perhaps suggests that echopraxia (and its counterpart in the speech domain, echolalia), imitation behavior, and utilization behavior all differ along a dimension of automaticity, occurring in the context of varying degrees of awareness and/or voluntary control.

Environmental Dependency Syndrome

A more complicated pathological behavioral sign also seen in some frontally-lesioned patients, lends support to the provoked versus incidental distinction made by Shallice et al. in regard to utilization behavior. In the "Environmental Dependency Syndrome" (Lhermitte, 1986) patients' behavior appears automatically triggered by the environment in which they find themselves. Thus whole strings of behavior patterns are elicited by the patient's environment. Lhermitte describes how one patient, when being shown around the experimenter's flat, entered the bedroom and on seeing the bed prepared, the patient immediately began to get undressed, got into bed, pulled the sheet up to his neck, and prepared to go to sleep. Another patient, having been shown a syringe, picked up the needle, soaked a cotton ball in antiseptic, and bent down to give the experimenter an injection! This form of behavior appears without provocation from the experimenter. It would appear that the environmental cues alone are sufficient a trigger. We have already discussed this type of behavior in normals under the label of capture errors. The first example was given by William James (1890). He went upstairs with the intention of changing his clothes and the next thing he knew was that he found himself in bed. Usually these sorts of mistakes in behavior occur when one is either preoccupied with some thought, tired, or under stress (Broadbent et al., 1982; Reason, 1993). Shallice (1988, p. 328) gives another example. He found that at one time, he walked into a room he knew well and suddenly noticed that he was making a pulling movement with his arm, which he did not understand. He remarks, "I eventually realized what was at some level 'known' – that the light switch in that room was controlled by a cord, which had got hooked up in a cupboard door." Shallice maintains that the fact that his behavior was so mystifying at the time indicates that initiation and execution of the action was not normally controlled by a conscious intention to execute it. In other words, the behavior was automatically triggered by the incoming perceptual triggers. That these kinds of mistakes are relatively uncommon in normals (and tend to occur only under certain circumstances), but may be shown by certain brain-damaged patients to a pathological degree, suggests that in the normally functioning individual certain cognitive systems exist which prevent such inappropriate behavior. What then are these systems?

Level 4: Where Automatic Behavioral Routines are not Sufficient: the Integration of Action and Cognition

So far, this chapter has described patients who have deficits in producing appropriate actions in the context of intact motor abilities (ideational apraxia); those that can produce these actions but are unable to stop them (perseveration); those that can produce actions, and stop them, but where actions are performed in the wrong sequence (action disorganization syndrome); and, finally, those that can produce correct behavioral sequences, but where they are activated (apparently) without voluntary control. If it were the case that automatic scheduling of behavior were sufficient for all situations in everyday life, then members of this last category of patients would never show their problem. It was the incongruity of their behavior with the social or environmental demands that made the behavior abnormal. Given then that no two situations in a lifetime are exactly identical, and some may even be completely novel, there must be some processes which enable adaptation of behavior in situations where automatic processing alone would be insufficient. What might these situations be?

Norman and Shallice (1980, 1986; see also Shallice and Burgess, 1991a) outline five types of situations where routine automatic activation of behavior would not be sufficient for optimal performance:

1 Ones that involve planning or decision making.
2 Ones that involve error correction or trouble-shooting.
3 Ones where responses are not well learned or contain novel sequences of actions.
4 Ones judged to be dangerous or technically difficult.
5 Ones that require the overcoming of a strong habitual response or of resisting temptation.

These five types can be essentially reduced to two broader characterizations. The first would be where an incorrect response has been, or is liable to be, produced by automatic behavioral routines. The second type is where no routine procedure is available to produce an appropriate response, that is the situation is in some essential respect novel. This latter category must apply in particular to situations in which the delayed activation of an intention is involved – that is where one intends to do something later. This is because rarely do individuals intend to carry out exactly the same sets of behaviors in exactly the same circumstances as they have done many times before. Sadly, real-world life is not usually that straightforward. Can neuropsychology, then, provide examples of patients whose problems show up only in these types of situations?

The Strategy Application Disorder

One of the earliest examples of a patient who has a relatively isolated deficit in higher control of action is reported by Penfield and Evans (1935). The report was actually about the author's (Penfield) own sister, who was unfortunate enough to have developed a brain tumor in her right frontal lobe. This case is perhaps especially relevant to the scenario being considered in this book. Penfield and Evans write: "She had planned to get a simple supper for one guest (Penfield himself) and four members of her family. She looked forward to it with pleasure and had the whole day for preparation. When the appointed hour arrived she was in the kitchen, the food was all there, one or two things were on the stove, but the salad was not ready, the meat had not been started and she was distressed and confused by her long continued effort alone." But why should this situation be so difficult for this patient? The answer is that organizing and cooking a meal is a surprisingly complex task (surprising, perhaps, only for those who rarely undertake it): planning has to be carried out and preparatory purchases made. Different considerations – costs, tastes, available time, etc. – may have to weighed against each other. Many minor decisions need to be made and typically many behavioral sequences have to be activated at the correct time and in the correct sequence (requiring **prospective memory**, the activation of an intention at delay).

Deficits in the ability to carry out actions at delay, or to prioritize action sequences, can have a dramatic effect on one's effectiveness in the real world. For instance, Eslinger and Damasio (1985) reported the case of an accountant, EVR, who six years after an operation for the removal of a large orbitofrontal meningioma had an IQ of over 130 (in the top few percent of the population) and no problems with his language, memory or perceptual functions. However his ability to organize his life was grossly impaired. He was dismissed from a series of jobs even though his basic skills, manner, and temper were all appropriate; he went bankrupt and was involved in two divorces in two years. Relatively simple matters would take him hours: thus, to go out to dinner required that he consider the seating plan, menu, atmosphere, and management of each restaurant and he might even drive to see how busy each of them was, but still he would be unable to come to a decision.

Quite obviously these patients' deficits are not just at the level of pure action. They are at the level of correct formulation of a goal (which includes planning, problem solving, and priority setting: see Duncan, 1986), maintenance of that goal in the face of environmental distractors, and activation of intentions at a time remote from their creation. Analysis of behavior at this level of complexity requires consideration of the interaction between thoughts, intentions and actions, and so descriptions of the patients' behavior at this kind of level needs to specify some bridge processes between internal cognitive states and external demands.

SAQ 12.3 How might you determine whether someone has a deficit at the level of strategy application?

Shallice and Burgess (1991b) were interested in identifying the exact locus of impairment that leads to deficits such as those described in EVR or in Penfield's sister. The hope was that these patients might show patterns of dissociation which would further our understanding of the stages which are involved in very complex behavior. They described three patients who all had evidence (from **brain imaging**) of frontal lobe damage following head injuries, but who had been left with little or no neuropsychological deficits in language, memory, or perceptual abilities. In addition they were all, like EVR, quite intellectually gifted. Despite this, all three had attempted to make returns to their former employment which had been unsuccessful. Their friends, employers, and relatives complained that they were disorganized, erratic, and showed poor judgment and decision making. Burgess and Shallice gave these patients two tests which were designed to tap the kinds of processes required in **non-routine** situations. In the first, they had to carry out six quite easy, but open-ended tasks in 15 minutes; two involved dictating routes, two involved the carrying out of a series of fairly simple sums, and two involved writing down the names of objects drawn on cards. However, to complete all six tasks would take much longer than the limited time available. The patients had to judge how much time to devote to each task so as to optimize their performance given some simple rules, of which the most important was that the scoring system guaranteed diminishing returns on a task as it continued to be carried out. The second task was undertaken in a shopping precinct, and again, although the actual task items themselves were easy (such as "buy a birthday card"), they had to follow certain arbitrary rules in carrying out the task, and some of the items involved some problem solving (e.g., find out where the coldest place in Britain was yesterday).

All three patients made significantly more mistakes than a matched control group, and these mistakes could be divided into four categories: Inefficiencies – where a course of action did not break task constraints, but could have been more efficient; Rule Breaks – where the patient explicitly broke a task rule, or a social convention (such as not paying for an item in a shop); Interpretation Failure – where the patient misunderstood some aspect of the task; and Task Failure – where a task was not completed or was performed incorrectly. For all types of mistake except Interpretation Failure at least two of the patients were significantly poorer than controls. The appearance of these types of error in everyday life in the context of an otherwise intact neuropsychological state was termed the "Strategy Application Disorder".

Processes enabling the pursuit of goals

Shallice and Burgess (1991a; 1993) argue, with reference to research findings in other areas of cognitive science, that complex non-routine behavior requires the interleaved operation of three broad processing stages. The first is Plan Formulation and Modification (where a person is first deciding what they should do in an unfamiliar situation), considerable examination of which appears in the literature (see, e.g., de Groot, 1965). The second stage, Marker Creation and Triggering, is newer to cognitive science, however. This refers to the operation of the creation and maintenance of goals and intentions, and their realization at appropriate times (i.e., prospective memory). A marker is basically a message that some future behavior or event should not be treated as routine and instead, some particular aspect of the situation should be viewed as especially relevant for action. If the behavior or event does occur later, then the marker would be triggered and this would lead to inhibition of the activity being carried out, the reassessment of the situation, and so, potentially, the switching in or out of a particular course of action linked to the marker. There are probably many different kinds of markers – for instance ones that respond to social convention cues, and temporal cues as well as situation-specific ones. The third set of processes in the organization of complex behavior is Evaluation and Goal Articulation (see, e.g., Duncan, 1986). This is were someone is assessing, after carrying out a set of actions, how close they now are to their initial goal state. Sometimes their actions, for whatever reason, have failed to achieve all that they intended, and a new goal (Goal Articulation) has to be decided upon.

SAQ 12.4 Lucy mentions something that will require the creation of a temporal marker by one of the family. What does Lucy say and who is she talking to?

Shallice and Burgess (1991a, b) provide a schematic representation of how the above three processes might relate to each other (see figure 12.2), and which emphasizes the divisions made in this chapter between automatic behavioral routines and those which are utilized in non-routine situations. In this figure, the vertical lines indicate the succession of the stages involved, and the horizontal lines indicate the flow of information between non-routine processing and automatic processing at a particular stage.

Recent evidence (Burgess and Shallice, 1994) broadly supports the characterization of the stages shown in figure 12.2: individual patients can be found who show striking dissociations in ability at different stages. However, any complex behavior will consist of complex interactions between the products of these processing stages, and this suggests that the analysis of very complex behavioral sequences will require an understanding of how one process can affect a second which in turn can affect the first. This may require the use of methodologies which, though common in other areas of cognitive science, have not as yet been applied in cognitive neuropsychology.

Two Models of the Control of Action

Having outlined the neuropsychological evidence for the mechanisms which are involved in the control of thought and action, we are now in a position to consider some of the computational models that have been developed within this domain. This evidence should, as in other areas of cognitive science, provide constraints on theorizing, and any adequate account of complex behavior should be informed by it (see, e.g., Bullinaria, 1994). Nevertheless, it is currently unrealistic to expect any one theory to account for all of the phenomena. A good account must therefore explain some of them, without contradicting those that it does not address.

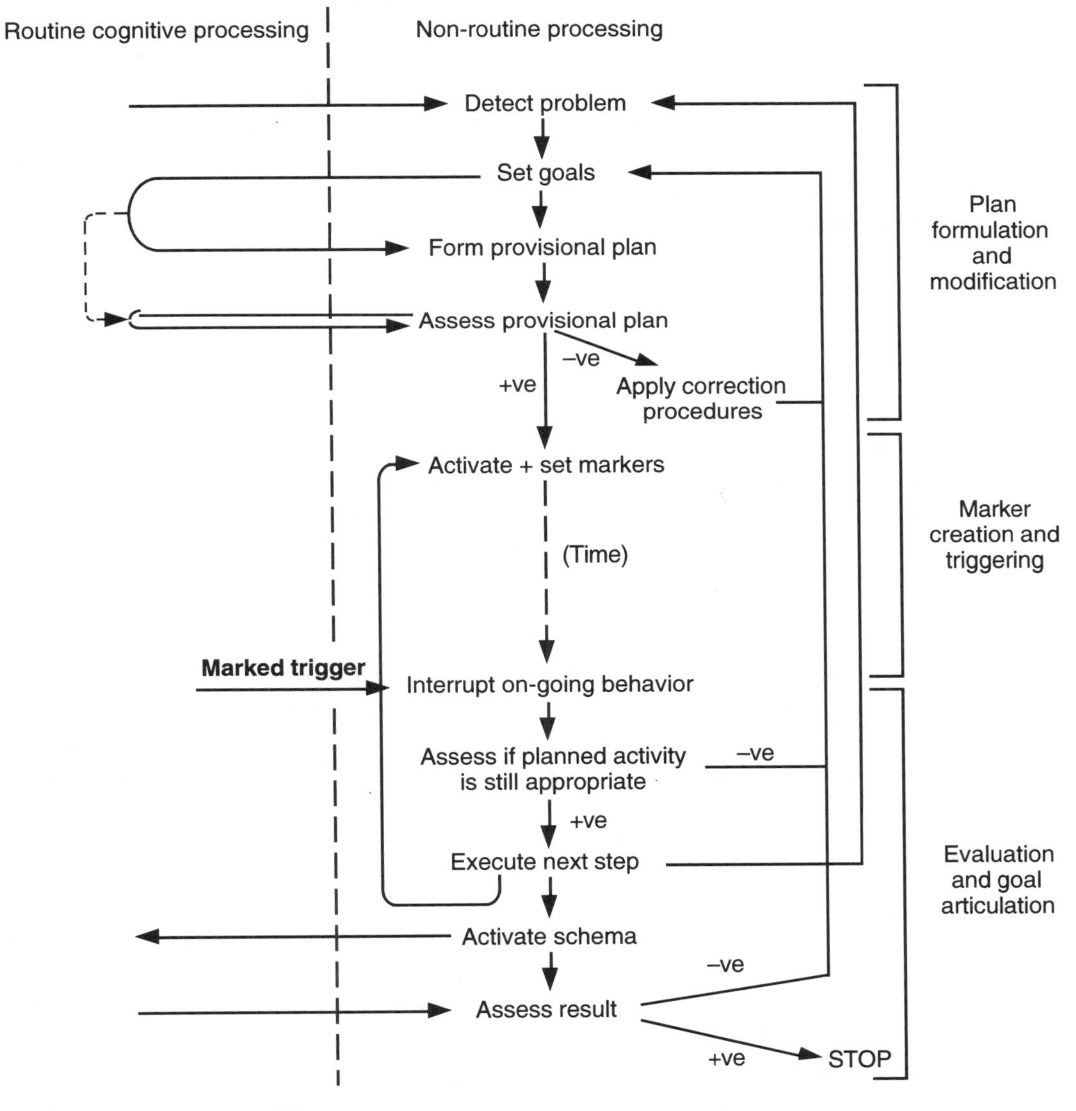

12.2 Outline of the interaction between routine and non-routine processing stages which are involved in coping with a novel situation requiring the delayed realization of an intention (adapted from Shallice and Burgess, 1991a). Vertical lines indicate the succession of stages, and horizontal lines indicate the flow of information between non-routine and routine processing.

We may view the neuropsychological evidence as, at minimum, suggesting that there are separate cognitive systems for the operation of routine and non-routine behavior. Routine behavior is the province of levels 1 and 2 (the ability to make object-appropriate actions and the ability to schedule those actions into behavioral routines). Non-routine behavior concerns levels 3 and 4 (voluntary control over behavioral routines and the integration of action and cognition). Some computational models of thought and action reflect this via separate systems for routine and non-routine processing. Others must find some other means to account for everyday action lapses and the dissociations found in impaired individuals. In this section we consider in detail two computational models of control processes. In each case, the assumptions underlying the model are presented, followed by a discussion of how processing is effected within the model, and an analysis of how the error patterns of intact and impaired individuals might be explained within the model.

The Norman and Shallice Model

Overview

Norman and Shallice (1980, 1986) proposed a theory of attention based on a fundamental distinction between automatic and controlled processing. They argued that separate systems were responsible for routine/automatic and non-routine/controlled processing and behavior. Contention Scheduling (CS), the system responsible for routine processing/behavior, is claimed to have direct control over effector systems (such as hands) as is required for routine behavior and **cognitive processing resources** (such as special purpose **modules**; see chapter 3) as required by routine processing. In situations where automatic control is insufficient, such as those involving planning, trouble-shooting, and novelty, the affect of CS is modulated by a second system, the Supervisory Attentional System (SAS). In the terms used above, this modulation may involve level 3 processes (voluntary control over behavioral routines) or level 4 processes (the integration of cognition and action).

If one assumes that attentional resources are limited, as is evidenced by our inability to attend to more than one thing at a time, then there are clear evolutionary advantages in having separate systems for controlled and automatic processing. In particular, delegating mundane, over-learned activities to an automatic system leaves the attentional system free to work on other, more cognitively demanding tasks. For example, when attempting to escape from a predator, the attentional system can plan an escape route while the automatic system controls the running action of the arms and legs.

In the complete model, behavior results from the interaction of four processing levels (see figure 12.3): (1) cognitive units; (2) schemas; (3) CS; and (4) SAS. We discuss each of these in turn.

Cognitive units are basic cognitive abilities that relate to specific neu-

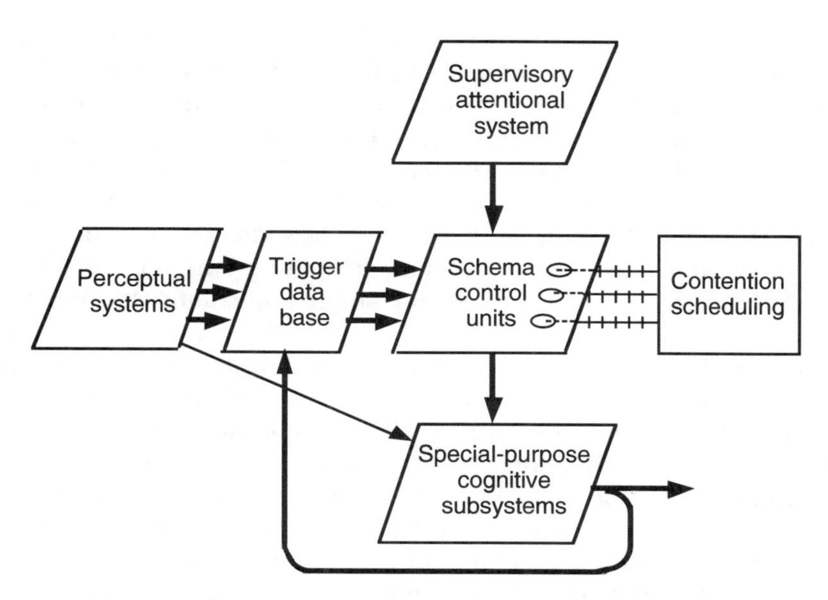

12.3 The Norman & Shallice model of the control of thought and action. The dotted lines refer to control information, the thin continuous line to specific information required by particular special-purpose cognitive subsystems, and the hatched lines to the primarily inhibitory interaction between the activation levels of schemas in contention scheduling (adapted from Shallice et al., 1989).

roanatomical or neuropsychological systems. These include perceptual abilities, memory, language, and so forth.

Schemas are strings of over-learned actions or thought sequences which usually occur together. As examples, Shallice (1982) cites drinking from a container, doing long division, making breakfast, or finding one's way home from work. Schemas can be activated in two ways. They may be triggered either by perceptual input or by the output of other schemas. As we will see, this first form of triggering (from perceptual input) is held to account for capture errors and environmental dependency syndrome. The second form (from other schemas) is due to the hierarchical structuring of schemas. Burgess and Alderman (1990) use the analogy of driving a car as an example of such hierarchical organization. A low-level schema might be a very basic behavioral routine, for example, the control of the movement of the head and eyes to look in the mirror. A higher-level schema might be a whole complex action routine, such as one's actions on approach to some traffic lights. This would include slowing down, indicating, and so forth, and would be triggered by the sight of the lights turning red. An example of the triggering of one schema by another might be that when this "approach traffic lights" schema has been activated, this might activate other related sub-schemas – for instance gear changes, checking the mirror, and looking at road signs.

Contention scheduling is a process that has the function of selecting the most efficient set of schemas which will fulfill the task demands. This is a routine function and is designed to deal with familiar situations. It works by selecting a limited number of compatible schemas so that they can successfully control the available resources until the goal is achieved, or alternatively, until a higher-priority schema is activated (as would be the case in the situation where one is doing something fairly routine and suddenly some unexpected event happens which requires one's immediate attention). In short, CS is a rapid triggering and selection process for familiar situations with clear rules and guidelines.

The supervisory attentional system has the job of dealing with unfamiliar or novel situations where routine schema selection is not sufficient to fulfill the task demands (i.e., level 4: the integration of action and cognition). This it does in a much slower and more deliberate fashion than CS. Critically, it works only indirectly by altering the likelihood of certain schemas being selected. Its most common function is that of monitoring. In the situation where CS fails (because there is no known appropriate behavioral routine or because schemas are so weakly activated that they cannot trigger other schemas) the SAS takes charge. Shallice (1982) characterizes his model as an information-processing analogue of Luria's (1973) unit for the programming, regulation, and verification of behavior and predicts that damage to this system (which is held to be part of the functioning of the frontal lobes) will lead to disturbances of attention, concreteness, lack of flexibility, perseveration, and so forth.

The most well-developed aspect of the theory is Contention Scheduling. The heart of CS is the hierarchical network of schemas. These operate at level 2 (scheduling actions into behavioral routines). Each schema in the network has an activation value, a number corresponding to the activity of the schema which varies depending on the degree of excitation or inhibition to which the schema is subject. Activation values are influenced by a number of sources, including environmental triggering (objects or situations in the environment which excite schemas) and attentional control (based on the direct influence of the SAS). In addition, schema activations interact via an interactive activation network (see chapter 2, "Interactive Activation Networks"). Schemas which compete for resource requirements (including cognitive and effective resources) inhibit each other, and this inhibition is partially balanced by each schema's self activation.

As noted above, the purpose of CS is to select a set of compatible schemas to control behavior. The selection of schemas is based on their activation level, such that all schemas whose activation rises above a certain threshold are automatically selected. Selection either alters the dynamics of activation flow in the network (for high-level schemas) or directly controls the allocation of cognitive or effector resources (for low-level schemas). In the former case, selection enables the flow of activation from a schema to its component subschemas. Thus, selection of a schema, such as that corresponding to "make coffee," will enable the flow of activation to the component sub-schemas, including that for "add sugar from sachet." This will boost the activation of the sub-schemas, possibly leading to their selection. In the latter case, selection enables the use of resources by low-level schemas operating at level 1 (i.e., at the level of object-appropriate actions). Thus, selection of the schema for grasp will result in a grasping action. Schemas remain selected (even if their activation level falls) until either the schema attains its goal, it is actively inhibited by the SAS, or the activation of a competing schema exceeds the selection threshold.

Learning within the CS/SAS model involves building new schemas by extending the schema hierarchy within the CS module. This requires the creation of new nodes in the schema hierarchy, the association of a sequence of sub-schemas within this schema, and the association (and subsequent refinement) of triggering conditions with the new nodes. However, the theory of learning remains underdeveloped. Thus, although the CS/SAS model accounts for the distinction between controlled and automatic processing, it remains to be demonstrated how it can account for the shift from controlled to automatic behavior, which appears to occur as individuals learn any routine task.

SAQ 12.5 Below we discuss how some of the sorts of errors mentioned in the introduction to this chapter may be accounted for within the Norman and Shallice model. Can you work out how such errors might arise within the model?

Processing within the Norman and Shallice model

The result of the CS processes described above is a volatile network that can work in an automatic fashion if the appropriate high-level schemas are activated by the SAS. The network is, however, subject to environmental influences, and hence capture errors. Situations trigger schemas and, if there is insufficient attentional control (in the form of insufficient activation of appropriate schemas by the SAS), this triggering can control behavior. The resulting behavior is, nevertheless, still constrained by the schemas available in the CS system, and as such will be routine in its nature. Utilization behavior can similarly be explained within the CS/SAS framework if attention is diverted or supervisory control is impaired. A third class of errors, perseverative errors, can be accounted for within the framework by assuming a breakdown in the monitoring by the SAS of the attainment of schemas' goals.

The original accounts of the CS/SAS systems of Norman and Shallice (1980, 1986) and Shallice (1988) did not attempt to account for the selection of objects and/or places (collectively known as arguments) to which schemas are to be applied. Thus, the theory did not specify processes which might determine which hand to use when picking up the sugar sachet, or the container into which to pour its contents. As such, an explanation of object substitution errors and place substitution errors was beyond the scope of the theory. More recently, however, Cooper, Shallice, and Farringdon (1995) have produced a computational implementation of the CS theory, and in the process have augmented the theory to allow the selection of arguments. The augmentation involves the explicit inclusion of a representation of the environment, with activation values attached to objects within that representation. The objects participate in an interactive activation network (analogous to the schema network) with objects effectively competing to act as arguments for selected schemas. Active schemas excite the representations of their arguments, and active objects excite (via triggering conditions) schemas which may use them. In this way, feedback ensures that when a schema is selected, the appropriate arguments are most active. Nevertheless, simulations have shown that without supervisory control, object substitution errors and place substitution errors can arise if the objects or places are sufficiently similar.

The CS/SAS system is therefore able to account for each of the major error classes described at the beginning of this chapter. This is possible because of the relatively loose relationship between the two systems involved. The CS system can function with minimal supervisory control producing automatic behavior, but even when supervisory attention is engaged, the SAS can only modulate behavior by exciting appropriate low-level schemas. In no circumstances does the SAS have complete control. Its modulatory affect can always be overridden with sufficient triggering from the environment.

Soar

The Norman and Shallice theory is really only well developed for behavior at the lower levels in our classification. Newell and colleagues (Laird et al., 1987; Newell, 1990; Rosenbloom, Laird, Newell, and McCarl, 1991) have developed a theory, Soar (see also chapter 11, "Expertise"), that is well developed at the higher levels.

Overview

The philosophical roots of Soar date back to 1973, when Newell argued that experimental psychology tended to focus on what he called "micro-theories": specific theories of individual tasks or small domains. Irrespective of how well such theories are able to capture the empirical results within their domain, the concentration on isolated domains can tell us very little about how tasks fit together in a complete theory of cognition. To this end, Newell has argued for the development of unified theories of cognition – theories of the complete cognitive architecture – and in 1990 Newell advanced Soar as one such unified theory (see chapter 2, "Architectures" and chapter 3 for background).

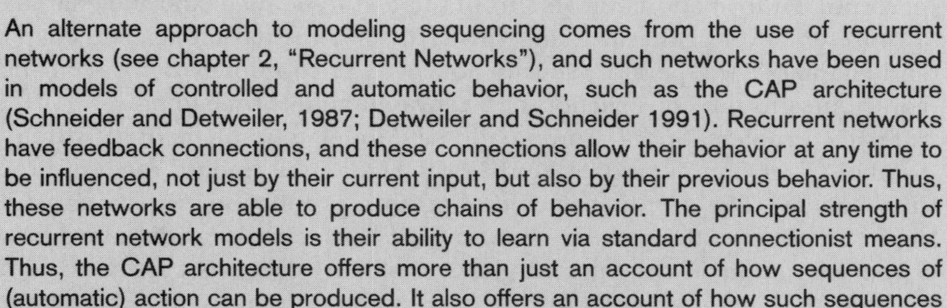

Devil's Advocate Box 12.1 Recurrent network approaches to sequencing

An alternate approach to modeling sequencing comes from the use of recurrent networks (see chapter 2, "Recurrent Networks"), and such networks have been used in models of controlled and automatic behavior, such as the CAP architecture (Schneider and Detweiler, 1987; Detweiler and Schneider 1991). Recurrent networks have feedback connections, and these connections allow their behavior at any time to be influenced, not just by their current input, but also by their previous behavior. Thus, these networks are able to produce chains of behavior. The principal strength of recurrent network models is their ability to learn via standard connectionist means. Thus, the CAP architecture offers more than just an account of how sequences of (automatic) action can be produced. It also offers an account of how such sequences can be learned.

One difficulty with the chaining approach to sequenced behavior, however, is that it is unclear how some types of action slips may be accounted for within such a model. In particular, it is difficult to explain how anticipation and omission errors might arise, given that these require the breaking of the "chain" of action.

At one level, Soar is a theory of the organization of high-level cognitive operations such as those involved in problem solving. This side of the theory is known as the Problem Space Computational Model. At another level, however, Soar is a theory of the implementation of those operations within a (symbolic) production system (see chapter 2, "Production Systems"). This aspect is known as the Symbol Level Computational Model. We describe each of these in turn.

Central to the Problem Space Computational Model is the notion of a state – a representation of a snap-shot of the problem domain. A state contains all relevant information that is known about a particular problem. Within the domain of chess-playing, for example, the state at a particular time would effectively be a snapshot of the board (including the positions of all pieces) at that time, together with an indication of whose move is next. Rules which specify the possible transitions from one state to another are known as operators. In the chess domain, operators correspond to legal moves, and specify how the state changes when the corresponding move is made. Processing consists of stepping from an initial state to a goal state, via various intermediate states, by the application of a series of operators. The key to intelligent behavior lies in selecting the appropriate operators at the appropriate times. The states and operators for any particular representation of a domain form a problem space.

The theoretical roots of Soar lie in Newell's early work on problem solving (e.g., Newell and Simon, 1972), and it is easy to see how various puzzles and games can be expressed in terms of problem spaces (see chapter 11, "Expertise"). Soar achieves its status as a unified theory of cognition by treating all cognitive tasks in terms of traversing a state space. All cognitive tasks are specified in terms of initial and goal states, together with a set of operators. In a task such as reading, for example, the initial state is a representation of the perceptual information available from the text to be read, and the goal state is a representation of the meaning (in terms of a mental model: see chapter 8 and chapter 11). Operators specify how perceptual information systematically contributes to the construction of a representation of meaning.

Soar's Problem Space Computational Model is implemented within the Symbol Level Computational Model, which is a nonstandard form of production system. The working memory of this production system contains a representation of the current goal (and all higher goals, as discussed below), a representation of the current goal's problem space, its current state, and possibly the operator which is to be next applied to that state. The long-term memory contains rules (i.e., productions) for selecting problem spaces, states, and operators, and rules for evaluating and applying operators. Processing is cyclic and each cycle consists of two phases. In the first, all long-term knowledge (in the form of productions) is brought to bear on the current representation of the task (i.e., the contents of working memory). This yields a set of potential working memory modifications (including altering the problem space, current state, or next operator), each tagged with an indication of its appropriateness. In the second phase, which is effectively a delayed conflict resolution phase, the most appropriate of these modifications is selected and working memory is modified accordingly. In this way, the ideas of operator selection and application, within the Problem Space Computational Model, are mapped onto the more familiar production system operations.

Soar has a further mode of operation, however. If Soar is unable to choose an operator or state (because, several operators are equally good, or because no states are appropriate, for example), the system is said to have reached an impasse. Soar automatically responds to such situations by setting itself a new goal: the goal of resolving the impasse. So in the case of solving Lucy's problem with the broken toaster, where there is no direct method available to cook the toast, Soar would set itself the goal of evaluating its alternatives (this presupposes that Soar knows what alternative ways of cooking toast are available). Once it had done this, it would select the best alternative and continue as before. Problem solving proceeds within the subgoal as usual, and Soar is normally able to resolve such impasses (though possibly only after a series of subgoals have been created) and return to its original goal. A typical progression of states (including the raising and resolving of an impasse) is shown in figure 12.4.

The Problem Space Computational Model and the Symbol Level Computational Model effectively correspond to "central cognition." When Newell (1990) advanced Soar as a unified theory of cognition, it was necessary to develop an account within that architecture of the interaction between central cognition and perception and action. Newell took perception and action to be

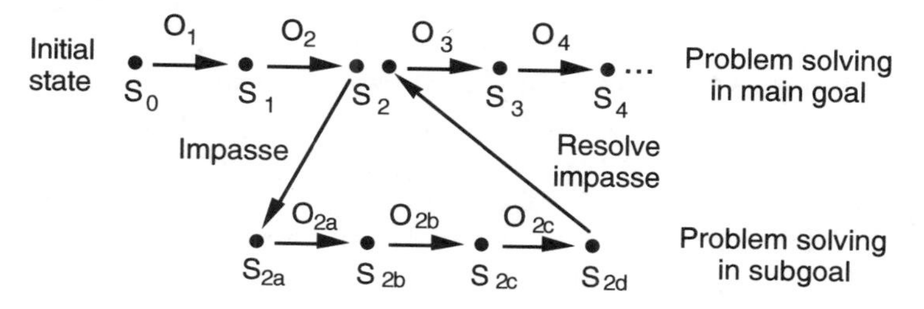

12.4 Processing and subgoaling with Soar. States (marked S) are indicated by black circles. Operators (marked O) are indicates by arrows between states.

simple encoding and decoding operations respectively. Special "perceptual" productions map perceived stimuli into working memory elements (forming part of the current state). "Motor" productions are triggered by other working memory elements, mapping them to muscle movements.

Deliberate processing and immediate behavior

The distinction between controlled and automatic processing as it occurs in the Norman and Shallice model may be equated within Soar with the distinction between processing that does and does not involve the selection of further states or operators between perceptual and motor productions. That is, automatic behavior may be defined as behavior in which the outputs of perceptual productions directly trigger motor productions. However, Soar is also equipped with a learning mechanism (as discussed below), and because of the way this works, it is more appropriate to distinguish between **deliberate processing**, in which impasses and subgoals are invoked, and immediate behavior, where subgoaling is unnecessary, and behavior is determined by the direct selection of a sequence of states and operators between perceptual and motor processes. If we make this distinction, then immediate behavior includes automatic processing.

SAQ 12.6 How do the four levels outlined in the beginning of this chapter map onto processing within Soar?

Deliberate processing in Soar corresponds to level 4 processing (the integration of action and cognition) as discussed above, and typically involves the comparison of several alternate possibilities, possibly by "looking ahead" and comparing the results of the various alternatives. In this way, Soar can, for example, compare the results of two or more hypothetical action sequences, and then select the operator that will trigger the first action in the most appropriate sequence. This form of processing is apparent when Soar is actively solving problems, such as solving AI puzzles or controlling the non-routine movement of a robot arm.

Immediate behavior might be equated with processing at levels 2 (the ability to schedule actions into routines) and 3 (voluntary control over behavioral routines), although strictly speaking there is no notion of voluntary control within Soar. Immediate behavior involves no impasses. This occurs when at each stage in processing there is either a single appropriate state to select or a single appropriate operator to apply. All processing in this case must occur in what Newell (1990) terms the Base Level Problem Space. Such behavior is characterized by the chaining of productions, resulting in fluent action without deliberation. Newell (1990) provides sketches of Soar's behavior in several immediate behavior tasks, ranging from the single choice reaction task (which requires a button to be pressed when a light comes on) to skilled typing (see Cooper and Shallice, 1995, for criticisms of this work). Processing at level 1 (the ability to make object appropriate actions) is not directly addressed by Soar.

SAQ 12.7 How might our ability to handle interruptions be accounted for within Soar? Can you see how some of the sorts of errors mentioned in the introduction to this chapter might be accounted for within Soar?

Unlike the Norman and Shallice theory, the development of Soar was not influenced by any of the neuropsychological evidence presented above. As such, Soar is less able to provide convincing accounts of action slips and the behavior of patients. Nevertheless, as a unified theory of cognition, these phenomena are within Soar's remit. Let us consider, first, how Soar might explain interruptions, such as that experienced by Mom when called to open the parcel while cooking the toast. As Soar's perceptual productions function independently of its cognitive productions, situations in the world may add elements to Soar's working memory at any time. Such elements may, however, trigger cognitive productions allowing automatic or deliberate switching of tasks. It is conceivable that action lapses, such as capture errors and utilization behavior might be given similar explanations, possibly involving the malfunctioning of the mechanism which selects the appropriate state or operator. Malfunctioning of the system is, however, a prerequisite of such behavior. There is at present no concept of attention, and as such these errors cannot be explained in terms of Soar being subject to some distraction. Other forms of action lapses, however, such as place and object substitutions, perseverations, or omissions, cannot be explained so simply, and it is unclear how such slips might be accounted for within the Soar architecture.

Learning within Soar

One of the distinctive features of Soar as a production system is its mechanism for learning. Soar is able to learn from experience by forming new productions as problem solving within a subgoal proceeds. The conditions of such productions are given by the circumstances which caused the impasse leading to the subgoal, and the actions are derived from the results of the subgoal. Thus, if a subgoal arose because the toaster was broken, and the result of the subgoal was to try to cook the bread in the frying pan, Soar would learn that (all other things being equal) when the toaster is broken (and the goal is to make toast) try to cook the bread in a frying pan. Productions learned in this way may apply in later problem solving when a situation similar to that which previously led to a subgoal arises, thus preventing the need for subgoaling. As Soar gains experience, learning can result in deliberate processing being reduced to immediate behavior through the elimination of all impasses in favor of appropriate productions. Further learning is not possible, however, as there is no scope for immediate behavior to be reduced to automatic behavior.

There is some empirical support for Soar's learning mechanism. In particular, the improvement that Soar shows with practice is comparable to that shown by humans in the sense that both exhibit the so-called power law of practice (Snoddy, 1926; Seibel, 1963; Newell and Rosenbloom, 1981) – in both cases the relationship between the speed up in performance on a specified task (expressed in Soar in terms of the number of cycles, and in humans by the number of milliseconds required to complete a given task) is related to the number of attempts at that task according to a power function.

Towards a Theoretical Integration

The CS/SAS theory and Soar are both very different. They each have strengths and weaknesses, but the criterion of assessment – that of potentially accounting for neuropsychological deficits – is tough. How do the theories compare, and how might research on the control of thought and action be advanced?

The Norman and Shallice model is theoretically able to explain the full gamut of the data we have considered. All the major error classes have been shown to arise when supervisory attention is lacking, but the SAS remains under-specified. Until this component has been further specified (and appropriate simulations have been carried out), the explanations of aspects of high-level control (such as the performance degradation observed when attempting multiple tasks simultaneously) and learning (i.e., the transfer of non-routine to routine behavior) within the model must remain speculative. It should be emphasized, however, that the success of the Norman and Shallice model with respect to the neuropsychological data is in part due to its origin. It was developed specifically with much of these data in mind.

Soar was developed to model problem solving. As such, it is in its element in accounting for controlled high-level thought and behavior. It is unclear, however, how the undamaged system might account for everyday action lapses, and the behavior of the damaged system (with respect to the neuropsychological evidence) remains to be demonstrated. Soar does have a story about learning, but the learning mechanism is limited to the transfer of deliberate processing to immediate behavior. This means that there is no scope for learning beyond the stage of immediate processing: learning cannot occur within action or thought sequences that do not involve impasses.

Thus, the models are most successful in different areas. The Norman and Shallice model (or at least that part of the model which has been implemented: contention scheduling) addresses the domain of routine action. Soar addresses both routine and non-routine behavior, but it is most successful in its account of non-routine behavior. The challenge for cognitive science, then, is to develop a single principled model which ties together these domains through a principled account of learning.

There is a further area, however, that we have not discussed, but which both models must measure up to. This concerns the interpretation of simulation results (see chapter 2, "How does Simulation Assist Explanation?"). As in much simulation work, interpretive assumptions are necessary in order to compare the behavior of these computational models of thought and action with human experimental data. At the very least then, we require not just integration of the theories, but also integration of the interpretive assumptions used to judge the adequacy of their computational instantiations.

Summing-up

We began this chapter by consider the sorts of errors which are common in everyday action. Such errors are often seen in exaggerated frequencies in patients with various neurological disorders, but the data reveals a number of distinct dimensions along which the control of action can break down. Dissociations between such breakdowns give us glimpses of the functional architecture of the mind, and hence criteria for judging the adequacy of computational models of control processes. In the second half of the chapter we presented two such models, Norman and Shallice's CS/SAS model and Newell and colleagues' Soar model, and considered the extent to which each of these models was supported by the evidence.

Further Reading

The reader interested in the considerable debate about how inferences are made in neuropsychology, and their legitimacy for theorizing, is directed to the opening chapters of *From Neuropsychology to Mental Structure*, by T. Shallice (1988), Cambridge University Press, Cambridge. The full variety of error in human action is detailed in *Human Error*, by J. Reason (1990), Cambridge University Press, Cambridge, UK.

More complete descriptions of the computational models discussed can be found in the articles referred to in the text. The most complete description of Soar is given in *Unified Theories of Cognition*, by A. Newell (1990), Harvard University Press, Cambridge, MA. The most thorough description of the Norman and Shallice model can be found in *From*

Neuropsychology to Mental Structure, by T. Shallice (1988), Cambridge University Press, Cambridge, UK (see especially chapter 14), which also contains a great deal of argumentation in support of the model.

Exercises

1. Attempt to explain your own action lapses (those listed in response to SAQ 1), within all four levels, according to, firstly, the Norman and Shallice theory, and then Soar.
2. Given the complementary strengths and weaknesses of the two models discussed in this chapter, consider the possibility of developing a hybrid model, drawing on each of the submodels for different domains. What problems may arise in such a venture?

Glossary

2½D sketch: a description of a surface in terms of surface distance and local surface orientation from a specific viewpoint.

accommodation: one of the three domain-general processes of Piaget's constructivist view of cognitive development; it is the process of modification of the existing knowledge structures to adapt to new input (see **assimilation**).

action lapse: the performance of an action that was not intended, usually when one's thoughts are distracted.

activation: a number, often between 0 and 1, which represents the relevance of an object to current processing. As processing progresses, the number rises and falls as the relevance of the object changes (see **interactive activation** network).

activation vector: an ordered set of **activations**, usually corresponding to a layer of nodes in a connectionist network.

addressee: person to whom an utterance is directed (as opposed to someone who merely overhears).

affordance (of objects): the potential use(s) of an object that is held by Gibson to be directly perceivable.

agnosia: the selective loss or impairment of visual perception or recognition following brain damage.

agreement: two constituents agree on a feature if they have compatible values for that feature. For example: *he is, they are, *she are.*

algorithm: a computational procedure, usually involving a fixed series of steps, for arriving at the solution to some problem (compare **heuristic**).

amnesia: the selective loss or impairment of memory following brain damage.

analogical problem solving: process of mapping the conceptual structure of a known or familiar domain on to that of an unfamiliar domain.

anterograde amnesia: the inability to acquire new memories.

aphasia: the selective loss or impairment of language abilities following brain damage.

apraxia: the selective loss or impairment of voluntary action following brain damage.

argument: a phrase acting as a "player" in the "scenario" specified by a verb. For instance, the verb *sleep* specifies a scenario in which some person is doing the sleeping. In the sentence *John sleeps, John* is an argument because he acts as a player in the scenario specified by *sleep.*

artificial intelligence: a branch of computer science which aims to make machines do the sorts of things that are done by human minds; its importance for cognitive science is in showing the considerable complexity and computational power involved in the most ordinary everyday achievements of the human mind.

assembled: a system which has been "put together" from some stock of more basic elements or processes.

assimilation: one of the three domain-general processes of Piaget's constructivist view of cognitive development; it is the process of modification of perceptual inputs by existing knowledge structures (see **accommodation**).

associationism: the view that complex mental activity, such as perceiving, remembering, and thinking, can be explained in terms of some sort of hooking together of simple elements; in the case of classical associationism these basic atoms are ideas or concepts; in the case of **behaviorism** they are stimuli and responses.

associative network: a connectionist network which stores a set of patterns and is able to use this information to complete a pattern given a partial or corrupted input pattern.

Behaviorism: a doctrine in psychology which insists that the goal of the subject is to formulate laws linking stimulation to observable behavior, and that because they are unobservable, mental events and states should not figure in these laws (see also **reflexes**).

bottom-up: processing of an input that relies on extracting information from the sensory stimulus, as opposed to being guided by higher-level knowledge or contextually generated expectations (compare **top-down**).

box/arrow diagram: a diagram showing the functional components which interact in the production of some aspect of behavior.

brain imaging: various techniques for producing a picture of the brain. These may vary from relatively simple techniques such as Computerized Tomography (CT), which is like an X-ray of the brain, to complex methods like Positron Emission Tomography (PET), which measures actual activity of the brain (e.g., glucose metabolism).

categorical perception: perception in which identification and discrimination performance of continuous (usually acoustic) dimensions is limited by the categories of the response labels provided. This mode of perception was initially thought to have been peculiar to speech.

causal modeling: a specification of the sequence of deficiencies at the biological and cognitive levels that lead to particular developmental abnormalities or insufficiencies.

central systems: those cognitive systems which take the deliverances of the **input systems** and integrate them with existing general knowledge in updating beliefs and making decisions.

classical conditioning: a learning procedure in which as a result of pairing a neutral conditioned stimulus (CS), such as a tone, with a motivationally-significant unconditioned stimulus (US), such as food or shock, the CS comes to elicit a conditioned response (CR), which is often similar to the unconditioned response (UR) evoked by the US on its own.

coarticulation: the influence on a segment of speech of the other segments around it.

cognitive architecture: the organization of the mind in terms of a set of **functional components** which support all aspects of cognition.

cognitive effects: within **relevance theory**, the kind of interaction which a newly processed stimulus must have with information already existing in the cognitive system in order for it to be relevant to the system. There are three types of cognitive effect it may have: (a) supporting and so strengthening existing assumptions, (b) contradicting and eliminating assumptions, and (c) interacting with them to produce new conclusions.

cognitive neuropsychology: the building of models of how mental processes are organized by observing patterns of impairment in brain-damaged individuals.

cognitive penetration: the involvement of higher-level (**central**) knowledge (beliefs, judgments, expectations) in lower-level perceptual processes (compare **informational encapsulation**).

cognitive processing resource: a theoretical process or ability which may be invoked by other processors or abilities in the production of behavior.

cognitive science: the interdisciplinary scientific discipline that studies mind as an information processing system.

contextual effects: see **cognitive effects**.

coherence: the general property of the connectedness or relevance of utterances in discourse or text.

complement: the sister of a **head**.

compositionality: the property of a representation that its meaning is a function of the meaning of its part and their manner of combination.

computation: a mental process that maps one representation or set of representations into another; for instance, the processes that map the acoustic representation of an utterance into its semantic representation (see **representational view of mind**).

computational model: a computer program which implements a theory of some aspect of cognition.

concepts: mental representations that guide the process of categorization and provide the basic units of thoughts, memories, and language.

conceptual knowledge: knowledge built around concepts, with several concepts being combined together into a higher-order structure.

conceptually-driven processing: top-down processing of a stimulus that is guided by prior knowledge and expectations.

conditioning: the behaviorist theory of learning, according to which organisms acquire behavioral patterns through experiencing a number of pairings of a stimulus or action and

reinforcer. In classical, or Pavlovian, conditioning the organism develops an **association** between a signal (say, a tone or light) and the reinforcer (food or some other reward); in instrumental, or operant, conditioning the association is between the organism's own action or behavior and the reinforcer.

connected curve: a curve which returns to its starting point.

connectionism: a branch of computational modeling which rejects the explicit processing of symbols in favor of processing based on the interaction of numerous simple processing devices (units).

consolidation: the process by which the brain gradually fixes an event in memory.

constructivism: a view of cognitive development in humans, associated with Jean Piaget, according to which there is little innate knowledge, and development takes place through the interaction of three general processes (**assimilation, accommodation**, and **equilibration**) with information in the environment.

context: as used in pragmatics, in a loose sense, the total linguistic and non-linguistic background to a text or utterance; more narrowly, as used in **relevance theory**, the set of mentally represented assumptions that interact with an utterance to give rise to **contextual effects**.

contrast: if two segments occur in the same place in two otherwise identical words, and there is a change in meaning, then the sounds are used contrastively in phonology, and each segment is assigned a separate phonemic identity.

contribution model: a model of conversation in which speakers and addressees both contribute to establishing that they have mutually understood one another to a degree suited to their conversational goal.

conversational implicature: a term derived from the philosopher Grice (1975) that refers to the implications which are inferred from an utterance on the basis of the maxims of conversation.

cooperative principle: a term derived from the philosopher Grice (1975) that refers to the assumption speakers and listeners make about the accepted goals and purpose of a conversation (see maxims).

coordination: a process by which individuals reach an agreed way of describing an object or situation.

cross-modal: the property of a process involving more than one of the five basic sensory modalities; for instance, speech perception takes a particular type of acoustic stimulus (language sounds) and a particular visual stimulus (movement of vocal organs) as the basis for its analysis.

data-driven processing: bottom-up processing of a stimulus that is predominantly guided by perceptual input.

declarative memory: memory for facts and personal episodes, assumed to be available to consciousness.

deixis: those features of language whose meaning is relative to the personal, temporal, or locational characteristics of the situation in which the utterance takes place.

depth of processing: the depth to which a stimulus is processed at encoding can have a significant impact on retention. Deep processing, in which the meaning of the stimulus is extracted, generally leads to better memory than shallow, perceptual processing.

derivation: broadly, a sequence of operations carried out by a computational system. Within the **Minimalist Program:** the sequence of steps carried out by the computational procedure of the **language faculty** in producing two interface representations from a selection of lexical items, or the process of applying a formal rule to derive a conclusion.

development contingency modeling: a specification of the skills and knowledge which it is necessary for a child to achieve before another target skill can develop.

discipline: an area of human investigation that explores problems of a certain type and involves specific practices and knowledge.

dissociation: a pattern of impaired performance on one task coupled with spared performance on another in any one individual (see also **double dissociation**).

distal stimulus (or distal layout): the object or scene in the world which is the ultimate cause of the proximal stimulations at the sensory receptors (the image on the retina, the vibrations at the ear, etc.), and which the perceptual processes of the organism are geared to form a representation of.

domain-general: knowledge or procedures which apply across a wide range of the stimulus and knowledge domains that the organism's overall cognitive system processes (compare **domain-specific**).

domain-specific: knowledge or procedures which apply to some delimited subset of the stimuli and knowledge that the organism processes; for example, knowledge of grammar, the principles involved in shape perception, and those mental maps which enable us to navigate our way about the world are three distinct and probably independent domains (compare **domain-general**).

double dissociation: a frequently used method in cognitive neuropsychology to demonstrate that two cognitive abilities are independent from each other and perhaps modular; one person is shown to be well able at task A but significantly impaired at task B, and another person is shown to be the opposite; for instance, there are cases of people with impaired language in the presence of normal intellectual ability (see, e.g., **aphasia**), and those with great linguistic ability but severe cognitive deficit (e.g., people with Williams Syndrome).

dyslexia: systematic, cognitive impairments of reading and spelling brought on by brain damage in later life (acquired dyslexia) or by genetically influenced developmental abnormalities.

economy: within **generative grammar**: the idea that representations of the **language faculty** contain no superfluous items and that **derivations** of the language faculty contain no superfluous steps.

effector system: a component which translates internal motor commands to actions in the world.

empiricism: the view that (in general) knowledge/ideas/concepts are not innately specified but are based on experience, so that the structure of human cognition derives from the structure of the environment which impinges on it.

encoding operations: the initial analytic processes carried out on a stimulus which creates a representation suitable for memorizing.

encoding specificity principle: the claim that a stimulus will only be effective as a retrieval cue if it was present at encoding and consequently formed part of the memory trace.

epigenesis: a view of physical and/or cognitive development, according to which three factors interact: (a) genetic endowment, (b) environment (both social and physical), and (c) self-regulatory processes within the organism, (see **constructivism**).

episodic memory: memory for events and episodes that are bound up with a specific context.

equilibration: the process whereby a structure settles into a relatively stable state so that it can accept and adapt to varied input without any essential change; it is the result of the two interacting opposed processes of **accommodation** and **assimilation**.

ethology: a discipline whose central domain is the biological study of the behavior of animals in their natural environment, but which goes beyond this to look at how and why a particular pattern of behavior has evolved, how it develops in the lifetime of an individual, and what internal mechanisms (biological and/or cognitive) and external factors govern its occurrence.

exemplar effects: observations that categorization responses are biased by specific category members that have been presented.

explicit knowledge: knowledge which is actually represented in a system as opposed to being merely inferable, or in some other way extractable from it (compare implicit knowledge).

explicit memory tests: tests which specifically instruct subjects to recall information from some prior episode.

extinction: the gradual reduction in the magnitude or frequency of a conditioned response when the conditioned stimulus is no longer followed by the unconditioned stimulus.

feature: within **generative grammar:** a structure consisting of a name and a value associated with a syntactic category.

feature: within vision, an elemental part of an object.

feature checking: in the **Minimalist Program**: the elimination of uninterpretable **features** during a **derivation** through specifier-**head agreement**.

feed-forward network: a type of connectionist network consisting of a series of "layers" of units, and in which processing involves the flow of **activation** from one layer to the next.

figurative meaning: the use of language as in **metaphor** and **irony** where the **intended meaning** of an utterance does not include its **literal meaning**.

finite state automaton: a simple computing device which is characterized by its internal state (which may be any one of a finite set) and its state transition function (which determines how the device's state changes given its current state and its current input).

first-order representation: a mental representation of some aspect of the world.

flow chart: a diagram showing the sequence of steps and branches in an algorithm.

foreshortening: the effect on the image of an object produced by rotating the object in depth.

forgetting: loss of information from memory, either because the information has been unlearned or because it has become inaccessible.

formants: regions of energy concentration in a speech signal around particular frequencies. The formants are numbered in terms of ascending frequency position (referred to as the first, second, third formant, etc.).

frame problem: a problem faced by a cognitive system, which can, in principle, draw on all the knowledge or information available to it in forming a new belief or making a decision; that is, the problem of when to stop thinking (and start acting).

functional approach: theories of linguistics that treat the notion of the function of an utterance (e.g., the function of declaring, requesting, permitting, etc.) as central.

functional component: a component of a processing device which is defined in terms of its function rather than in terms of its physical manifestation.

generalized cone: a simple volume constructed by sweeping a **connected curve** along an axis, sometimes referred to as a generalized cylinder.

generative grammar: any approach to the study of human language is a form of **generative grammar** if it adopts the requirement that a **grammar** has to be perfectly explicit. The Chomskyan approach to linguistics represents a form of generative grammar.

global shape: the shape of an object that can be influenced by a parameter the scope of which includes the whole object, for example, the global shape of the face changes when you prepare to sneeze.

grammar: the theory of the mature state of the **language faculty**, a mentally represented system of rules which underlies our ability to produce and comprehend language.

grapheme–phoneme conversion: The means of converting letter sequences to sound without the use of a lexicon.

grounding: the process by which parties to a conversation establish the mutual belief that the **addressee** has a satisfactory understanding of the speaker's meaning.

head: the most important word of a phrase.

heuristic: a strategy or procedure for solving a problem which has high utility (it may be quick and economical) but is not foolproof; a trial and error approach (compare **algorithm**).

hidden layer: intermediate layer of units in a connectionist network that fit between the input layer and the output layer.

hippocampus: a structure in the medial temporal lobe that is intimately involved in the consolidation of memories.

homunculus: literally "little man"; usually applied to circular accounts of psychological processes which involve postulating an internal device or entity which has the very properties it was introduced to explain; for instance, a theory of vision that postulates a device which scans or inspects images on the retina and decides what they are.

hybrid model: a **computational model** which employs techniques drawn from both **connectionism** and **symbol systems**.

image processing pipeline: an architecture for image processing in which operations on an image give rise to a new image which can then be used as input for a subsequent process.

immediate behavior: in Soar, behavior which is determined by the direct selection of a sequence of states and operators between perception and action, in contrast to behavior requiring the resolution of subgoals.

implicit knowledge: knowledge which is not **explicit**ly represented in a system, but which is, in principle at least, extractable from knowledge which is explicitly represented; for instance, a system which forms the plural of nouns by having a procedure for each individual noun has implicitly within it the rule that the plural of most nouns is formed by adding "s."

implicit memory tests: tests which reveal the influence of prior learning without asking the subjects specifically to recall information from the learning episode.

inference: a systematic process (which may be conscious or unconscious) of drawing a conclusion from premises or evidence. Deductive (or valid) inferences are those which are foolproof in that, provided their premises are true, so must their conclusion be. Non-demonstrative inferences, on the other hand, give rise to conclusions which, though supported by available evidence, remain hypotheses which are open to further confirmation or to disconfirmation.

information flow: the representation of mental operations involved in a particular task, structured in terms of their sequence (see **flow chart**).

information processing approach: a way of describing mental operations by analogy with the operation of a computer or other physical system (contrasts with **information flow**).

informational encapsulation: the property of a cognitive subsystem, which does not have free access to information present in the overall system of which it is a part; for example, low-level perception (of shape or sound) is said to be immune to general knowledge and expectations of the wider cognitive system.

input layer: the set of units in a connectionist network which are directly activated by the input.

input systems: the perceptual systems, whose function is to deliver to the **cen-**

tral systems information about the outside world (the **distal layout**).

instrumental conditioning: a change in the probability or strength of a response as a result of it being paired with reinforcement or punishment.

intended meaning: the meaning the speaker seeks to convey by an utterance. An assertion such as "It's hot" can be an indirect request to open a door or to turn on the air conditioning. The **literal meaning** is about temperature, but the intended meaning is a request.

interactive activation network: a network consisting of a set of nodes with **activation** levels. Nodes excite and inhibit each other, leading to a situation in which, as processing continues, the activation of one node rises above all others.

interpolation: the process of finding an intermediate point between two given points.

invariant: a quality or quantity which remains unchanged under transformation.

inverse problems: problems related to finding multiple causes from singular effects, for example, attempting to work out the influence of surface colour, illumination, and surface orientation on the brightness of a point on the surface.

irony: the use of language by which a speaker communicates implicitly a mocking attitude to a thought or view which the speaker implicitly attributes to some other person at the time of the utterance.

language acquisition: the process by which the initial state of the **language faculty** develops to a mature, steady state.

language faculty: the language module of the mind.

larynx: the cartilaginous structure at the bottom of the vocal tract, part of which is the elastic vocal folds.

learning: the acquisition of information, knowledge, or skills.

levels of description: intelligent human activity can be described at three levels: behavioral, the cognitive, and biological.

lexical decision task: an experimental technique where a string of letters is presented to the subject who has to decide whether or not it corresponds to a word.

lexicon: the stored mental representation of words.

literal meaning: the linguistic meaning of an utterance that is invariant over different contexts of use, or the proposition expressed by the utterance (see **propositional representation**).

Logical Form: the interface of the **language faculty** which specifies linguistically determined aspects of meaning; also logical form as a term is used to refer to what is extracted from a sentence in **rule theory**, permitting abstract formal rules to derive a conclusion, if any, that must then be translated back into the content of the initial sentence (contrasts here with the theory of **mental models**).

logogen system: a type of **lexicon** which operates by the accumulation of evidence.

logographic stage: the earliest mental representation of printed words in an alphabetic script wherein letter order is not crucial.

long-term memory: a memory system that stores information over intervals longer than a few seconds.

long-term potentiation: a mechanism for learning at the neural level that has been studied in the hippocampus.

manner of production: the magnitude of the maximum constriction in the vocal tract. This varies from complete closure in the case of plosive stop consonants through to an open vocal tract in the case of vowels.

maxims of conversation: a term derived from the work of the philosopher Grice (1975). Maxims are general principles which are thought to underlie the use of language (i.e., the maxims of quality, quantity, relevance, and manner).

means-end analysis: the operation of factoring a problem into a set of subproblems whose solution leads to the goal in question.

memory: the storage and retention of information, knowledge, or skills.

mental image: the perceptual form of a **mental model**.

mental model: a mental representation which has a similar structure to an actual or imagined situation in contrast with a **propositional representation**. Each entity (object or person) in the world is represented by a token in the model and properties of entities are represented by corresponding properties of tokens. Relations among entities are represented by relations among tokens.

mental representation: a mental structure which stands in some correspondence to the world.

Merge: within the **Minimalist Program:** an operation of the computational procedure of the **language faculty** which combines two categories and **projects** one of them.

metalinguistic tasks: tasks which depend on the subject's ability to reflect upon their linguistic knowledge and representations, such as grammaticality judgments and **phoneme awareness tasks**.

metaphor: a figure of speech in which a word or phrase is applied to an object or action that it does not literally denote in order to convey some of its properties from the speaker's view point.

metarepresentation: a representation which represents another representation (as opposed to some aspect of the external world); for instance, a mental representation of an utterance (which is a public representation) or of someone else's belief or intention (which are themselves mental representations).

Minimalist Program: a research program for the study of language, which seeks to establish how "perfect" human language is, the leading idea being that both representations and computations of the **language faculty** satisfy **economy** conditions in an optimal way.

modularity: a term used of mental systems which have a cluster of properties including **domain-specificity**, innate specification, hard-wiredness, and relative autonomy within the overall cognitive system.

module: a set of processes that don't draw on other processes in performing their function (see also **information encapsulation, modularity**).

Move: within the **Minimalist Program**: an operation of the computational procedure of the **language faculty** responsible for displacing categories from the position in which they are interpreted.

nativism: the view that, to a substantial extent, the mind is innately endowed with structure and knowledge, and that psychological development is a preprogrammed unfolding of innate properties. Two kinds of nativism may be distinguished: (a) Cartesian nativism (named after the philosopher Descartes) where some kinds of knowledge or content are innately given; and (b) architectural nativism where the mind is innately modular (to some interesting degree).

natural selection: the main driving force

accounting for the evolution of ever more complex species: among the variants in a particular population some are better able to cope than others with the problems posed by the environment; these survive (or leave more offspring) so that their features become permanent adaptations or traits of the population while the other variants die out.

non-routine processing/behavior: effortful processing/behavior, usually conceived of as involving conscious intent, such as concentrating hard when solving a novel problem (see also **routine processing/behavior**).

object constancy: the problem of recognizing an object as being the same from different **viewpoints**.

object substitution error: an appropriate action performed with the wrong object.

omission error: leaving out one or more components of a sequence of actions.

ontogenesis: the sequence of events involved in the development of the individual, both physical and cognitive (compare **phylogenesis**).

optimal relevance: a property that an utterance has, on a given interpretation, when (a) it has enough **contextual effects** to be worth the addressee's attention, and (b) it puts the addressee to no gratuitous **processing effort** in achieving these effects.

output layer: the last layer of units in a connectionist network. These are interpreted directly as the output of the network.

parallel distributed processing: the branch of **connectionism** which employs units corresponding to "micro-features", such that representations of complete entities are distributed across a set of units.

parsing: the process of building a representation of the structure of a sentence, thereby checking that the sentence is grammatical.

perseveration: the unintended repetition of an action or thought.

phoneme: the minimal contrastive unit used within a language.

phoneme awareness tasks: tasks that require subjects to manipulate their internal phonological representations of words.

Phonetic Form: the interface of the language faculty to the articulatory-perceptual systems of the mind.

phonology: the systematic description of the structure of speech, including phonemes, stress, and intonation.

phylogenesis: the sequence of events involved in the development (or evolution) of a species (compare **ontogenesis**).

place of articulation: the position in the vocal tract where maximum constriction occurs.

place substitution error: the proper execution of some intended action with the intended objects, but with an incorrect spatial location.

pose: the orientation of a 3D object in space, usually used with reference to a rotation about a vertical axis.

pragmatics: the study of language usage in context.

prelinguistic communication: in language acquisition, the period immediately preceding the emergence of words.

presupposition: certain pragmatic inferences or assumptions that seem to be built into linguistic expressions and which can be isolated using specific linguistic tests (e.g., constancy under negation).

primacy effect: superior recall of the early items on a list relative to the middle items.

priming: a universal psychological principle whereby processing a particular stimulus makes it easier shortly afterwards to process the same or a related stimulus.

primitives: an elemental unit of a description.

problem space: the set of states that can be reached by applying a set of operators or actions in a particular problem domain.

procedural memory: memory that underlies skills and habits, and is assumed to be unavailable for conscious reflection.

processing effort: as used within **relevance theory**, the effort which a cognitive system must expend in order to arrive at a satisfactory interpretation of incoming information (involving factors such as the retrieval of relevant information from memory and the inferential work involved in integrating new information).

production: as used in a **production system**, a rule consisting of a condition (or goal) and an action.

production system: a type of **symbol system** consisting of a working memory and a rule base. Processing involves the repeated application of rules from the rule base to the contents of working memory.

projection: the higher-level category formed when two syntactic categories A and B combine, one of which passes up its properties to the new category.

propositional attitude: a mental attitude such as believing, desiring, expecting, intending, etc. which holds between an individual and a mental representation.

propositional representation: an explicit and abstract representation of (conceptual) objects and relations. Relations are typically viewed as predicates (e.g., ON), and objects as arguments as in the propositional representation of the statement "The toast is on the table": ON(TOAST,TABLE).

prospective memory: the realization of a delayed intention (i.e., "remembering to remember" to do something).

prototype: a mental representation of a categories, assumed to be correspond to an average of the category members that have been presented.

proximal stimulus: the stimulation at the sensory receptor; for instance, the two-dimensional, inverted, ever-fluctuating image on the retina caused by looking at a table (compare **distal stimulus**).

push down automaton: a computing device with a finite set of states, a state transition function, and a **stack** structured memory. On input, the device changes state (with the new state being dependent on the current state, the input, and the current stack) and (optionally) alters its stack.

range map: an image in which each point specifies the distance to a surface.

recall: the ability to retrieve episodic information from memory given an appropriate cue.

recency effect: superior recall of the final items on a list relative to the middle items.

recognition: the ability to judge whether an item has been seen before, either in an absolute sense or with respect to a given learning episode.

recurrent network: a type of layered connectionist network with feedback connections from one layer to a previous layer, allowing the behavior of the network to be dependent not just on its current input, but also on its previous state.

recursion: a procedure that invokes itself (usually on a subpart of the original input).

referential communication: a conversation or task dialogue that refers to concrete objects or events.

reflex: an automatic reaction to a stimulus impinging on the body or within it; examples of physical reflexes are the knee jerk and the sneeze. Some people believe that some mental processes, such as low-level perceptual processes are reflexes.

reinforcer: a motivationally-significant event, such as food, that is capable of changing the probability or strength of an instrumental response that precedes and produces it.

relevance theory: a cognitively-based pragmatic theory which accounts for how individuals reach an appropriate interpretation of an utterance. It is based on two principles: a cognitive principle which says that human cognition is geared towards the maximization of **relevance** (i.e., the achievement of as many **contextual effects** as possible for as little **processing effort** as possible) and a communicative principle, which says that every utterance creates an expectation or presumption of its own **optimal relevance**.

representational redescription: cognitive processes which take some system of knowledge or procedures already functioning within the mind and reformulate it at some higher level so that it is more explicit, more systematic, and more available to other systems within the mind.

representational system: a mapping between one set of objects and relations (the represented objects) and another set of objects and relations (their representations), such that manipulation of the representations mirrors manipulation of the represented objects.

representational theory of mind: the view that mental activity involves the manipulation of representations, which may consist of images or of a structured string of symbols. Particular mental systems may construct and transform several levels of representation; for instance, the language system may involve a level of phonetic representation, various levels of syntactic representation, and a level of logical or semantic representation.

retrieval cue: a stimulus or event that is capable of reactivating an episodic memory.

retroactive interference: an explanation of forgetting which attributes it to the interfering effects of subsequent learning.

retrograde amnesia: inability to remember events that were learned prior to the brain trauma.

routine processing/behavior: automatic and fairly effortless processing/behavior that requires minimal conscious control, such as very well-rehearsed motor movements like running (see also **non-routine processing/behavior**).

rule theory: the claim that reasoning is the application of mental rules of **inference** on purely arbitrary symbols.

satisficing: finding a solutuion to a problem which is good enough for the purposes in hand, but not necessarily the best.

savant: a person with an outstanding ability in a particular cognitive domain, usually accompanied by deficiency in a wide range of other areas of cognitive processing (hence the term "idiot-savant").

schema: strings of over-learned actions or thought sequences which usually occur together (plural schemas or schemata).

second-order representation: a mental representation of a **first-order representation**, for example, a belief.

semantic memory: memory for facts, knowledge of word meanings, etc.

semantic network: a **symbol system** structured as a network in which nodes represent symbols and semantically-related nodes are connected with links.

short-term memory: a memory system that stores information over intervals of a few seconds.

similarity: the concept of proximity in a psychological space. Two stimuli that are close together in psychological space are likely to be responded to in similar ways.

simulation: the execution of a computational model in an attempt to replicate some aspect of behavior or cognition.

source and filter: a representation which can be applied to the process of producing speech. The vocal tract is the filter and the source one of several types of excitation, such as voiced excitation or whisper.

specifier: the first phrase to be **Merge**d with a **projection** after the **complement** has been **Merge**d.

spectrogram: a visual depiction of speech. The three dimensions depicted are energy, frequency, and time. Time and frequency are on the x- and y-axes. Energy at a particular frequency and time is represented by the darkness at that point (the darker a point is, the more energy occurs at that frequency).

speech act: the uttering of a sentence in order to express an intention (e.g., to declare, ask a question, etc.) the speaker wants the listener to recognize.

speech event: a culturally recognized social activity in which language plays a specific, and often rather specialized role (e.g., teaching in a classroom).

stack: a form of data store with a top-most element that permits just two operations: elements may be "pushed" onto the stack (thus obscuring the previous top element and making the new element the top-most element) or "popped" from the stack (thus revealing the previous top-most element).

structural description: a description of an object in terms of the relative locations and dimensions of its component parts.

subconcepts: conceptual primitives that make up the mental representation of concepts.

surface geometry: the study of the characterization and measurement of surfaces.

surface normal: a **vector** of unit length orthogonal to the surface.

surface orientation: the local orientation of a surface usually specified by the surface normal.

syllogism: an argument that consists of two premises and a conclusion.

symbol system: a cognitive model in which objects are represented as symbols and in which processing involves the manipulation of those symbols.

synesthesia: the experience of a sensation or perception in a modality other than, and in addition to, the one which was actually stimulated; for example, the experience of seeing a color or shape when hearing a sound.

systematicity: a property of a representation that allows well-formed representations to be generated by the replacement of subparts of the original representation by other representations of an appropriate type.

tacit knowledge: knowledge within a cognitive system which is not accessible to consciousness and not able to be inferentially integrated with other knowledge in the system; it includes all **implicit** procedural knowledge and those elements of explicit knowledge that cannot be brought to consciousness.

teleological: an explanation that looks to considerations of purpose or design in accounting for the structure of some physical or cognitive system; for instance, the reflex-like nature of visual perception may be explained by pointing out that its purpose is to enable quick reactions such as attacking (of prey) or avoiding (or predators).

templates: stored images which are used for recognition by matching corresponding points on the input representation.

theory of mind: This term is used in two ways: (1) to refer to the study of children's developing understanding of the mind, and (2) to refer to a domain of knowledge that all mentally intact humans have about the minds of others; that is, the capacity to attribute **propositional attitudes**, such as beliefs, desires, and intentions, at various levels of complexity (see **metarepresentation**).

think-aloud protocols: transcripts of individuals talking aloud about what they are attending to or thinking about as they solve a problem.

top-down: processing of an input that is driven, or guided, by higher-level information; for instance, visual or linguistic processing which is affected by general knowledge or contextual considerations (compare **bottom-up**).

transducer: as used in cognitive psychology, sensory receptors which convert (or translate) one form of energy, say light or sound, into another form of energy, which is interpretable at the cognitive level as computational.

transfer-appropriate processing: the idea that memory benefits from an overlap between the mental operations performed at study and recall.

truth-conditional semantics: the view that the meaning of a sentence is the set of conditions that must hold in the world for the sentence to be true.

Turing machine: a computing device (devised by Alan Turing) with an infinite tape, which it may read from or write to. The device is characterized by its state (which may be any one of a finite set) and its state transition function (which determines how the state changes when the machine reads a symbol and how, if at all, the machine should alter the tape).

Universal Grammar (UG): the theory of the initial state of the **Language faculty**.

utilization behavior: the unintended and inappropriate use of an object.

vector: a mathematical object which has a direction and a length.

viewpoint: a position in space from which a scene is viewed.

voiced excitation: brought about when the vocal folds vibrate quasi-periodically, resulting in a buzz entering the vocal tract.

volumetric primitives: elemental parts of a description which are defined in terms of 3D volumes.

working memory: the name of a modular memory system for the storage and manipulation of short-term linguistic and visuo-spatial information.

References

Abbeduto, L. and Rosenberg, S. (1985). Children's knowledge of the presuppositions of know and other cognitive verbs. *Journal of Child Language*, 12, 621–41.

Abbs, J.H. (1973). The influence of the gamma motor system on jaw movements during speech: A theoretical framework and some preliminary observations. *Journal of Speech and Hearing Research*, 16, 175–200.

Ackerman, B.P. (1978). Children's understanding of presuppositions of 'know' and other cognitive verbs. *Child Development*, 49, 311–18.

Adelson, B. (1984). When novices surpass experts: The difficulty of a task may increase with expertise. *Journal of Experimental Psychology: Learning, Memory and Cognition*, 10, 483–95.

Aitkenhead, A.M. and Slack, J.M. (eds) (1985). *Issues in Cognitive Modeling.* Hove: Lawrence Erlbaum Associates.

Altmann, G. and Steedman, M. (1988). Interaction with context during human sentence processing. *Cognition*, 30, 191–238.

Anderson, J.R. (1976). *Language, Memory, and Thought.* Hillsdale, NJ: Lawrence Erlbaum Associates.

Anderson, J.R. (1983). *The Architecture of Cognition.* Cambridge, MA: Harvard University Press.

Anderson, J.R. (1990). *The Adaptive Character of Thought.* Hillsdale, NJ: Lawrence Erlbaum Associates. Cambridge, MA: Harvard University Press.

Anderson, J.R. (1993). *Rules of the Mind.* Hillsdale, NJ: Lawrence Erlbaum Associates.

Anderson, J.R. and Schooler, L.J. (1991). Reflections of the environment in memory. *Psychological Science*, 2, 396–408.

Anselmi, D., Tomasello, M., and Acuzo, M. (1986). Young children's responses to neutral and specific contingent queries. *Journal of Child Language*, 13, 135–44.

Anzai, Y. and Simon, H.A. (1979). The theory of learning by doing. *Psychological Review*, 86, 124–40.

Armstrong, D.F., Stokoe W.C., and Wilcox S.E. (1995). *Gesture and the Nature of Language.* Cambridge: Cambridge University Press.

Astington, J.W. (1988). Children's production of commissive speech acts. *Journal of Child Language*, 15, 411–23.

Astington, J.W., Harris, P.L., and Olson, D.R. (eds) (1988). *Developing Theories of Mind.* Cambridge: Cambridge University Press.

Atwood, M.E. and Polson, P.G. (1976). A process model for water jug problems. *Cognitive Psychology*, 8, 191–216.

Austin, J.L. (1962). *How to do Things with Words.* Oxford: Clarendon Press.

Awbery, G.M. (1976). *The Syntax of Welsh.* Cambridge: Cambridge University Press.

Baddeley, A.D. and Hitch, G. (1974). Working memory. In G. H. Bower (ed.), *The Psychology of Learning and Motivation* (vol. 8, pp. 47–90). New York: Academic Press.

Baddeley, A.D. and Warrington, E.K. (1970). Amnesia and the distinction between long- and short-term memory. *Journal of Verbal Learning and Verbal Behavior*, 9, 176–89.

Baron-Cohen, S. (1988). Social and pragmatic deficits in autism: Cognitive or affective? *Journal of Autism and Developmental Disorders*, 18, 379–402.

Baron-Cohen, S. (1989). The autistic child's theory of mind: A case of specific developmental delay. *Journal of Child Psychology and Psychiatry*, 30, 285–97.

Baron-Cohen, S. (1993). From attention-goal psychology to belief-desire psychology: The development of a theory of mind, and its dysfunction. In S. Baron-Cohen, H. Tager-Flusberg and D.J. Cohen (eds), *Understanding other Minds: Perspectives from Autism.* Oxford: Oxford University Press.

Baron-Cohen, S., Harrison, J., Goldstein, L., and Wyke, L. (1993). Coloured speech perception: Is synaesthesia what happens when modularity breaks down? *Perception*, 22, 419–26.

Baron-Cohen, S., Leslie, A., and Frith, U. (1985). Does the autistic child have a "theory of mind"? *Cognition*, 21, 37–46.

Baron-Cohen, S., Leslie, A.M., and Frith, U. (1986). Mechanical, behavioral and intentional understanding of picture stories in autistic children. *British Journal of Developmental Psychology*, 4, 113–25.

Baron-Cohen, S., Tager-Flusberg, H., and Cohen, D. J. (1993). *Understanding other Minds: Perspectives from Autism.* Oxford: Oxford University Press.

Bates, E. (1976). *Language and Context: The Acquisition of Pragmatics.* New York: Academic.

Bates, E., Benigni, L., Bretherton, I., Camaioni, L., and Volterra, V. (1979). Cognition and communication from 9–13 months: Correlational findings. In E. Bates, L. Benigni, I. Bretherton, L. Camaioni and V. Volterra (eds), *The Emergence of Symbols: Cognition and Communication in Infancy* (pp. 69–131). New York: Academic.

Bauman, R. and Sherzer, J. (eds). (1974). *Explorations in the Ethnography of Speaking.* Cambridge: Cambridge University Press.

Baumgartner, P. and Payr, S. (1995). *Speaking Minds.* Princeton, N.J.: Princeton University Press.

Bechtel, W. and Abrahamsen, A. (1991). *Connectionism and the Mind.* Basil Blackwell, Oxford, UK.

Beiderman, I. (1987). Recognition-by-components: A theory of human image understanding. *Psychological Review,* 94, 115–47.

Benson, P.J. and Perrett, D.I. (1994). Visual processing of facial distinctiveness. *Perception,* 23, 75–93.

Binford, T.O. (1971). Visual perception by computer. *IEEE Conference on Systems and Control.* Miami.

Blakemore, D. (1992). *Understanding Utterances.* Oxford: Blackwell.

Blank, M., Gessner, M., and Esposito, A. (1979). Language without communication: A case study. *Journal of Child Language,* 6, 329–52.

Bloom, L., Rocissano, L., and Hood, L. (1976). Adult-child discourse: developmental interaction between information processing and linguistic knowledge. *Cognitive Psychology,* 8, 521–52.

Bottini, G., Corcoran, R., Sterzi, R., Paulesu, E., Schenone, P., Scarpa, P., Frackowiak, R.S.J., and Frith, C.D. (1994). The role of the right hemisphere in the interpretation of figurative aspects of language: Positron emission tomography activation study. *Brain,* 117, 1241–53.

Bradley, L.L. and Bryant, P.E. (1983). Categorizing sounds and learning to read: A causal connection. *Nature,* 301, 419.

Bradley, L.L. and Bryant, P.E. (1985). *Rhyme and Reason in Reading and Spelling.* Ann Arbor: University of Michigan Press.

Braine, M.D.S., and O'Brien, D.P. (1991). A theory of If: A lexical entry, reasoning program and pragmatic principles. *Psychological Review,* 98, 182–203.

Brennan, S.E. (1985). The caricature generator. *Leonardo,* 18, 170–8.

Bretherton, I. and Bates, E. (1979). The emergence of intentional communication. In I.C. Uzgiris (ed.), *Social Interaction and Communication during Infancy.* San Francisco, CA: Jossey Bass.

Bretherton, I., McNew, S., and Beeghly-Smith, M. (1981). Early person knowledge as expressed in gestural and verbal communication: When do infants acquire a "theory of mind"? In M.E. Lamb and L.R. Sherod (eds), *Infant Social Cognition* (pp. 333–73). Hillsdale, NJ: Lawrence Erlbaum.

Broadbent, D.E. (1958). *Perception and Communication.* London: Pergamon.

Broadbent, D.E. (ed.) (1993). *The Simulation of Human Intelligence.* Basil Blackwell, Oxford, UK.

Broadbent, D.E., Cooper, P.J., Fitzgerald, P.F., and Parkes, K.R. (1982). The Cognitive Failures Questionnaire (CFQ) and it's correlates. *British Journal of Clinical Psychology, 21,* 1–16.

Brooks, R. (1991). Intelligence without representation. *Artificial Intelligence,* 47, 139–59.

Brown, T.L. and Carr, T.H. (1989). Automaticity in skill acquisition: Mechanisms for reducing interference in concurrent performance. *Journal of Experimental Psychology: Human Perception and Performance,* 15, 686–700.

Bruce, V. (1988). *Recognising Faces.* Hove: Lawrence Erlbaum.

Bruce, V., Hanna, E., Dench, N., Healey, E., Mason, O., Coombes, A., Fright, R., and Linney, A. (1993). Sex discrimination: How do we tell the difference between male and female faces? *Perception,* 22, 131–52.

Bruner, J. (1957). On perceptual readiness. *Psychological Review,* 64, 123–52.

Bruner, J. and Klein, G. (1960). The functions of perceiving: New look retrospective. In Kaplan, B. and Wapner, S. (eds), *Perspectives in Psychological Theory.* International Universities Press.

Bruner, J., Roy, C., and Ratner, N. (1982). The beginnings of request. In K.E. Nelson (ed.), *Children's Language* (vol. 3 pp. 91–35). New York: Garden Press.

Brunswick, E. (1956). *Perception and the Representative Design of Psychological Experiments.* Berkeley, CA: University of California Press.

Bryant, P. and Bradley, L. (1985) *Children's Reading Problems.* Oxford: Blackwell.

Bullinaria, J.A. (1994). Simulating nonlocal systems: Rules of the game. *Behavioral and Brain Sciences,* 17, 61–2.

Bülthoff, H.H. and Edelman, S. (1992). Psychophysical support for a two-dimensional view interpolation theory of object recognition. *Proceedings of the National Academy of Science of the USA,* 89, 60–4.

Bülthogy, H.H. and Edelman, S.S. (1993). Evaluating object recognition theories by computer graphics psychophysics. In D. Glaser and T. Poggio (eds), *Exploring Brain Functions: Models in Neuroscience* (pp.

139–64) Chichester/England: John Wiley and Sons.

Burgess, P.W. and Alderman, N. (1990). Rehabilitation of dyscontrol syndromes following frontal lobe damage: A cognitive neuropsychological approach. In R.L Wood and I. Fussey (eds), *Cognitive Rehabilitation in Perspective* (ch. 9, pp. 183–203).London, UK: Taylor and Francis.

Burgess, P.W. and Shallice, T. (1994). Fractionnement du syndrome frontal. *Revue de Neuropsychologie, 4*, 345–70.

Butterworth, G., Harris, P., Leslie, A., and Frith, U. (1991). *Perspectives on the Child's Theory of Mind.* Oxford: Oxford University Press/British Psychological Society.

Byrne, R.M.J. and Johnson-Laird, P.N. (1989). Spatial reasoning. *Journal of Memory and Language*, 28, 564–75.

Cann, R. (1993). *Formal Semantics.* Cambridge: Cambridge University Press.

Carston, R. (1988). Implicature, explicature and truth-conditional semantics. In R. Kempson (ed.), *Mental Representation and the Interface between Language and Reality.* Cambridge: Cambridge University Press.

Carston, R. (1993). Syntax and pragmatics. In *The Encyclopedia of Language and Linguistics.* Pergamon and Aberdeen University Press.

Chandler, C.C. (1993). Accessing related events increases retroactive interference in a matching recognition test. *Journal of Experimental Psychology: Learning, Memory, and Cognition*, 19, 967–74.

Charness, N. (1989). Expertise in chess and bridge. In D. Klahr and K. Kotovsky (eds), *Complex Information Processing: The Impact of Herbert Simon* (pp. 183–208). Hillsdale, NJ: Lawrence Erlbaum Associates.

Charney, R. (1979). The comprehension of "here" and "there". *Journal of Child Language*, 6, 69–80.

Charniak, E. (1972). Towards a model of children's story comprehension. *MIT Artificial Intelligence Laboratory Monographs*, No. 226. Cambridge, MA.

Chase, W.G. and Ericsson, K.A. (1982). Skill and working memory. In G.H. Bower (ed.), *The Psychology of Learning and Motivation* (vol. 16, pp. 1–58). New York: Academic Press.

Chase, W.G. and Simon, H.A. (1973). Perception in chess. *Cognitive Psychology*, 4, 55–81.

Cheng, P.W. (1985). Restructuring versus automaticity: Alternative accounts of skill acquisition. *Psychological Review*, 92, 414–23.

Chiat, S. (1986). Personal pronouns. In P. Fletcher and M. Garman (eds), *Language Acquisition* (2nd edn) (pp. 339–55). Cambridge: Cambridge University Press.

Chomsky, N. (1965). *Aspects of the Theory of Syntax.* Cambridge: MIT Press.

Chomsky, N. (1977). On Wh-Movement. In P. Culicover, T. Wasow and A. Akmajian (eds), *Formal Syntax.* New York: Academic Press.

Chomsky, N. (1981). *Lectures on Government and Binding.* Dordrecht: Foris.

Chomsky, N. (1986a). *Knowledge of Language*, New York: Praeger.

Chomsky, N. (1986b). *Barriers.* Cambridge, MA: MIT Press.

Chomsky, N. (1988). *Language and Problems of Knowedge.* Cambridge MA: MIT Press.

Chomsky, N. (1993). A minimalist program for linguistic theory. In K. Hale and S. Keyser (eds), *The View from Building 20.* Cambridge, MA: MIT Press.

Churchland, P. (1988). Perceptual plasticity and theoretical neutrality: A reply to Jerry Fodor. *Philosophy of Science*, 55, 167–87.

Clark, A., and Karmiloff-Smith, A. (1993). The cognizer's innards: A psychological and philosophical perspective on the development of thought. *Mind and Language*, 8, 487–519.

Clark, E.V. (1978). From gesture to word: On the natural history of deixis in language acquisition. In J. Bruner and A. Garton (eds), *Human Growth and Development: Wolfson College Lectures 1976* (pp. 85–120). Oxford: Oxford University Press.

Clark, H.H. (1992). *Arenas of Language Use.* Chicago: University of Chicago Press.

Clark, H.H. and Brennan, S.E. (1991). Grounding in communication. In L.B. Resnick., J.M. Levine and S.D. Teasley (eds), *Perspectives on Socially Shared Cognition.* Washington: American Psychological Association.

Clark, H.H. and Clark, E.V. (1977). *Psychology and Language.* New York: Harcourt Brace Jovanovich.

Clark, H.H. and Gerrig, R.J. (1984). On the pretence theory of irony. *Journal of Experimental Psychology: General*, 113, 121–6.

Clark, H.H. and Lucy, P. (1975). Understanding what is meant from what is said: A study in conversationally conveyed requests. *Journal of Verbal Learning and Verbal Behavior*, 14, 56–72.

Clark, H.H. and Marshall, C.R. (1981). Definite reference and mutual knowledge. In A.K. Joshi, B.L. Webber, and I.A. Sag (eds), *Elements of Discourse Understanding* (pp. 10–63). Cambridge: Cambridge University Press.

Clark, H.H. and Schaefer, E.E. (1987). Collaborating

on contributions to conversations. *Language and Cognitive Processes*, 2, 19–41.

Clark, H.H. and Schaefer, E.E. (1989). Contributing to discourse. *Cognitive Science*, 13, 259–94.

Clark, H.H. and Wilkes-Gibbs, D. (1986). Referring as a collaborative process. *Cognition*, 22, 1–39.

Clarke, R. and Morton, J. (1983). Cross modality facilitation in tachistoscopic word recognition. *Quarterly Journal of Experimental Psychology*, 35A, 79–96.

Clement, J. (1991). Nonformal reasoning in experts and in science students: The use of analogies, extreme cases, and physical intuition. In J. Voss, D. Perkins and J. Siegel (eds), *Informal Reasoning and Education* (pp. 345–62). Hillsdale, NJ: Lawrence Erlbaum Associates.

Cohen, N.J. and Squire, L.R. (1980). Preserved learning and retention of pattern-analyzing skill in amnesia: Dissociation of knowing how and knowing that. *Science*, 21, 207–10.

Cole, M. and Scribner, S. (1974). *Culture and Thought: A Psychological Introduction.* New York: John Wiley and Sons.

Coltheart, M. (1980). Deep dyslexia: A right-hemisphere hypothesis. In M. Coltheart, K.E. Patterson and J.C. Marshall (eds), *Deep Dyslexia.* London: Routledge and Kegan Paul.

Coltheart, M., Curtis, B., Atkins, P., and Haller, M. (1993). Models of reading aloud: Dual-route and parallel-distributed-processing approaches. *Psychological Review*, 100, 589–608.

Coltheart, M. and Rastle, K. (1994). Serial processing in reading aloud: Evidence for dual-route models of reading. *Journal of Experimental Psychology: Human Performance and Perception*, 20, 1197–211.

Coltheart, M., Sartori, G., and Job, R. (eds) (1986). *The Cognitive Neuropsychology of Language.* London: Lawrence Erlbaum Associates.

Cooper, R. and Shallice, T. (1995). Soar and the case for Unified Theories of Cognition. *Cognition*, 55, 115–49.

Cooper, R., Shallice, T., and Farringdon, J. (1995). Symbolic and continuous processes in the automatic selection of actions. In J. Hallam (ed.), *Hybrid Problems, Hybrid Solutions* (pp. 27–37). Amsterdam: IOS Press.

Corkin, S. (1984). Lasting consequences of bilateral medial temporal lobectomy: Clinical course and experimental findings in H.M. *Seminars in Neurology*, 4, 249–59.

Cosmides, L. and Tooby, J. (1994a). Beyond intuition and instinct blindness: Toward an evolutionarily rigorous cognitive science. *Cognition*, 50, 41–77.

Cosmides, L. and Tooby, J. (1994b). Origins of domain specificity: The evolution of functional organization. In L. Hirschfeld, and S. Gelman, (eds), *Mapping the Mind* (1994), 85–116.

Cosmides, L., Tooby, J., and Barkow, J.H. (1992). Introduction: evolutionary psychology and conceptual integration. In J.H. Barkow, L. Cosmides and J. Tooby (eds), *The Adapted Mind: Evolutionary Psychology and the Generation of Culture.* New York: Oxford University Press.

Cossu, G., Rossini, F. and Marshall, J.C. (1993). When reading is acquired but phonemic awareness is not: A study of literacy in Down's syndrome. *Cognition*, 46, 129–38.

Craik, F.I.M. and Tulving, E. (1975). Depth of processing and the retention of words in episodic memory. *Journal of Experimental Psychology: General*, 104, 268–94.

Craik, K. (1943). *The Nature of Explanation.* Cambridge: Cambridge University Press.

Crain, S. (1991). Language acquisition in the absence of experience. *Behavioral and Brain Sciences*, 14, 597–650.

Curcio, F. (1978). Sensorimotor functioning and communication in mute autistic children. *Journal of Autism and Childhood Schizophrenia*, 5, 281–92.

Cutting, J.E. and Rosner, B.S. (1974). Categories and boundaries in speech and music. *Perception and Psychophysics*, 16, 564–70.

Cytowic, R. (1989). *Synesthesia: A Union of the Senses.* Berlin: Springer.

D'Andrade, R.G. (1989). Cultural cognition. In M.I. Posner (ed.), *Foundations of Cognitive Science* (ch. 20, pp. 795–830). Cambridge, MA: The MIT Press.

Daniloff, R.G. and Hammarberg, R.E. (1973). On defining coarticulation. *Journal of Phonetics*, 1, 239–48.

Daniloff, R. and Moll, K. (1968). Coarticulation of lip rounding. *Journal of Speech and Hearing Research*, 11, 707–21.

Dawkins, R. 1986. *The Blind Watchmaker.* New York: Norton.

De Groot, A. (1965/1946). *Thought and Choice in Chess.* The Hague: Mouton (orig. published in Dutch).

DeHart, G. and Maratsos, M. (1984). Children's acquisition of presuppositional usuages. In R.L. Schiefelbusch and J. Pickar (eds), *The Acquisition of Communicative Competence* (pp. 237–93). Baltimore, MD: University Park Press.

Delattre, P.C., Liberman, A.M., and Cooper, F.S. (1955). Acoustic loci and transitional cues for consonants. *Journal of the Acoustical Society of America*, 27, 769–73.

Dell, G.S. (1986). A spreading-activation theory of retrieval in sentence production. *Psychological Review*, 93, 83–321.

Denes, P.B. and Pinson, E.N. (1973). The speech chain: The physics and biology of spoken language. *Anchor Science Study Series*.

Detweiler, M. and Schneider, W. (1991). Modelling the acquisition of dual-task skill in a connectionist/control architecture. In Damos, D.L. (ed.), *Multiple-Task Performance*, pp. 69–99. Taylor and Francis, London, UK.

Deutsch, D. (1975). Musical illusions. *Scientific American*, 233, 92–104.

Deutsch, D. (1983). Auditory illusions and audio. Special issue of the *Journal of the Audio Engineering Society*, 31, 9.

Diamond, R. and Carey, S. (1986). Why faces are not special: An effect of expertise. *Journal of Experimental Psychology: General*, 115, 107–17.

Dorman, M.F., Studdert-Kennedy, M., and Raphael, L.J. (1977). Stop consonant recognition: Release bursts and formant transitions as functionally equivalent, context-dependent cues. *Perception and Psychophysics*, 22, 109–22.

Dromard, G. (1905). Erude psychologique et clinique sur l'echopraxie. *Journal Psychologie (Paris)*, 2, 385–403.

Dunbar, R. (1993). Coevolution of neocortical size, group size and language in humans. *Behavioral and Brain Sciences*, 16, 681–735.

Duncan, J. (1986). Disorganization of behavior after frontal lobe damage. *Cognitive Neuropsychology*, 3, 273–90.

Duncker, K. (1945/1935). On problem solving. *Psychological Monographs*, 58 (whole number 270), pp. 1–113 (orig. pubd in German in 1935).

Ebbinghaus, H. (1885). *Memory: A Contribution to Experimental Psychology*. New York: Dover.

Edelman, G.M. (1992). *Bright Air, Brilliant Fire: On The Matter of the Mind*. UK: Penguin Books.

Egan, D.W. and Greeno, J.G. (1974). Theories of rule induction: Knowledge acquired in concept learning, serial pattern learning, and problem solving. In L.W. Gregg (ed.), *Knowledge and Cognition* (pp. 43–104). New York: John Wiley and Sons.

Eisenberg, A. (1985). Learning to describe past experiences in conversation. *Discourse Processes*, 8, 177–204.

Ekman, P. and Friesen, W.V. (1978). *Facial Action Coding System (FACS): A Technique for the Measurement of Facial Action*. Palo Alto, CA: Consulting Psychologists Press.

Ellis, A. (1984) *Reading, Writing and Dyslexia*. London: Erlbaum.

Elman, J. L. (1990). Finding structure in time. *Cognitive Science*, *14*, 179–211.

Elman, J.L. and Zipser, D. (1988). Learning the hidden structure of speech. *Journal of the Acoustical Society of America*, 83, 1615–26.

Emslie, H.C. and Stevenson, R.J. (1981). Pre-school children's use of the articles in definite and indefinite referring expressions. *Journal of Child Language*, 8, 313–28.

Eslinger, P.J. and Damasio, A.R. (1985). Severe disturbance of higher cognition after bilateral frontal lobe ablation: Patient EVR. *Neurology*, *35*, 1731–41.

Evans, J.St.B.T., Barston, J.L., and Pollard, P. (1983). On the conflict between logic and belief in syllogistic reasoning. *Memory and Cognition*, 11, 295–306.

Evans, J.St.B.T., Newstead, S.E., and Byrne, R.M.J. (1993). *Human Reasoning: The Psychology of Deduction*. Hillsdale, NJ: Lawrence Erlbaum Associates.

Evans, J.St.B.T., Over, D.E., and Manktelow, K.I. (1993). Reasoning, decision-making and rationality. *Cognition*, 49, 165–87.

Falkenhainer, B., Forbus, K.D., and Gentner, D. (1989). Structure-mapping engine. *Artificial Intelligence*, 41, 1–63.

Farah, M. (1994). Neuropsychological inference with an inter-active brain: A critique of the "locality" assumption. *Behavioral and Brain Sciences* 17, 43–104.

Fay, W.H. and Schuler, A.L. (1980). *Emerging Language in Autistic Children*. London: Edward Arnold.

Feagans, L. and Short, E. (1984). Developmental differences in the comprehension and production of narratives by reading disabled and normally achieving children. *Child Development*, 55, 1727–36.

Ferreira, F and Clifton, C. (1986). The independence of syntactic processing. *Journal of Memory and Language*, 25, 348–68.

Flanagan, O. (1993). *The Science of the Mind* (2nd edn). Cambridge, MA: MIT Press

Flavell, J.H., Botkin, P.T., Fry, C.L., Wright, J.W., and Jarvis, P.E. (1968). *The Development of Role-taking and Communication Skills in Children*. New York: Wiley.

Fodor, J. (1983). *The Modularity of Mind*. Cambridge: MIT Press.

Fodor, J. (1985). Precis of The Modularity of Mind. *Behavioral and Brain Sciences*, 8, 1–5.

Fodor, J. (1986). The modularity of mind. In Z. Pylyshyn and W. Demopoulos (eds), *Meaning and*

Cognitive Structure, 3–18 and 129–35. Norwood, NJ: Ablex.

Fodor, J. (1987). Modules, frames, fridgeons, sleeping dogs and the music of the spheres. In Z. Pylyshyn (ed.), *The Robot's Dilemma: The Frame Problem in Artificial Intelligence.* Norwood, NJ: Ablex.

Fodor, J. (1988). A reply to Churchland's "perceptual plasticity and theoretical neutrality." *Philosophy of Science* 55, 188–98.

Fodor, J. (1989). Why should the mind be modular? In A. George (ed.), *Reflections on Chomsky*, 1–22. Oxford: Basil Blackwell. Reprinted in J. Fodor (1990). *A Theory of Content and Other Essays*, 207–30. Cambridge MA: MIT Press.

Fodor, J., Garrett, M., Walker, E., and Parkes, C. (1980). Against definitions. *Cognition*, 8, 263–367.

Fodor, J. and Pylyshyn, Z. (1988). Connectionism and cognitive architecture: A critical analysis. *Cognition*, 28, 3–71.

Foley, J.D., van Dam, A., Feiner, S.K., and Hughes, J.F. (1990). *Computer Graphics* (2nd edn). MA: Addison Wesley.

Ford, M. (1994). Two modes of representation and problem solution in syllogistic reasoning. *Cognition*, 54, 1–71.

Foster, S.H. (1990). *The Communicative Competence of Young Children.* New York: Longman.

Fowler, C.A. (1986). An event approach to the study of speech perception from a direct-realist approach. *Journal of Phonetics*, 14, 175–207.

Fowler, C.A., Rubin, P., Remez, R.E,. and Turvey, M.T. (1980). Implications for speech production of a general theory of action. In B. Butterworth, *Language Production* (vol. 1). Academic Press, London and New York.

Fowler, R.L. and Kimmel, H.D. (1962). Operant conditioning of the GSR. *Journal of Experimental Psychology*, 63, 563–67.

Frith, U. (1985). Beneath the surface of developmental dyslexia. In K. Patterson, J. Marshall and M. Coltheart (eds), *Surface Dyslexia, Neuropsychological and Cognitive Studies of Phonological Reading.* London: Erlbaum.

Frith, U. (1986). A developmental framework for developmental dyslexia. *Annals of Dyslexia*, 36, 69–81.

Fujimura, O. and Lindqvist, J. (1970). Sweep-tone measurements of vocal-tract characteristics. *Journal of the Acoustical Society of America*, 49, 142–6.

Fuster, J.M. (1980). *The Prefrontal Cortex: Anatomy, Physiology and Neuropsychology of the Frontal Lobe.* New York, NY: Raven Press.

Gallistel, C. and Cheng, K. (1985). A modular sense of place? *Behavioral and Brain Sciences*, 8, 11–12.

Gardner, H. (1985). *The Mind's New Science.* New York: Basic Books Inc.

Gardner, R.A. and Gardner, B.T. (1969). Teaching sign language to a Chimpanzee. *Science*, 165, 664–72.

Garfield, J. (ed.) (1987). *Modularity in Knowledge Representation and Natural Language Understanding.* Cambridge, MA: MIT Press.

Garnham, A. and Oakhill, J. (1992). Discourse processing and text representation from a "mental model" perspective. *Language and Cognitive Processes*, 7, 193–204.

Garnham, A., and Oakhill, J. (1994). *Thinking and Reasoning.* Oxford: Blackwell.

Garrod, S. and Anderson, A. (1987). Saying what you mean in dialogue: A study in conceptual and semantic coordination. *Cognition*, 27, 181–218.

Garrod, S. and Doherty, D. (1994). Conversation, coordination and convention: An empirical investigation into how groups establish linguistic conventions. *Cognition*, 53, 181–215.

Garvey, C. (1975). Requests and responses in children's speech. *Journal of Child Language*, 2, 41–63.

Gay, T. and Hirose, H. (1973). Effect of speaking rate on labial consonant production. *Phonetica*, 27, 44–56.

Gay, T., Ushijima, T., Hirose, H., and Cooper, F.S. (1974). Effect of speaking rate on labial consonant-vowel articulation. *Journal of Phonetics*, 2, 47–63.

Gazdar, G., Klein, E. Pullum G.K. and Sag I. (1985). *Generalized Phrase Structure Grammar.* Harvard University Press.

Gazzaniga, M. (1989). Organization of the human brain. *Science*, 245, 947–52.

Gentner, D. (1983). Structure-mapping: A theoretical framework for analogy. *Cognitive Science*, 7, 155–70.

Gentner, D. (1989). Mechanisms of analogical learning. In S. Vosniadou and A. Ortony (eds), *Similarity and Analogical Reasoning.* Cambridge: Cambridge University Press.

Gentner, D. and Forbus, K.D. (1991). MAC/FAC: A model of similarity-based retrieval. Thirteenth Annual Conference of the Cognitive Science Society. Hillsdale, NJ: Lawrence Erlbaum Associates.

Gentner, D. and Gentner, D.R. (1983). Flowing waters or teeming crowds: Mental models of electricity. In D. Gentner and A.I. Stevens (eds), *Mental Models* (pp. 99–129). Hillsdale, NJ: Lawrence Erlbaum Associates.

Gibbs, R.W. (1984). Literal meaning and psychological theory. *Cognitive Science*, 8, 275–304.

Gibbs, R.W. (1986). On the psycholinguistics of sarcasm. *Journal of Experimental Psychology: General*, 115, 3–15.

Gibson, J.J. (1977). The theory of affordances. In R.E. Shaw and J. Bransford (eds), *Perceiving, Acting, and Knowing*. Hillsdale, NJ: Lawrence Erlbaum Associates.

Gibson, J.J. (1979). *The Ecological Approach to Visual Perception*. Boston: Houghton Mifflin.

Gick, M.L. and Holyoak, K.J. (1980). Analogical problem-solving. *Cognitive Psychology*, 12, 306–55.

Gick, M.L., and Holyoak, K.J. (1983). Schema induction in analogical transfer. *Cognitive Psychology*, 15, 1–38.

Gilbert, D.T. (1991). How mental systems believe. *American Psychologist*, 46, 107–19.

Gilhooly, K.J., Logie, R.H., Wetherick, N.E. Wynn, V. (1993). Working memory and strategies in syllogistic-reasoning tasks. *Memory and Cognition*, 21, 115–24.

Glanzer, M. and Cunitz, A. R. (1966). Two storage mechanisms in free recall. *Journal of Verbal Learning and Verbal Behavior*, 5, 351–60.

Gluck, M.A. and Bower, G.H. (1988). From conditioning to category learning: An adaptive network model. *Journal of Experimental Psychology: General*, 117, 225–44.

Glucksberg, S. and Danks, J.H. (1975). *Experimental Psycholinguistics: An Introduction*. Hillsdale, NJ: Lawrence Erlbaum Associates.

Glucksberg, S., Gildea, P., and Bookin, H. (1982). On understanding non-literal speech: Can people ignore metaphors? *Journal of Verbal Learning and Verbal Behavior*, 21, 85–98.

Glucksberg, S. and Keysar, B. (1990). Understanding metaphorical comparisons: Beyond similarity. *Psychological Review*, 97, 3–18.

Goldin-Meadow, S. and Feldman, H. (1979). The development of language-like communication without a language model. *Science*, 197, 401–3.

Golinkoff, R.M. (1983). The preverbal negotiation of failed messages: Insights into the transition period. In R.M. Golinkoff (ed.), *The Transition from Prelinguistic to Linguistic Communication*. Hillsdale, NJ: Lawrence Erlbaum.

Golinkoff, R.M. (1993). When is communication a "meeting of minds"? *Journal of Child Language*, 20, 199–207.

Golinkoff, R.M. (1986). "I beg your pardon?" The preverbal negotiation of failed messages. *Journal of Child Language*, 13, 455–76.

Gopnik, A. (1982). Words and plans: Early language and the development of intelligent action. *Journal of Child Language*, 9, 303–18.

Gopnik, A. and Meltzhoff, A.N. (1986). Relations between semantic and cognitive development in the one-word-stage: The specificity hypothesis. *Child Development*, 57, 1040–53.

Goren, C.C., Sarty, M., and Wu, P.Y.K. (1975). Visual following and pattern discrimination of face-like stimuli by newborn infants. *Pediatrics*, 56, 544–9.

Graf, P., Shimamura, A.P., and Squire, L.R. (1985). Priming across modalities and priming across category levels: Extending the domain of preserved function in amnesia. *Journal of Experimental Psychology: Learning, Memory, and Cognition*, 11, 386–96.

Greenfield, P.M. and Smith, J. (1976). *The Structure of Communication in Early Language Development*. New York: Academic.

Gregory, R. (1970). *The Intelligent Eye*. McGraw-Hill.

Grice, H.P. (1968). Utterer's meaning, sentence-meaning, and word-meaning. *Foundations of Language*, 4, 1–18.

Grice, H.P. (1975). Logic and conversation. In P. Cole and J.L. Morgan (eds), *Syntax and Semantics, Vol. 3: Speech acts* (pp. 225–42). New York: Academic Press.

Grimm, H. (1975, September). Analysis of short-term dialogues in 5–7 year olds: Encoding of intentions and modifications of speech acts as a function of negative feedback. In Paper presented at the Third International Child Language Symposium, London.

Gruber, H.E. (1981). *Darwin on Man: A Psychological Study of Scientific Creativity* (2nd edn). Chicago: University of Chicago Press.

Hagert, G. (1984). Modeling mental models: Experiments in cognitive modeling of spatial reasoning. In T. O'Shea (ed.), *Advances in Artificial Intelligence*. Amsterdam: North-Holland.

Halliday, M.A.K. (1975). *Learning How to Mean*. London: Edward Arnold.

Happ, F.G. (1993). Communicative competence and theory of mind in autism: A test of relevance theory. *Cognition*, 48, 101–19.

Harnad, S. (1987). *Categorical Perception*. Cambridge: Cambridge University Press.

Harnad, S. (1990). The symbol grounding problem, Physica D, 42, 335–46.

Harris, J. (1994). *English Sound Structure*. Oxford: Blackwell.

Hayes, J.R. and Simon, H.A. (1977). Psychological differences among problem isomorphs. In N.J. Castellan, D.B. Pisoni and G.R. Potts (eds), *Cognitive*

Theory (vol. 2). Hillsdale, New Jersey: Lawrence Erlbaum Associates.

Hebb, D.O. (1949). *The Organization of Behavior: A Neuropsychological Theory.* New York: John Wiley and Sons.

Heilmen, K.M. and Rothi, L.J. (1985). Apraxia. In K.M. Heilman and E. Valenstein (eds), *Clinical Neuropsychology* pp. 131–50. New York: Oxford University Press.

Henderson, L. (1982). *Orthography and Word Recognition in Reading.* London: Academic Press.

Henke, W.L. (1966). Dynamic articulatory model of speech production using computer simulation. Doctoral dissertation, MIT.

Hertz, J.A., Krogh, A.S., and Palmer, R.G. (1991). *Introduction to the Theory of Neural Computation.* Santa Fe Institute Studies in the Sciences of Complexity. Reading, MA: Addison-Wesley.

Hesse, M.B. (1966). *Models and Analogies in Science.* Notre Dame, IN: Notre Dame University Press.

Hinton, G.E. and Shallice, T. (1991). Lesioning an attractor network: Investigations of acquired dyslexia. *Psychological Review,* 98, 74–95.

Hirschfeld, L. and Gelman, S. (1994). Towards a topography of mind: An introduction to domain specificity. In L. Hirschfeld, and S. Gelman (eds), 3–35.

Hirschfeld, L. and Gelman, S. (eds) (1994). *Mapping the Mind: Domain Specificity in Cognition and Culture.* Cambridge: Cambridge University Press.

Hirst, W., Spelke, E.S., Reaves, C.C., Caharack, G., and Neisser, U. (1980). Dividing attention without alternation or automaticity. *Journal of Experimental Psychology: General,* 109, 98–117.

Hofmann, Th.R. (1993). *Realms of Meaning.* London: Longman.

Holding, D.H. (1985). *The Psychology of Chess Skill.* Hillsdale, NJ: Lawrence Erlbaum Associates.

Holding, D.H. and Reynolds, R.J. (1982). Recall or evaluation of chess positions as determinants of chess skill. *Memory and Cognition,* 10, 237–42.

Holland, J.H., Holyoak, K.J., Nisbett, R.E., and Thagard, P.R. (1989). *Induction: Processes of Inference, Learning and Discovery.* Cambridge, MA: MIT Press.

Holyoak, K.J. (1991). Symbolic connectionism: Toward third generation theories of expertise. In K.A. Ericsson and J. Smith (eds), *Toward a General Theory of Expertise: Prospects and Limits.* Cambridge: Cambridge University Press.

Holyoak, K.J. and Koh, K. (1987). Surface and structural similarity in analogical transfer. *Memory and Cognition,* 15, 332–40.

Holyoak, K.J. and Thagard, P.R. (1989). Analogical mapping by constraint satisfaction. *Cognitive Science,* 13, 295–355.

Homa, D., Sterling, S., and Trepel, L. (1981). Limitations of exemplar-based generalization and the abstraction of categorical information. *Journal of Experimental Psychology: Human Learning and Memory,* 7, 418–39.

Howell, P. (1981), Velar coarticulation in the speech of French–English bilinguals. In F. Ferrero (ed.), 4th FASE Symp. Acoustics and Speech, Edizione Scientifiche Associate: Rome.

Howell, P. (1983). The extent of coarticulatory effects: Implications for models of speech recognition. *Speech Communication,* 2, 159–63.

Howell, P. and Harvey, N. (1983). Perceptual equivalence and motor equivalence in speech. In B. Butterworth (ed.), *Language Production,* (vol. 2), 203–24.

Hudson, J.A. (1993). Reminiscing with mothers and others: Autobiographical memory in young two-year-olds. *Journal of Narrative and Life History,* 3, 1–32.

Humphreys, G.W. and Riddoch, M.J. (1987). *To See or Not to See.* Hove: Lawrence Erlbaum.

Hunt, M. (1982). *The Universe Within.* Brighton: Harvester Press.

Huttenlocher, J. (1974). The origins of language comprehension. In R. Solso (ed.), *Theories in Cogntive Psychology.* Hillsdale, NJ: Lawrence Erlbaum.

Hymes, D. (1972). Models of the interaction of language and social life. In J.J. Gumperz and D.H. Hymes (eds), *Directions in Sociolinguistics* (pp. 35–71). New York: Holt, Rinehart and Winston.

Isaacs, E.A. and Clark, H.H. (1987). References in conversation between experts and novices. *Journal of Experimental Psychology,* 116, 26–37.

Isaacs, E.A. and Clark, H.H. (1990). Ostensible invitations. *Language in Society,* 19, 493–509.

Jackendoff, R. (1983). *Semantics and Cognition.* Cambridge, MA: MIT Press.

Jackendoff, R. (1991). *Semantic Structures.* Cambridge, MA: MIT Press.

Jackendoff, R. (1993). *Patterns in the Mind.* New York: Harvester Wheatsheaf.

Jacob, F. (1988), *The Statue Within: An Autobiography.* New York: Basic Books.

James, W. (1890). *The Principles of Psychology.* New York, NY: Holt.

Johnson, C.N. and Wellman, H.M. (1980). Children's developing understanding of mental verbs: Remember, know and guess. *Child Development,* 51, 1095–102.

Johnson, M.H. and Morton, J. (1991). *Biology and Cognitive Development: The Case of Face Recognition.* Oxford: Blackwell.

Johnson-Laird, P.N. (1983). *Mental Models: Towards a cognitive science of language, inference and consciousness.* Cambridge: Cambridge University Press.

Johnson-Laird, P.N. (1988). *The Computer and the Mind.* London: Fontana Press.

Johnson-Laird, P.N. (1989). Analogy and the exercise of creativity. In S. Vosniadou and A. Ortony (eds), *Similarity and Analogical Reasoning.* Cambridge: Cambridge University Press.

Johnson-Laird, P.N. (1993). *Human and Machine Thinking.* Hillsdale, NJ: Lawrence Erlbaum Associates.

Johnson-Laird, P.N. and Byrne, R.M.J. (1991). *Deduction.* Hillsdale, NJ: Lawrence Erlbaum Associates.

Johnson-Laird, P.N., Byrne, R.M.J., and Schaeken, W. (1992). Propositional reasoning by model. *Psychological Review,* 99, 418–39.

Johnson-Laird, P.N. and Wason, P.C. (1970). A theoretical analysis of insight into a reasoning task. *Cognitive Psychology,* 1, 134–48.

Johnston, A. (1992). Object constancy in face processing: Intermediate representations and object forms. *Irish Journal of Psychology,* 13, 425–38.

Johnston, A., Hill, H. and Carman, N. (1992). Recognising faces: effects of lighting direction, inversion and brightness reversal. *Perception,* 21, 356–76.

Johnston, A. and Passmore, P.J. (1994). Independent encoding of surface orientation and surface curvature. *Vision Research,* 34, 3005–12.

Jordan, M.I. (1986). Serial order: A Parallel Distributed Processing Approach. ICI Report 8604. Institute for Cognitive Science, University of California at San Diego.

Jorgensen, J., Miller, G.A., and Sperber, D. (1984). Test of the mention theory of irony. *Journal of Experimental Psychology: General,* 113, 112–20.

Karmiloff-Smith, A. (1992). *Beyond Modularity: A Developmental Perspective on Cognitive Science.* Cambridge, MA: MIT Press.

Karmiloff-Smith, A. (1994). Precis of *beyond modularity: A developmental perspective on cognitive science. Behavioral and Brain Sciences,* 17, 693–706.

Karttunen, L. (1971). Implicative verbs. *Language,* 47, 340–58.

Kaye, J. (1989). *Phonology: A Cognitive View.* Hillsdale, NJ: Lawrence Erlbaum.

Keane, M.T. (1990). Incremental analogising: Theory and model. In K.J. Gilhooly, M.T. Keane, R. Logie and G. Erdos (eds), *Lines of Thinking: Reflections on the Psychology of Thought,* (vol 1). Chichester: John Wiley.

Keane, M.T., Ledgeway, T., and Duff, S. (1994). Constraints on analogical mapping: A comparison of three models. *Cognitive Science,* 18, 387–438.

Keenan, E.O. and Schieffelin, B. B. (1976). Topic as a discourse notion: A study of topic in the conversations of children and adults. In C.N. Ni (ed.), *Subject and Topic* (pp. 335–84). New York: Academic Press.

Kempton, W. (1986). Two theories of home heat control. *Cognitive Science,* 10, 75–90.

Kent, R.D., Carney, P.J., and Severeid, L.R. (1974). Velar coarticulation and timing: Evaluation of a model for binary control. *Journal of Speech and Hearing Research,* 17, 470–88.

Kintsch, W., Welsch, D., Schmalhofer, F., and Zimny, S. (1990). Surface memory: A theoretical analysis. *Journal of Memory and Language,* 29, 133–59.

Kiparsky, P. and Kiparsky, C. (1970). Fact. In M. Bierwisch and K. Heidolph (eds), *Progress in Linguistics.* The Hague: Mouton.

Koenderink, J. J. (1990). *Solid Shape.* Cambridge, MA: MIT Press.

Korsakoff, S. S. (1889). Psychic disorder in conjunction with multiple neuritis. Translated by M. Victor and P.I. Yakovlev and reprinted in *Neurology,* 1955, 5, 394–406.

Kosslyn, S.M. (1980). *Image and Mind.* Cambridge, MA.: Harvard University Press.

Krauss, R.M. and Weinheimer, S. (1964). Changes in reference phrases as a function of frequency of usage in social interactions: A preliminary study. *Psychonomic Science,* 1, 113–14.

Kreuz, R.J. and Glucksberg, S. (1989). How to be sarcastic: The echoic reminder theory of verbal irony. *Journal of Experimental Psychology: General,* 118, 374–86.

Kuhl, P.K. (1981). Discrimination of speech by non-human animals: Basic auditory sensitivities conducive to the perception of speech–sound categories. *Journal of the Acoustical Society of America,* 70, 340–9.

Kuhl, P.K. and Miller, J.D. (1975). Speech perception by the chinchilla: Voiced–voiceless distinction in alveolar plosive consonants. *Science,* 190, 69–72.

Kuhl, P.K. and Meltzoff, A. (1982). The bimodal perception of speech in infancy. *Science,* 218, 1138–41.

Kuhn, T.S. (1970). *The Structure of Scientific Revolutions* (2nd edn). Chicago: University of Chicago Press.

Ladefoged, P. (1975). *A Course in Phonetics.* New York: Harcourt, Brace, Jovanovich.

Laird, J.E., Newell, A., and Rosenbloom, P.S. (1987). SOAR: An architecture for general intelligence. *Artificial Intelligence*, 33, 1–64.

Lashley, K.S. (1950). In search of the engram. *Symposia of the Society for Experimental Biology*, 4, 454–82.

Lesgold, A.M. (1988). Problem solving. In R.J. Sternberg and E.E. Smith (eds). *The Psychology of Human Thought.* Cambridge/New York: Cambridge University Press.

Leslie, A.M. (1987). Pretence and representation: the origins of "theory of mind." *Psychological Review*, 94, 412–26.

Leslie, A.M. (1988). The necessity of illusion: Perception and thought in infancy. In L. Weiskrantz, (ed.), *Thought Without Language*, 185–210. Oxford: Clarendon Press.

Leslie, A.M. (1994). ToMM, ToBY, and Agency: Core architecture and domain specificity. In L. Hirschfeld and S. Gelman (eds), *Mapping the Mind* (1994), 119–48.

Levelt, W.J.M. (1989). *Speaking: From Intention to Articulation.* Bradford Book, MIT Press, Cambridge, MA.

Levin, E. A. and Rubin, K.H. (1983). Getting others to do what you want them to do: The development of children's requestive strategies. In K.E. Nelson (ed.), *Children's Language.* Hillsdale, NJ: Lawrence Erlbaum.

Levinson, S.C. (1979). Pragmatics and social deixis. *Proceedings of the Fifth Annual Meeting of the Berkeley Linguistic Society*, 206–23.

Levinson, S.C. (1983). *Pragmatics.* Cambridge: Cambridge University Press.

Lhermitte, F. (1983). Utilization behaviour and its relation to lesions of the frontal lobe. *Brain*, 106, 55.

Lhermitte, F. (1986). Human autonomy and the frontal lobes. Part II: Patient behavior in complex and social situations: The "Environmental Dependency Syndrome". *Annals of Neurology*, 19, 335–43.

Liberman, A.M., Cooper, F.S., Shankweiler, D.P., and Studdert-Kennedy, M. (1967). Perception of the speech code. *Psychological Review*, 74, 431–61.

Liberman, A.M., Delattre, P.C., and Cooper, F.S. (1952). The role of selected stimulus variables in the perception of the unvoiced stop consonants. *American Journal of Psychology*, 65, 497–516.

Liberman, I.Y., Shankweiler, D., Liberman, A.M., Fowler, C., and Fisher, W.F. (1977). Phonetic segmentation and recoding in the beginning reader. In A.S. Reber and D.L. Scarborough (eds), *Toward a Psychology of Reading* (pp. 207–25). Hillsdale, NJ: Erlbaum.

Liles, B. (1985). Cohesion in the narratives of normal and language-disordered children. *Journal of Speech and Hearing Research*, 28, 123–33.

Lindblom, B. (1963). Spectrographic study of vowel reduction. *Journal of the Acoustical Society of America*, 35, 1773–81.

Lindblom, B. (1990). Explaining phonetic variation: A sketch of the HandH theory. In W.J. Hardcastle and A. Marchal (eds), *Speech Production and Speech Modelling.* Dordrecht: Kluwer Academic Publishers.

Lindenmayer, A. (1968). Mathematical models for cellular interactions in development. Parts I and II. *Journal of Theoretical Biology*, 18, 280–315.

Loftus, E.F. (1993). The reality of repressed memories. *American Psychologist*, 48, 518–37.

Loftus, E.F. and Loftus, G.R. (1980). On the permanence of stored information in the human brain. *American Psychologist*, 35, 49–72.

Long, J. and Dowell, J. (1989). Conceptions of the discipline of HCI: Craft, applied science and engineering. In A. Sutcliffe and L. Macaulay (eds), *People and Computers V.* Cambridge: Cambridge University Press.

Loveland, K.A. and Landry, S. (1986). Joint attention and language in autism and developmental language delay. *Journal of Autism and Developmental Disorders*, 16, 335–49.

Lovibond, P.F. (1993). Conditioning and cognitive-behaviour therapy. *Behaviour Change*, 10, 119–30.

Lukatela, G. and Turvey, M.T. (1994). Visual lexical access is initially phonological: I. Evidence from associative priming by words, homophones, and pseudohomophones. *Journal of Experimental Psychology: General*, 123, 107–28.

Luria, A.R. (1968). *The Mind of a Mnemonist.* Cambridge, MA: Harvard University Press.

Luria, A.R. (1971). Towards the problem of the historical nature of psychological processes. *International Journal of Psychology*, 6, 259–72.

Luria, A.R. (1973). *The Working Brain: An Introduction to Neuropsychology.* The Penguin Press, Allen Lane.

McCarthy, R. and Warrington, E. (1990). *Cognitive Neuropsychology: A Clinical Introduction.* San Diego, CA: Academic Press Inc.

McClelland, J.L. and Rumelhart, D.E. (1981). An interactive activation model of context effects in letter perception: Part 1. An account of basic findings. *Psychological Review*, 88, 375–407.

McClelland, J.L. and Rumelhart, D.E. (1985). Distributed memory and the representation of general and specific information. *Journal of Experimental Psychology: General*, 114, 159–88.

McClelland, J.L. and Rumelhart, D.E. (1986). *Parallel Distributed Processing* (vol. 2). Cambridge, MA: MIT Press.

McClelland, J. and Seidenburg, M. (1988). In D.A. Galaburda, (ed.), *From Neurons to Reading*. Cambridge, MA: MIT Press.

McCulloch, W.S. and Pitts, W. (1943). A logical calculus of the ideas immanent in nervous activity. *Bulletin of Mathematical Biophysics*, 5, 115–33.

MacFarlane, D.A. (1930). The role of kinesthesis in maze learning. *University of California Publications in Psychology*, 4, 277–305.

McGurk, H. and MacDonald, J. (1976). Hearing lips and seeing voices. *Nature*, 264, 746–8.

MacKay, D.G. (1987). *The Organization of Perception and Action: A Theory for Language and other Cognitive Skills*. New York: Springer Verlag.

Maclean, M., Bryant, P.E., and Bradley, L. (1987). Rhymes, nursery rhymes and reading in early childhood. *Merill-Palmer Quarterly*, 33, 255–82.

McNeil, J.E. and Warrington, E. K. (1993). Prosopagnosia: A face specific disorder. *Quarterly Journal of Experimental Psychology*, 46 (A), 1–10.

McTear, M. (1985). *Children's Conversation*. Oxford: Basil Blackwell.

McTear, M. (1989). Semantic–pragmatic disability: A disorder of thought? In D.M. Topping, D.C. Crowell and V.N. Kobayashi (eds), *Thinking Across Cultures: The Third International Conference on Thinking*. Hillsdale, NJ: Lawrence Erlbaum.

McTear, M.F. and Conti-Ramsden, G. (1992). *Pragmatic Disability in Children*. London: Whurr Publishers Ltd.

Mandler, G. (1975). Consciousness: Respectable, useful and probably necessary. In R. Solso (ed.), *Information Processing and Cognition: The Loyola Symposium* (pp. 229–54).

Markman, E.M. (1977). Realizing that you don't understand: A preliminary investigation. *Child Development*, 48, 986–92.

Marr, D. (1976). Early processing of visual information. *Phil. Trans. R. Soc. Lond. B*, 275, 483–524.

Marr, D. (1982). *Vision*. San Fransisco: Freeman.

Marr, D. and Nishihara, H.K. (1978). Representation and recognition of the spatial organisation of three-dimensional shape. *Proceedings of the Royal Society of London B*, 200, 269–94.

Marsh, G., Friedman, M., Welch, V., and Desberg, P. (1980). The development of strategies in spelling. In U. Frith (ed.), *Cognitive Processes in Spelling*. London: Academic Press.

Marshall, J.C. (1984). Multiple perspectives on modularity. *Cognition*, 17, 209–42.

Marshall, J.C. and Newcombe, F. (1973). Patterns of paralexia: A psycholinguistic approach. *Journal of Psycholinguistic Research*, 2, 175–99.

Marslen-Wilson, W.D. and Teuber, H.-L. (1975). Memory for remote events in anterograde amnesia: Recognition of public figures from news photographs. *Neuropsychologia*, 13, 353–64.

Marslen-Wilson, W.D. and Tyler, L.K. (1987). Against Modularity. In J. Garfield (ed.), *Modularity in Knowledge Representation and Natural Language Understanding*. Cambridge, MA: MIT Press.

Marslen-Wilson, W. D. and Welsh, A. (1978). Processing interactions and lexical access during word recognition in continuous speech. *Cognitive Psychology*, 10, 29–63.

Martone, M., Butters, N., Payne, M., Becker, J.T., and Sax, D.S. (1984). Dissociations between skill learning and verbal recognition in amnesia and dementia. *Archives of Neurology*, 41, 965–70.

Massaro, D.W. and Cowan, N. (1993). Information processing models: Microscopes of the mind. *Annual Review of Psychology*, 44, 383–425.

Matzel, L.D., Held, F.P., and Miller, R.R. (1988). Information and expression of simultaneous and backward associations: Implications for contiguity theory. *Learning and Motivation*, 19, 317–44.

Medin, D.L. (1989). Concepts and conceptual structure. *American Psychologist*, 44, 1469–81.

Mehler, J., Morton, J., and Juszuck, P.W. (1984). On reducing language to biology. *Cognitive Neuropsychology*, 1, 83–116.

Menig-Peterson, C.L. (1975). The modification of communicative behavior in preschool-aged children as a function of the listener's perspective. *Child Development*, 46, 1015–18.

Messr, D.J. (1995). *The Development of Communication: From Social Interactions to Language*. Chichester: John Wiley and Sons.

Miikkulainen, R. (1993). *Subsymbolic Natural Language Processing*. Cambridge, MA: MIT Press.

Miller, G.A. (1956). The magical number seven, plus or minus two: Some limits on our capacity for processing information. *Psychological Review*, 63, 81–97.

Miller, G.A., Galanter, E., and Pribram, K. (1960). *Plans and the Structure of Behavior*. New York: Holt, Rinehart and Winston.

Miller, G.A., and Johnson-Laird, P.N. (1976). *Language and Perception*. Cambridge: Cambridge University Press.

Miller, J.D., Wier, C.C., Pastore, R.E., Kelly, W.J., and Dooling, R.J. (1976). Discrimination and labelling of noise-buzz sequences with varying noise lead times: An example of categorical perception. *Journal of the Acoustical Society of America*, 60, 410–17.

Miller, P.H., Kessel, F.S., and Flavell, J.M. (1970). Thinking about people thinking about people thinking about . . . A study of social cognitive development. *Child Development*, 41, 613–23.

Miller, P.J. and Sperry, L.L. (1988). Early talk about the past: The origins of conversational stories of personal experience. *Journal of Child Language*, 15, 293–316.

Minsky, M. (1975). A framework for representing knowledge. In P. Winston (ed.), *The Psychology of Computer Vision* (pp. 211–77). New York: McGraw-Hill.

Minksy, M. and Papert, S. (1988/first pub. 1969). *Perceptrons: An Introduction to Computational Geometry.* Cambridge, MA: MIT Press.

Miscione, J.L., Marvin, R.S., O'Brien, R.G., and Greenberg, M.T. (1978). A developmental study of preschool children's understanding of the words "know" and "guess". *Child Development*, 49, 1107–13.

Moll, K.L. and Daniloff, R.G. (1971). Investigations of the timing of velar movements. *Journal of the Acoustical Society of America*, 50, 678–84.

Moore, C., Bryant, D., and Furrow, D. (1989). Mental terms and the development of certainty. *Child Development*, 60, 167–71.

Moore, C. and Davidge, J. (1989). The development of mental terms: Pragmatics of semantics? *Journal of Child Language*, 16, 433–41.

Morais, J., Alegria, J., and Content, A. (1987). The relationship between segmental analysis and alphabetic literacy: An interactive view. *Cahiers de Psychologie Cognitive*, 7, 415–38.

Morais, J., Cary, L., Alegria, J., and Bertelson, P. (1979). Does awareness of speech as a sequence of phones arise spontaneously? *Cognition*, 7, 323–31.

Morris (1994). The neural basis of learning with particular reference to the role of synaptic plasticity: Where are we a century after Cajal's speculations? In N.J. Mackintosh (ed.), *Animal Learning and Cognition* (pp. 135–83). San Diego, CA: Academic Press.

Morton, J. (1969). Interaction of information in word recognition. *Psychological Review*, 76, 165–78.

Morton, J. (1970). A functional model for memory. In D.A. Norman (ed.), *Models of Human Memory.* New York: Academic Press.

Morton, J. (1988). An information account of reading acquisition. In D.A. Galaburda (ed.), *Neurons to Reading.* Cambridge, MA: MIT Press.

Morton, J. and Frith, U. (1995). Causal modelling: A structural approach to developmental psychopathology. In D. Cicchetti and D.J. Cohen (eds), *Manual of Developmental Psychopathology.* (pp. 357–90). New York: Wiley.

Morton, J. and Patterson, K.E. (1980). A new attempt at an interpretation, or, an attempt at a new interpretation. In M. Coltheart, K.E. Patterson and J.C. Marshall (eds), *Deep Dyslexia.* London: Routledge and Kegan Paul.

Moscovitch, M. (1992). Memory and working-with-memory: A component process model based on modules and central systems. *Journal of Cognitive Neuroscience*, 4, 257–67.

Moss, H. and Marslen-Wilson, W. (1993). Access to word meanings during spoken language comprehension: Effects of sentential semantic context. *Journal of Experimental Psychology: Learning, Memory and Cognition*, 19, 1254–76.

Mundy, P., Sigman, M., Ungerer, J. A., and Sherman, T. (1986). Defining the social deficit of autism: The contribution of non-verbal communication measures. *Journal of Child Psychology and Psychiatry*, 27, 657–69.

Neisser, U. (ed.) (1987). *Concepts and Conceptual Development: Ecological and Intellectual Factors in Categorization.* Cambridge: Cambridge University Press.

Nelson, K. (1981). Social cognition in a script framework. In J.H. Flavell and L. Ross (eds), *Social Cognitive Development.* New York: Cambridge University Press.

Nelson, K. and Gruendel, J. (1979). At morning it's lunchtime: A scriptal view of children's dialogues. *Discourse Processes*, 2, 73–94.

Nelson, K. and Seidman, S. (1984). Playing with scripts. In I. Bretherton (ed.), *Symbolic Play.* New York: Harcourt Brace.

Newell, A. (1973). You can't play 20 questions with nature and win. In W.G. Chase (ed.), *Visual Information Processing* (pp. 283–308). San Diego, CA: Academic Press.

Newell, A. (1990). *Unified Theories of Cognition.* Cambridge, Mass.: Harvard University Press.

Newell, A. (1992). SOAR as a unified theory of cognition: Issues and explanations. *Behavioral and Brain Sciences*, 15, 464–92.

Newell, A. and Rosenbloom, P. (1981). Mechanisms of skill acquisition and the law of practice. In Anderson, J.R. (ed.), *Cognitive Skills and their Acquisition.* Hillsdale, NJ: Lawrence Erlbaum Associates.

Newell, A. and Simon, H. (1972). *Human Problem Solving*. Englewood Cliffs, NJ: Prentice-Hall.

Norman, D.A. (1981). Categorization of action slips. *Psychological Review*, 88, 1–15.

Norman, D.A. and Shallice, T. (1980). Attention to action: Willed and automatic control of behavior. Chip report 99, University of California, San Diego, CA.

Norman, D.A. and Shallice, T. (1986). Attention to action: Willed and automatic control of behavior. In R. Davidson, G. Schwartz, and D. Shapiro (eds), *Consciousness and Self Regulation: Advances in Research and Theory* (vol.4, pp. 1–18). New York: Plenum.

Novick, L.R. and Holyoak, K.J. (1991). Mathematical problem-solving by analogy. *Journal of Experimental Psychology: Learning, Memory and Cognition*, 17, 398–415.

Ohlsson, S. (1984). Induced strategy shifts in spatial reasoning. *Acta Psychologica*, 57, 46–67.

O'Neill, D. K. (in press). Two-year-old children's sensitivity to a parent's knowledge state when making requests. *Child Development*.

Partee, B., ter Meulen, A., and Wall, R. (1990). *Mathematical Methods in Linguistics*. Dordrecht: Kluwer Academic Publishers.

Patterson, F.G. (1978). The gestures of a gorilla: Language acquisition in another pongid. *Brain and Language*, 5, 56–71.

Patterson, K.E. and Morton, J. (1985), From orthography to phonology: An attempt at an old interpretation. In K.E. Patterson, J.C. Marshall and M. Coltheart (eds), *Surface Dyslexia*. London: Erlbaum.

Pavlov, I.P. (1927). *Conditioned Reflexes*. Oxford: Oxford University Press.

Payne, S.J. (1991). Display-based action at the user interface. *International Journal of Man-Machine Studies*, 35, 275–89.

Penfield, W. and Evans, J. (1935). The frontal lobe in man: A clinical study of maximum removals. *Brain*, 58, 115–33.

Penfield, W. and Perot, P. (1963). The brain's record of auditory and visual experience. *Brain*, 86, 595–696.

Penry, J. (1971). *Looking at Faces and Remembering Them*. London: Elek Books.

Pentland, A. (1986). Perceptual organisation and the representation of natural form. *Artificial Intelligence*, 28, 293–331.

Perner, J. (1991). *Understanding the Representational Mind*. Cambridge, MA: Bradford Books/MIT Press.

Perner, J. and Leekam, S.R. (1986). Belief and quantity: Three-year-olds' adaptation to listener's knowledge. *Journal of Child Language*, 13, 305–15.

Perruchet, P. and Amorim, M.-A. (1992). Conscious knowledge and changes in performance in sequence learning: Evidence against dissociation. *Journal of Experimental Psychology: Learning, Memory, and Cognition*, 18, 785–800.

Piaget, J. (1926). *The Language and Thought of the Child*. New York: Harcourt Brace.

Pick, A. (1905). *Studien uber motorrische Apraxie und ihr nahestehende Erscheinungen*. Deuticke, Leipzig.

Pinker, S. (1994). *The Language Instinct: The New Science of Language and Mind*. London/New York: Allen Lane, The Penguin Press.

Plaut, D.C., McClelland, J.L., Seidenberg, M.S., and Patterson, K. (in press). Understanding normal and impaired word reading: Computational principles in quasi-regular domains. *Psychological Review*.

Plaut, D. and Shallice, T. (1993). Deep dyslexia: A case study of connectionist neuropsychology. *Cognitive Neuropsychology*, 10, 377–500.

Pollack, J.B. (1990). Recursive distributed representations. *Artificial Intelligence*, 46, 77–105.

Pollard, C. and Sag, I.A. (1987). Information-Based Syntax and Semantics Vol. 1: Fundamentals. CSLI Lecture Notes 13.

Pollard, C. and Sag, I.A. (1993). *Head-Driven Phrase Structure Grammar*. Chicago: University of Chicago Press.

Potter, J., Edwards, D., and Wetherwell, M. (1993) A model of discourse in action. *American Behavioral Scientist*, 36, 383–401.

Pylyshyn, Z. (ed.) (1987). *The Robot's Dilemma: The Frame Problem in Artificial Intelligence*. Norwood, NJ: Ablex.

Pylyshyn, Z. (1991). The role of cognitive architectures in theories of cognition. In K. VanLehn, (ed.), *Architectures for Intelligence* (pp. 189–224). Hillsdale, NJ: Lawrence Erlbaum.

Quinlan, P.T. (1991). *Connectionism and Psychology: A Psychological Perspective on New Connectionist Research*. London: Harvester Wheatsheaf.

Ramachandran, V.S. (1985). The neurobiology of perception. *Perception*, 14, 97–103.

Read, B.K. and Cherry, L.J. (1978). Preschool children's production of directive forms. *Discourse Processes*, 1, 233–45.

Reason, J.T. (1979). Actions not as planned: The price of automatization. In G. Underwood and R. Stevens (eds), *Aspects of Consciousness*, (vol. 1, ch. 4, pp. 67–89). London: Academic Press.

Reason, J.T. (1984). Lapses of attention in everyday life. In W. Parasuraman and R. Davies (eds), *Varieties of Attention* (ch. 14, pp. 515–49). San Diego, CA: Academic Press.

Reason, J.T. (1990). *Human Error.* Cambridge: Cambridge University Press.

Reason, J.T. (1993). Self-report questionnaires in cognitive psychology: Have they delivered the goods? In A. Baddeley and L. Weiskrantz (eds), *Attention: Selection, Awareness, and Control* (ch. 20, pp. 406–24). Oxford: Clarendon Press.

Récanati, F. (1995). Processing models for non-literal discourse. *Cognitive Science,* 19, 207–32.

Reif, F. and Larkin, J.H. (1991). Cognition in scientific and everyday domains: Comparison and learning implications. *Journal of Research in Science Teaching,* 28, 733–60.

Reinhart, T. (1981). Pragmatics and linguistics: An analysis of sentence topics. *Philosophica,* 27, 53–94.

Ribot, T. (1882). *Diseases of Memory.* New York: Appleton.

Rips, L. J. (1989). Similarity, typicality, and categorization. In S. Vosniadou and A. Ortony (eds), *Similarity and Analogical Reasoning.* New York: Cambridge University Press.

Rips, L.J. (1994). *The Psychology of Proof: Deductive Reasoning in Human Thinking.* Cambridge, MA: MIT.

Roberts, L.E., Williams, R.J., Marlin, R.G., Farrell, T., and Imiolo, D. (1984). Awareness of the response after feedback training for changes in heart rate and sudomotor laterality. *Journal of Experimental Psychology: General,* 113, 225–55.

Robinson, E.J. and Robinson, W.P. (1976a). Developmental changes in the child's explanation of communication failure. *Australian Journal of Psychology,* 28, 155–65.

Robinson, E. J., and Robinson, W. P. (1976b). The young child's understanding of communication. *Developmental Psychology,* 12, 328–33.

Rock, I. (1984). *Perception.* New York: Scientific American Library.

Roediger, H.L. (1993). Learning and memory: Progress and challenge. In D.E. Meyer and S. Kornblum (eds), *Attention and Performance XIV: Synergies in Experimental Psychology, Artificial Intelligence, and Cognitive Neuroscience* (pp. 509–28). Cambridge, MA: MIT Press.

Rosch, E., Simpson, C., and Miller, R.S. (1976). Structural bases of typicality effects. *Journal of Experimental Psychology: Human Learning and Memory,* 2, 491–502.

Rosen, S.M. and Howell, P. (1981). Plucks and bows are not categorically perceived. *Perception and Psychophysics,* 30, 156–68.

Rosen, S. and Howell, P. (1991). *Signals and Systems for Speech and Hearing.* London and San Diego: London and San Diego.

Rosenbloom, P.S., Laird, J.E., Newell, A., and McCarl, R. (1991). A preliminary analysis of the Soar architecture as a basis for general intelligence. *Artificial Intelligence,* 47, 289–325.

Rozin, P. and Gleitman, L.R. (1977). The structure and acquisition of reading. II: The reading process and the acquisition of the alphabetic principle. In A.S. Reber and D.L. Scarborough (eds), *Toward a Psychology of Reading.* Hillsdale, NJ: Erlbaum.

Rubenstein, B.B. (1972). On metaphor and related phenomena. In R.R. Holt and E. Peterfreund (eds), *Psychoanalysis and Contemporary Science* (vol. 1, pp. 70–108). London: Macmillan and Co.

Rumelhart, D.E., Hinton, G.E., and Williams, R.J. (1986). Learning internal representations by error propagation. In D.E. Rumelhart and J.L. McClelland (eds), *Parallel Distributed Processing, Vol. 1: Foundations,* 318–62. Cambridge, MA: MIT Press.

Rumelhart, D.E. and Norman, D.A. (1985). Representation of knowledge. In A.M. Aitkenhead, and J.M. Slack (eds), *Issues in Cognitive Modeling* (ch. 2 pp. 15–62). Hove: Lawrence Erlbaum Associates.

Rumelhart, D.E. and McClelland, J.L. (1986). *Parallel Distributed Processing, Vol 1 Foundations.* Cambridge, MA: MIT Press.

Rumelhart, D.E. and Ortony, A. (1977). The representation of knowledge in memory. In R.C. Anderson, R.J. Spiro and W. E. Montague (eds), *Schooling and the Acquisition of Knowledge.* Hillsdale, NJ: Lawrence Erlbaum.

Rundus, D. (1971). Analysis of rehearsal processes in free recall. *Journal of Experimental Psychology,* 89, 63–77.

Saariluoma, P. (1985). Chess players' intake of task-relevant cues. *Memory and Cognition,* 13, 385–91.

Sachs, J. (1983). Talking about the there and then: The emergence of displaced reference in parent-child discourse. In K.E. Nelson (ed.), *Children's Language* (pp. 1–28). Hillsdale, NJ: Lawrence Erlbaum.

Sacks, H., Schegloff, E.A., and Jefferson, G. (1974). A simplest systematics for the organization of turn-taking in conversation. *Language,* 50, 696–735.

Salkie, R. (1990). *The Chomsky Update.* London: Unwin Hyman.

Sandson, J. and Albert, M.L. (1984). Varieties of perseveration. *Neuropsychologia,* 6, 715–32.

Scardamalia, M. and Bereiter, C. (1991). Literate expertise. In K.A. Ericsson and J. Smith (eds), *Toward a General Theory of Expertise: Prospects and Limits.* Cambridge: Cambridge University Press.

Schank, R.C. and Abelson, R.P. (1977). *Scripts, Plans, Goals, and Understanding: An Inquiry into Human Understanding.* Hillsdale, NJ: Lawrence Erlbaum.

Schiefelbus, R.L. and Pickar, J. (eds) (1984) *The Acquistion of Communicative Competence.* Baltimore: University Press.

Schneider, W. and Detweiler, M. (1987). A connectionist/control architecture for working memory. In Bower, G.H. (ed.), *The Psychology of Learning and Motivation* (vol. 21, pp. 53–119). San Diego, CA: Academic Press.

Schneider, W. and Shiffrin, R.M. (1977). Controlled and automatic human information processing: I. Detection, search and attention. *Psychological Review,* 84, 1–66.

Schober, M.F. and Clark, H.H. (1989). Understanding by addressees and overhearers. *Cognitive Psychology,* 21, 211–32.

Schon, D.A. (1979). Generative metaphor: A perspective on problem-solving and social policy. In A. Ortony (ed.), *Metaphor and Thought.* Cambridge: Cambridge University Press.

Schwartz, B. and Reisberg, D. (1991). *Learning and Memory.* New York: Norton.

Schwartz, M.F., Reed, E.S., Montgomery, M., Palmer, C., and Mayer, N.H. (1991). The quantitative description of action disorganization after brain damage: A case study. *Cognitive Neuropsychology,* 8, 381–414.

Scoville, R.P. and Gordon, A.M. (1980). Children's understanding of factive presuppositions: An experiment and a review. *Journal of Child Language,* 7, 381–99.

Searle, J.R. (1969). *Speech Acts: An Essay in the Philosophy of Language.* Cambridge: Cambridge University Press.

Searle, J.R. (1976). The classification of illocutionary acts. *Language in Society,* 5, 1–24.

Searle, J. (1980). Minds, brains and programs. *Behavioral and Brain Sciences,* 3, 417–57.

Seibel, R. (1963). Discrimination reaction time for a 1023-alternative task. *Journal of Experimental Psychology,* 66, 215–26.

Seidenburg, M.S. and McClelland, J. (1989). A distributed, developmental model of word recognition and naming. *Psychological Review,* 96, 523–68.

Seidenberg, M.S., Tannenhaus, M., Leiman, J., and Bienkowski, M. (1982). Automatic access of the meaning of ambiguous words in context: Some limitations of knowledge-based processing. *Cognitive Psychology,* 14, 489–537.

Sellars, W. (1954). Presupposing. *Philosophical Review,* 63, 197–215.

Sells, P. (1985). Lectures on Contemporary Syntactic Theories. CSLI Lecture Notes 3.

Sergent, J. (1984). An investigation into component and configural processes underlying face recognition. *British Journal of Psychology,* 75, 221–42.

Shallice, T. (1972). Dual functions of consciousness. *Psychological Review,* 79, 383–93.

Shallice, T. (1982). Specific impairments of planning. *Philosophical Transactions of the Royal Society of London,* B298, 199–209.

Shallice, T. (1988). *From Neuropsychology to Mental Structure.* Cambridge: Cambridge University Press.

Shallice, T. (1991). The revival of consciousness in cognitive science. In W. Kessen, A. Ortony and F. Craik (eds), *Memories, Thoughts and Emotion: Essays in Honour of George Mandler.*

Shallice, T. and Burgess, P.W. (1991a). Deficits in strategy application following frontal lobe lesions. *Brain,* 114, 727–41.

Shallice, T. and Burgess, P.W. (1991b). Higher-order cognitive impairments and frontal lobe lesions in man. In H.S. Levin, H.M. Eisenberg and A.L. Benton (eds), *Frontal Lobe Function and Dysfunction* (ch. 6, pp. 125–38). Oxford: Oxford University Press.

Shallice, T. and Burgess, P.W. (1993). Supervisory control of action and thought selection. In A. Baddeley and L. Weiskrantz (eds), *Attention: Selection, Awareness, and Control* (ch. 9, pp. 171–87). Oxford: Clarendon Press.

Shallice, T., Burgess, P. W., Schon, F., and Baxter, D. M. (1989). The origins of utilisation behaviour. *Brain,* 112, 1587–98.

Shanks, D.R. and St. John, M.F. (1994). Characteristics of dissociable human learning systems. *Behavioral and Brain Sciences,* 17, 367–447.

Shatz, M. (1978). Children's comprehension of their mothers' question-directives. *Journal of Child Language,* 5, 39–46.

Shatz, M. (1994). *A Toddler's Life.* Oxford: Blackwell.

Shatz, M. and McClosky, L. (1984). Answering appropriately: A developmental perspective on conversational knowledge. In S.A.I. Kuczaj (ed.), *Discourse Development: Progress in Cognitive Developmental Research.* (pp. 19–36). New York: Springer Verlag.

Shatz, M. and O'Reilly, A.W. (1990). Conversational or communicative skill? A reassessment of two-year-olds' behavior in miscommunication episodes. *Journal of Child Language,* 17, 131–46.

Shepard, R.N. (1981). Psychophysical complementarity. In M. Kubovy and J. Pomerantz (eds), *Perceptual Organization* (pp. 279–341). Hillsdale, NJ: Lawrence Erlbaum Associates.

Shepherd, J.W. (1981). Social factors in face recognition. In G. Davies, H. Ellis and J.W. Shepherd (ed.), *Social factors in face recognition.* London: Academic Press.

Sherry, D. and Schacter, D. 1987. The evolution of multiple memory systems. *Psychological Review,* 94, 439–54.

Shiffrin, R.M. and Nosofsky, R.M. (1994). Seven plus or minus two: A commentary on capacity limitations. *Psychological Review,* 101, 357–61.

Shimamura, A.P. and Squire, L. R. (1987). A neuropsychological study of fact memory and source amnesia. *Journal of Experimental Psychology: Learning, Memory, and Cognition,* 13, 464–73.

Simon, H. (1962). The architecture of complexity. *Proceedings of American Philosophical Society,* 106, 467–82.

Simon, H.A. (1969). *The Sciences of the Artificial.* Cambridge, MA: MIT Press.

Simon, H.A. and Gilmartin, K.A. (1973). A simulation of memory for chess positions. *Cognitive Psychology,* 5, 29–46.

Slobin, D.I. (1970). Universals of grammatical development in children. In G.B.F. d'Arcais and W.J.M. Levelt (eds), *Advances in Psycholinguistics* (pp. 175–208). Amsterdam: North-Holland.

Sloman, A. (1993). The mind as a control system. In C. Hookway and D. Peterson (eds), *Philosophy and Cognitive Science.* Cambridge: The Press Syndicate, University of Cambridge.

Smith, B.R. and Leinonen, E. (1992). *Clinical Pragmatics.* London: Chapman and Hall.

Smith, N.V. (1989). *The Twitter Machine.* Oxford: Blackwell.

Smith, N. (1994). Review of Karmiloff-Smith (1992) "Beyond Modularity: A Developmental Perspective on Cognitive Science." *European Journal of Disorders in Communication,* 29, 95–105.

Smith, N. and Tsimpli, I-M. (1991). Linguistic modularity? A case of a savant-linguist. *Lingua,* 84, 315–51.

Smith, N. and Tsimpli, I-M. (1995). *The Mind of a Savant: Language Learning and Modularity.* Oxford: Blackwell.

Smolensky, P. (1988). On the proper treatment of connectionism. *Behavioral and Brain Sciences,* 11, 1–74.

Snoddy, G.S. (1926). Learning and stability. *Journal of Applied Psychology,* 10, 1–36.

Snowling, M. (1987). *Dyslexia: A Cognitive Developmental Perspective.* Oxford: Blackwell.

Solomon, R.L. and Turner, L.H. (1962). Discriminative classical conditioning in dogs paralyzed by curare can later control discriminative avoidance responses in the normal state. *Psychological Review,* 69, 202–19.

Sperber, D. (1994a). The modularity of thought and the epidemiology of representations. In L. Hirschfeld and S. Gelman (eds) *Mapping the Mind* (1994), 39–67.

Sperber, D. (1994b). Understanding verbal understanding. In J. Khalfa (ed.), *What is Intelligence?* Cambridge: Cambridge University Press.

Sperber, D., Cara, F. and Girotto, V. (1995). Relevance theory explains the selection task. *Cognition,* 57, 31–95.

Sperber, D. and Wilson, D. (1981). Irony and the use-mention distinction. In P. Cole (ed.), *Radical Pragmatics* (pp. 295–318). New York: Academic Press.

Sperber, D. and Wilson, D. (1986). *Relevance: Communication and Cognition.* Cambridge, MA: Harvard University.

Sperber, D. and Wilson, D. (1995). Fodor's frame problem and relevance theory. *Behavioral and Brain Sciences.*

Sperber, D. and Wilson, D. (forthcoming). *Relevance and Meaning.*

Squire, L.R. (1987). *Memory and Brain.* Oxford: Oxford University Press.

Squire, L.R. (1992). Memory and the hippocampus: A synthesis from findings with rats, monkeys, and humans. *Psychological Review,* 99, 195–231.

Squire, L.R., Slater, P.C., and Chace, P.M. (1975). Retrograde amnesia: Temporal gradient in very long-term memory following electroconvulsive therapy. *Science,* 187, 77–9.

Steedman, M. (1993). Categorial Grammar. *Lingua,* 90.3, 221–58.

Stenning, K. and Oberlander, J. (1995). A cognitive theory of graphical and linguistic reasoning: Logic and implementation. *Cognitive Science,* 19, 97–140.

Stevens, K.N. (1972). The quantal nature of speech: Evidence from articulatory-acoustic data. In P.B. Denes and E.E. David (eds), *Human Communication. A Unified View* (pp. 51–74). New York: McGraw-Hill.

Stevens, K. N. (1981). Constraints imposed by the auditory system on the properties of speech sounds: Data from phonology, acoustics and psychoacoustics. In T. Myers, J. Laver and J. Andersen (eds), *The Cognitive Representation of Speech* (pp. 61–74). Amsterdam: North-Holland.

Stevenson, R.J. (1993). *Language, Thought and Representation.* Chichester: Wiley.

Strawson, P.F. (1950). On referring. *Mind*, 59, 320–44.

Strawson, P.F. (1964). Intention and convention in speech acts. *Philosophical Review*, 73, 439–60.

Suchman, L.A. (1987). *Plans and Situated Actions: The problem of Human Machine Communication.* Cambridge: Cambridge University Press.

Svartvik, J. and Quirk, R. (1980). *A Corpus of English Conversation.* Sweden, Lund: Gleerup.

Swinney, D. (1979). Lexical access during sentence comprehension: (Re)consideration of context effects. *Journal of Verbal Learning and Verbal Behavior*, 18, 645–59.

Taft, M. (1991) *Reading and the Mental Lexicon.* London: Erlbaum.

Tager-Flusberg, H. (1981). On the nature of linguistic functioning in early infantile autism. *Journal of Autism and Developmental Disorders*, 11, 45–56.

Tanz, C. (1980). *Studies in the acquisition of deictic terms.* Cambridge: Cambridge University Press.

Thagard, P., Holyoak, K.J., and Gochfield, D. (1990). Analogue retrieval by constraint satisfaction. *Artificial Intelligence*, 46, 259–310.

Thorndike, E.L. (1911). *Animal Intelligence: Experimental Studies.* New York: Macmillan.

Touretzky, D.S. and Hinton, G.E. (1988). A distributed connectionist production system. *Cognitive Science*, 12, 423–66.

Tulving, E. and Osler, S. (1968). Effectiveness of retrieval cues in memory for words. *Journal of Experimental Psychology*, 77, 593–601.

Tulving, E., Schacter, D.L., McLachlan, D.R., and Moscovitch, M. (1988). Priming of semantic autobiographical knowledge: A case study of retrograde amnesia. *Brain and Cognition*, 8, 3–20.

Turing, A.M. (1950). Computing machinery and intelligence. *Mind*, LIX, 236.

Tweney, R. (1985). Faraday's discovery of induction: A cognitive approach. In D. Gooding and F. James (eds), *Faraday Rediscovered: Essays on the Life and Work of Michael Faraday* (pp. 189–209). New York: Stockton Press.

Ullman, S. (1989). Aligning pictorial descriptions: An approach to object recognition. *Cognition*, 32, 193–254.

Valentine, T. and Bruce, V. (1986a). The effect of race, inversion, and encoding activity upon face recognition. *Acta Psychologica*, 61, 259–73.

Valentine, T. and Bruce, V. (1986b). The effects of distinctiveness in recognising and classifying faces. *Perception*, 15, 525–36.

Valentine, T. and Bruce, V. (1986c). Recognising familiar faces: The role of distinctiveness and familiarity. *Canadian Journal of Psychology*, 40, 300–5.

van Lehn, K. (ed.) (1991). *Architectures for Intelligence.* Hove: Lawrence Erlbaum Associates.

Vera, A.H., Lewis, R.L., and Lerch, F.J. (1993). Situated decision-making and recognition-based learning: Applying symbolic theories to interactive tasks. *Proceedings of the 15th Annual Conference of the Cognitive Science Society.* University of Colorado (pp. 635–40). Hillsdale, NJ: Lawrence Erlbaum Associates.

Vera, A.H. and Simon, H.A. (1993). Situated action: A symbolic interpretation. *Cognitive Science*, 17, 7–48.

von Neuman, J. (1947). Preliminary discussion of the logical design of an electronic computing instrument. US Army Ordnance Report.

Voss, J.E. and Post, T.A. (1988). On the solving of ill-structured problems. In M.T.H. Chi, R. Glaser and M.J. Farr (eds), *The Nature of Expertise* (pp. 261–86). Hillsdale, NJ: Lawrence Erlbaum Associates.

Wallman, J. (1992). *Aping Language.* New York: Cambridge University Press.

Warrington, E.K. and Shallice, T. (1969). The selective impairment of auditory verbal short-term memory. *Brain*, 92, 885–96.

Warrington, E.K. and Weiskrantz, L. (1968). New method of testing long-term retention with special reference to amnesic patients. *Nature*, 217, 972–74.

Webelhuth, G. (ed.) (1995). *Government and Binding Theory and the Minimalist Program.* Oxford: Blackwell.

Weldon, M.S. and Roediger, H.L. (1987). Altering retrieval demands reverses the picture superiority effect. *Memory and Cognition*, 15, 269–80.

Wilcox, M.J. and Howse, P. (1982). Children's use of gestural and verbal behavior in communicative misunderstandings. *Applied Psycholinguistics*, 3, 15–27.

Wilcox, M.J. and Webster, E.J. (1980). Early discourse behavior: An analysis of children's responses to listener feed-back. *Child Development*, 51, 1120–5.

Wilson, D. (1994). Relevance and understanding. In G. Brown (ed.), *Language and Understanding.* Oxford: Oxford University Press.

Wilson, D. and Sperber, D. (1992). On verbal irony. *Lingua*, 87, 53–76.

Wimmer, H. and Perner, J. (1983). Beliefs about

beliefs: Representation and constraining function of wrong beliefs in young children's understanding of deception. *Cognition*, 13, 103–28.

Winograd, T. and Flores, F. (1986). *Understanding Computers and Cognition: A New Foundation for Design.* Norwood, NJ: Ablex.

Witkin, A.P. and Tenenbaum, J.M. (1983). On the role of structure in vision. In J. Beck, B. Hope and A. Rosenfeld (ed.), *Human and Machine Vision* (pp. 481–543). New York: Academic Press.

Wittgenstein, L. (1958). *Philosophical Investigations.* Oxford: Blackwell.

Wixted, J.T. and Ebbesen, E. B. (1991). On the form of forgetting. *Psychological Science*, 2, 409–15.

Yamada, J. (1990). *Laura: A Case for the Modularity of Language.* Cambridge, MA: MIT Press.

Yin, R.K. (1969). Looking at upside-down faces. *Journal of Experimental Psychology*, 81, 141–45.

Young, A.W., Hellawell, D., and Hay, D.C. (1987). Configural information in face perception. *Perception*, 16, 747–59.

Zeki, S. (1992). The visual image in mind and brain. *Scientific American*, September 1992, 42–50.

Zukow, P. G., Reilly, J., and Greenfield, P. M. (1982). Making the absent present: Facilitating the transition from sensorimotor to linguistic communication. In K. Nelson (ed.), *Children's Language* (vol. 3). Hillsdale, NJ: Lawrence Erlbaum.

Subject Index

Name Index